W9-DIO-497

Understanding and Using Statistics

Understanding and Using Statistics

Basic Concepts

Second Edition

Marty J. Schmidt

University of New Hampshire

D. C. HEATH AND COMPANY

Lexington, Massachusetts   Toronto

Preface

The second edition of *Understanding and Using Statistics: Basic Concepts* is designed as a text for the first course in statistics. Its goals are to give students a sound conceptual understanding of and computational facility with applied statistics. It is intended to make the subject approachable to those with a wide variety of backgrounds and abilities and to show that statistics can be interesting and useful.

After teaching introductory statistics for many years, I have become aware of an acute need for a textbook that will meet *several* needs of each student. The text should explain basic ideas clearly and thoroughly; it should illustrate computational procedures carefully; and, above all, it should stress proper use and interpretations of statistical results. The text should also supply a generous number of practice problems. Finally, it should include "real" examples of statistical usage to complement the more sterile and contrived teaching examples. My intent in preparing this edition was to meet all of these needs in a single, well-integrated textbook.

A central theme of this book is that statistics is not simply an unapproachable, esoteric study that should be left to scientists and insurance actuaries. A sound conceptual understanding of statistical ideas and statistical logic can lend power and precision to one's thinking and communication in nearly all facets of life. That point is stressed repeatedly, through examples drawn from psychology, medicine, sociology, education, business, and other fields.

Many students are initially apprehensive about the subject — especially those who have not taken mathematics courses for some time. For these students especially, this book emphasizes that formal statistical thinking is an extension and refinement of ideas that everyone understands before taking a statistics course. We use familiar statistical words like "average" and "uncertainty," for instance, without fear of misunderstanding. The same ideas have a way of becoming mysterious and threatening, however, when we bring them into a statistics class disguised as "central tendency" and "probability," quantify them, and represent them with symbols. For this reason, special care is given to explaining fundamental concepts and to giving students a conceptual framework into which new ideas can be assimilated. Each chapter focuses on a number of key concepts that appear in boldface within the text when they are first defined and are listed again at the end of the chapter. Most concepts or terms are explained, with illustrations or examples when possible, and then related to concepts the student should already know. The intent is to teach the basic idea first, show how it operates in statistical computations, and then return to the basic idea for interpretation of the statistic. Key concepts are also defined once more in the Glossary, with page references to the longer text definitions.

Symbols, formulas, and computations are also troublesome for many students. Thus, computational procedures are carefully explained, step by step, so that students may see that even the most awesome-looking formula is little more than a set of instructions for repeated addition, subtraction, multiplication or division. Special care is also taken to explain the use of subscripted variables and summation notation.

Many students will want to know if they will need an electronic calculator for the first statistics course. Access to a calculator is not necessary — all computations can be done "by hand." However, the use of a calculator can reduce computational time and effort, and students can work the same problem several times to check their computations. Thus I recommend to students who wish to utilize a calculator an inexpensive model that adds, subtracts, multiplies, divides, and extracts square roots. It is also useful, but not necessary, to have a calculator with memory functions, a squaring function, and an inverse function. Of course, one can buy more powerful calculators that automatically deliver statistics such as the mean, standard deviation, Pearson r, and regression coefficients without involving the user in the computational process; these latter "preprogrammed" statistical functions, however, are of little help in acquiring an understanding of these statistics.

All chapters of this second edition have been updated; many have been rewritten. The chapter on nonparametric statistics has been greatly expanded, and two new chapters on analysis of variance have been added. Computational methods are treated in greater depth through step-by-step illustrations. Computation of many statistics is illustrated in two forms: through definitional formulas and through computational ("raw score") formulas. The latter kind of computational illustration appears in optional computational sections marked $\boxed{c}$. Sections marked $\boxed{o}$ represent specific statistics that some instructors will consider optional and may be omitted without affecting the content of the chapter.

Other features include checklists of objectives, spotlights, and exercises. A checklist of objectives, which appears at the beginning of each chapter, provides a rough conceptual framework of the entire chapter before the student plunges into complex details. Spotlights, or high-interest readings, show the application of statistics in current research and other situations or illustrate significant historical information about certain statistics. In response to requests from users, the exercises that appear at the end of each chapter contain more problems than the exercises in the previous edition. Answers to problems that have numerical solutions are provided in Appendix 4.

In preparing the second edition of this book, I received valuable assistance from many people. Comments and suggestions submitted by instructors and students who used the first edition have contributed to virtually every part of the second, and I am very grateful to all of them. Special thanks are due to Dr. Ted W. Allen, University of California, Los Angeles; Dr. David Kieras, University of Arizona; and Dr. O. R. Holsti, Duke University, for extensive critiquing of an early manuscript for this edition; the text is very much improved for their efforts.

I am also grateful to the literary Executor of the late Sir Ronald Fisher, F.R.S., to Dr. Frank Yates, F.R.S., and to Longman Group Ltd., London, for permission to reprint Tables IIi, III, and IV from their book *Statistical Tables for Biological, Agricultural, and Medical Research* (6th edition, 1974).

M.J.S.

Contents

Introduction

After reading this chapter you should be able to do the following:

1. *Recognize uses of statistics and statistical ideas in several phases of daily life.*

2. *Define the term statistics.*

3. *Discuss three factors that influence the interpretation of any statistic (source of the data,* *choice of the correct statistic, and appropriate presentation of the statistic).*

4. *Explain the difference between inferential statistics and descriptive statistics.*

Statistics: Why Study Them?

Most of us have been using and consuming statistics almost as long as we have been speaking English. They turn up everywhere, every day, in statements such as these:

> "A recent Gallup Poll indicates that 52 percent of the American public approves of the President's performance in office."

> "Data from animal studies suggest that regular use of saccharin may increase the risk of cancer."

> "The Dow-Jones index was off another 10 points today."

> "Eight out of ten doctors surveyed recommended the ingredients of Brand A Pain Reliever."

A large part of what we call the news, for example, consists of statistics. The results of public-opinion polls are often news items and so are various economic indices, such as the Dow Jones averages and the consumer price index. Advertising accounts for another large part of our daily statistical intake. We learn that 67 out of 100 owners of luxury car A

1

prefer the smooth, soft ride of luxury car B, or that 90 percent of the public suffers from dandruff at one time or another. Apparently, advertisers believe that statistics lend dignity and a suggestion of scientific proof to their messages.

The news, advertising, and of course, sports, have made us quite familiar with statistics like averages, proportions, and percentages. We are also familiar with the statistical concept of probability, which is used in weather reports ("There's a 70 percent chance of rain tomorrow") and at the race track ("That horse is a 30-to-1 long shot"). Even third- and fourth-grade children learn the elements of making and graphing statistical distributions from a booklet published by the children's newspaper *My Weekly Reader*.[1] So it isn't surprising that many people feel they have enough understanding of basic statistics and that more sophisticated statistics should be left to scientists, mathematicians, and insurance actuaries. We are so comfortable with everyday statistics that we may doubt the need for learning any more about them.

The problem, however, is that we cannot understand or use statistics correctly until we interpret them, and all statistics — even simple, familiar ones — are subject to misuse and misinterpretation. As product consumers and idea consumers it is to our advantage to recognize misuse and avoid misinterpretation of the many statistics with which we come in contact. For example, how do you respond when spokesmen on opposite sides of controversial issues each cite statistics to support their own points of view? Today one hears statistics that say nuclear power is safe and others that say it is dangerous. In such a situation, we must decide which statistics have the most meaning.

The advantages of knowing basic statistical principles go far beyond the instances in which we use and evaluate statistics ourselves. It is an inescapable fact of life that statistical decisions made in government, science, and industry often have a very direct influence on our lives. Statistical evidence is responsible for the Surgeon General's warning on cigarette packages and the presence of seat belts in your car. Statistics determine which of tonight's television shows will still be running next year, and which experimental drugs will end up at the pharmacy and which will never be marketed. As a responsible citizen, how do you evaluate the results of these statistical decisions?

Statistics thus affect our lives in many ways. Some can be produced and interpreted by nearly everyone; others require the combined efforts of experts and computers. But the surprising thing to many people is that almost all applied statistics can be appreciated and understood — at least in principle — with an understanding of a few basic concepts.

This is the main reason statistics courses are included in many college psychology, political science, sociology, and education programs. Much of the content of these disciplines consists of experimental results rather

---

[1]*Introducing Table and Graph Skills* (Middletown, Conn.: American Education Publications, 1968).

than of rules and laws. Many "facts" in these areas are only tenuously established, so it is useful to study the experiments that produced them. Thus, the literature of these fields is filled with studies and experiments. Quite often there are two or more plausible sides to a disputed issue, and you need to look at the experimental evidence supporting each side. Understanding basic statistical ideas is an invaluable skill in this process. Sometimes you will even interpret a set of experimental results differently from the researcher who reports them.

A very important point should be emphasized here. Taking a one-semester course in statistics will not equip you with the skills you would need to produce all the statistics that affect you, but it should present you with the knowledge necessary to evaluate them. A first course in applied statistics should not require mastery of highly abstract mathematical ideas, nor should it present you with difficult computational problems. If you have looked ahead and discovered some of the symbol-heavy formulas in the later chapters, be assured that despite their formidable appearance the formulas are simply *instructions* for repeated addition, subtraction, multiplication, and division. Learning to use a formula means that you learn to read the instructions; it does not mean learning new computational skills. If you can handle elementary arithmetic and a very few other simple procedures (primarily finding square roots, squaring a number, and solving an algebraic equation for one unknown value), you should have no trouble with the mathematical aspects of basic applied statistics. The crucial ideas to be learned have to do with the interpretation and use of statistics. Because high-powered computers and calculators are now readily available to nearly everyone, the only justification for making you labor through tedious statistical calculations is that interpretation sometimes depends on computational factors that produce the statistic. Once you understand how a statistic is produced, there is little virtue in computing it "by hand" if a calculator or computer can do it for you more quickly.

Maybe you can begin to see why a basic understanding of statistics can be useful to nearly everyone. Its value should become clearer as we take a closer look at the nature of statistics and examine some ways in which they are used.

Statistics: What Are They?

By now you know that statistics involve numbers. Statistics, in fact, *are* numbers. A single **statistic** is a single number that can be used to summarize, analyze, or evaluate a body of information. Your bowling average for the year is such a number. It can be used to *summarize* your performance from a great many games, it certainly can be used to *evaluate* your ability, and compared with last year's average, it can be used to *analyze* your progress.

The study of statistics includes, of course, the numerical methods that produce those numbers. But computing statistics is only a small part of the story. A statistic is of no use until it is interpreted, and proper interpretation does not automatically follow proper computation. Statistics, as you may suspect, can be technically correct and still misrepresent reality or lead to invalid conclusions. Whoever coined the old saw "Figures don't lie, but liars can figure," was almost certainly thinking about statistics.

In order to interpret a statistic properly you must know something about its origin and production. The statistical process can be described as the business of taking data in numerical form, applying computational procedures, and coming up with a single number (or a few numbers) that tell you something about the group of data as a whole. So the statistical consumer cannot respond intelligently to a statistic unless he or she knows something about the numerical raw material, the computational procedures applied to it, and the proper interpretation of the numerical end product. This three-part process requires us to know something about all three parts: data acquisition, computation, and interpretation.

Data Acquisition

Every time you come in contact with statistics, you should ask an important question: Where did the original data come from? Asking this question is most important when someone uses statistics to sell you something or solicits your opinion on a disputed issue. You have probably seen advertisements in which an authoritative-looking figure in a lab coat says something like, "In a recent study eight out of ten doctors surveyed recommended the ingredients in A Brand Pain Reliever." That is a statistical statement, and it is loaded with qualifications that should raise questions in your mind about the original source of the data. Take the word surveyed. Years ago many such ads said simply, "Eight out of ten doctors recommend . . ." Recently, however, the public — and the law — has become more sophisticated with respect to advertising claims, and the word "surveyed" is included to protect A Brand, Inc. Technically, the corporation is not responsible for the opinions of doctors who did not take part in the survey mentioned in the statement.

Most ads like this do not tell us how the doctors were chosen for the survey, how many doctors were chosen in all, what kind of doctors they were, or even how many surveys were taken. Perhaps A Brand included only ten doctors in the survey, perhaps 100. Perhaps A Brand took several surveys before getting a favorable set of results; perhaps only one survey was taken. From the information usually given in these ads, we simply cannot answer the question, Where did the original data come from?

Incidentally, you may have noticed that the doctors did not specifically recommend A Brand Pain Reliever, only its ingredients. The same ingredients may or may not be contained in competing brands. It is interesting that some recent advertisements invite you to send for more complete descriptions of surveys and tests mentioned in the ads; this, again, probably reflects an awareness among advertisers that the general public is becoming more sophisticated about statistics.

We will consider the important factors in data acquisition in more depth in Chapter 1.

Computation and Choice of Statistics

When confronted with a statistic, you should also ask if it is the proper statistic for the purpose at hand. Usually, a researcher can communicate a given statistical message with any one of several available statistics. Sometimes, however, computational factors make one or more of the available statistics inappropriate for the message in ways that are not immediately obvious. To emphasize the need for basic computational understanding, consider a rather extreme example in which a statistic as simple and familiar as the arithmetic average can be misleading.

Suppose I tell you that the five houses on my block have an average value of $100,000. That's a pretty impressive statement, and unless I give you more information, you would probably have to conclude that I live in an expensive house on an exclusive block. This impression may be changed, however, if I mention that my house and three others are worth about $10,000 each and the millionaire down the street lives in a $460,000 mansion.

To compute the **arithmetic average** of these values, add the individual values and divide the total by the number of houses. Suppose these figures were as follows:

House 1	$ 10,800
House 2	9,900
House 3	10,000
House 4	9,300
House 5	460,000
Total	$500,000

Arithmetic average = $5\overline{)500,000}$ = 100,000

Here the statistic, the arithmetic average, is correctly determined to be $100,000, but it tells us little about the typical home in this particular group. In this case, it might be more appropriate to suggest the typical home value by using a different statistic, the **median.** When the houses are lined up in order of value from low to high, the median may be taken to be the middle value:

House 4	$ 9,300	lowest value
House 2	9,900	
House 3	10,000	middle value = median
House 1	10,800	
House 5	460,000	highest value

Here the median home value is $10,000. Specifying the median value of these houses gives a more accurate picture of the typical house than does specifying the arithmetic average, although you would never guess from the median that there is a large mansion in the group. For this particular group of five houses, the two statistics, arithmetic average and median, produce radically different pictures of typical homes. This is an extreme example, but it should suggest to you that proper interpretation of a statistic depends upon knowing something of the statistic's computational basis.

Finally, a statistic can be derived from good data, be the proper statistic for the purpose at hand, and still be *presented* in a manner that is misleading. We will use the familiar arithmetic average again to show one way this can happen.

Presentation and Interpretation

It is a fact that the average miles per gallon obtained by full-size American cars declined during the period 1964 to 1974. But was the decline substantial or trivial? Suppose that I want to convince you that the decline in fuel efficiency was substantial, that during these years cars were rapidly turning into gas guzzlers. How do I impress you with the significance of this trend?

I'll begin by graphing in Figure I.1 the average miles per gallon obtained by full-size American cars for each car-model year from 1964 to 1974. The vertical **axis,** or dimension, of the graph represents average miles per gallon, and the horizontal axis represents various car-model years. The average miles per gallon for each year's cars are shown by dots, and the dots are connected with a line to make the year-to-year trend clearer. You can tell from the downward slope of the line that fuel efficiency declined during the period shown.

The decline is noticeable, but not striking when graphed in this form. Remember that the intention is to make the trend toward lower fuel efficiency look particularly awesome. So I'll present a graph of the same averages in a slightly different form. First, I'll select only a part of the vertical axis — just the part that includes all the averages shown (Figure I.2). In this case, the averages range from 10.8 miles/gal obtained by 1974 cars to 12.9 miles/gal obtained by 1964 cars. Next I'll throw away the rest of the vertical axis and stretch out the small section so that it covers almost the same distance as the entire original vertical axis. Then I'll combine the new vertical axis with the old horizontal axis to obtain a new graph for the same averages (Figure I.3).

This is the same group of statistics, but changing the vertical axis drastically alters the appearance of the graph. At first glance, the average miles per gallon seem to be rapidly declining. And the change in the vertical axis is strictly legal. Notice the little break at the base of the new vertical axis in Figure I.3. It indicates that the axis has been cut.

Suppose I want the decline in fuel efficiency to look even more impres-

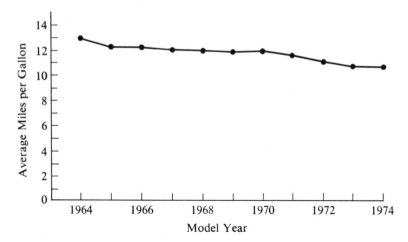

Figure I.1 Average miles per gallon obtained by cars weighing 3,850–4,250 lb (including two passengers and a full gas tank) for model years 1964 through 1974. The 1964 cars in this category averaged 12.9 miles/gal while comparable 1974 cars averaged only 10.8 miles/gal. The year-to-year trend toward lower fuel efficiency is suggested by the line connecting the dots. (Averages are from the Environmental Protection Agency.)

sive. This can be accomplished by compressing the horizontal axis and combining it with the new vertical axis (Figure I.4). This graph conveys exactly the same information as Figure I.1, but the two graphs appear quite different upon initial inspection. In Figure I.1 average gas mileages were represented by a line like (a) in Figure I.5. In Figure I.4 these averages were represented by a line like (b) in Figure I.5. Of course, you can look at the numbers on the vertical axis of either graph and see that the decrease has been about 2 miles/gal during this period. But it is well

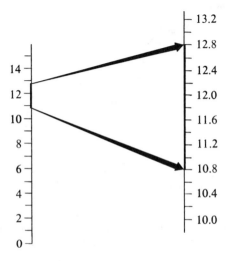

Figure I.2

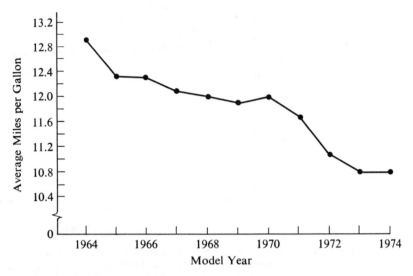

Figure I.3   Graph of the same average miles per gallon for different model-year cars shown in Figure I.1, presented with a different vertical axis. Notice that here the vertical axis covers a much smaller range of miles/gal values, producing a steeper slope in the line connecting yearly averages.

established by psychology that first impressions can have long-lasting and subtle effects on later interpretations, and the first impression created by Figure I.5 is more striking.

Take a close look at the graphical displays of statistics that appear in

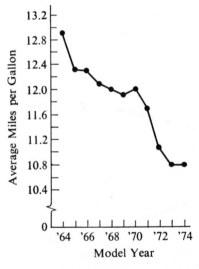

Figure I.4   The same averages that were shown in Figures I.1 and I.3. Here the year-to-year trend toward lower fuel efficiency is maximally accented by using a compressed horizontal axis and a cut and stretched vertical axis.

(a)

(b)

Figure I.5

news magazines and newspapers. Examples of such graph chopping and stretching are quite common. Sometimes the chopping and stretching add clarity; sometimes the altered graphs are genuinely misleading. But knowing what to look for can help you decide which is the case.

These examples were included simply to show that interpretation is an important part of the statistical process and that statistics do not interpret themselves. The examples should suggest that statistics are tools that must be handled properly in order to be useful. When you understand basic statistical concepts, they can lend power and precision to your thinking and communication. They are misleading only when generated and consumed in ignorance.

Statistics: How Are They Used?

Statistics were defined earlier as numbers that can be used to summarize, evaluate, or analyze a body of information. Although summarization, evaluation, and analysis are all involved to some extent whenever statistics are used, it is useful to distinguish two categories of statistics with different functions. In general terms, applied statistics can be classified in the following way:

Descriptive statistics — Describe and summarize data.

Inferential statistics — Make possible generalizations beyond the data at hand; evaluate differences among groups; estimate unknown values.

Both categories of statistics have many methods and techniques in common, such as means, standard deviations, and correlation coefficients. Some statistics, such as the standard deviation, are computed in slightly different ways for inferential and descriptive uses. However, the two basic categories of statistics differ not so much in computational methods as they do in use and interpretation. Briefly, the uses of each category (and the concerns of the two major divisions of this book) are as follows.

Descriptive Statistics

Descriptive statistics are used to describe or summarize information clearly and precisely. This information usually represents a collection of systematically made observations or measurements.

When our friends discuss the "statistics" from yesterday's football game, they are using descriptive statistics. Single numbers, such as the average yards per carry obtained by a certain running back, can convey a great deal of information in compact form. Almanacs and record books similarly consist mainly of large collections of descriptive statistics.

The important point to be made about descriptive statistics is that they apply only to the individuals or objects upon which the observations were actually made. If we use statistics to generalize beyond the immediate observations, then we are no longer using descriptive statistics.

Inferential Statistics

Sometimes we wish to do more than describe the observations at hand. We may want to study characteristics of large groups of people, for example, even though it is not actually possible to make an observation on each person in the group. In these cases, we select only some members of the group and study them. In statistical terms, this is known as choosing a **sample** (the people actually observed) from the **population** (the larger group in which we are ultimately interested). Drawing conclusions about the population based on observation of a sample is the concern of inferential statistics.

Inferential statistics are especially useful in evaluating the results of experiments, since experiments are performed on samples rather than on whole populations. Often researchers are interested in characteristics of very large populations. What proportion of the voting population favors candidate Smith? Does the population of smokers have a lower mean age at death than the population of nonsmokers? Since it is not possible to make an observation on each member of these populations, small groups of population members are selected for observation. Inferential statistics make possible generalizations about the population from sample data.

The crucial difference between descriptive and inferential statistics concerns the possibility of drawing erroneous conclusions with inferential statistics. Consider, for instance, the process of determining the mean (average) age at marriage of American women who married during 1950. The population of interest is specified clearly and it is quite large. In order to find the average using *descriptive* statistics, we would have to know the marriage age of every individual population member. Obviously, there are many practical reasons why we cannot collect all of that information.

Instead, we can select a sample from that population's members — say 1,000 women who married during 1950. By computing the mean age at which women in the sample married, we can *estimate* the population mean. In other words, we can make an inference about the population from sample observations. Using inferential statistics, however, we can never be absolutely certain that our estimate represents the true population mean. Many different samples of 1,000 women could be selected from that population, and each sample could have a different mean. How are we to decide whether or not the estimate is close to the real population mean? How much confidence should we place in the estimate? The special value of inferential statistics is that they let us make a probability statement about the likelihood that our inferences are in error. With inferential methods, we can make inferences *and* say something about the precision of those inferences.

The error problem also appears when inferential statistics are used to evaluate controlled experiments. Suppose that for a period of one month, one group of laboratory rats receives a food additive suspected of being a cancer-causing agent mixed with its regular food. A similar group of rats receives only its normal food allotment for the same month. Then, it is determined that the animals receiving the food additive have a higher incidence of cancerous tumors than do animals in the "food only" group. Did the food additive cause the difference between groups? Or, is it likely that other differences between animals in the two groups produced the different cancer rates? The experimenter must decide whether the food additive or "other" factors produced the difference, and inferential methods are designed to help make that kind of decision. Still, there is the possibility that different groups of animals would produce different results, and that the decision about cause is incorrect. Again, inferential methods allow the experimenter to state precisely the chances that his or her decisions are in error. The chance of error cannot be eliminated from inferential statistics, but it can be controlled and quantified.

SUMMARY

A statistic is a number used to summarize, evaluate, or analyze a body of information. The study of statistics concerns the numerical processes that produce these numbers as well as other important factors that influence their interpretation. In order to interpret a statistic properly, you must know something about the original data that produced it, characteristics of the individual statistic, and whether the statistic is presented appropriately.

Applied statistics can be divided into two general categories with different functions. Descriptive statistics are used to describe and summarize data clearly and precisely; they apply only to the individuals or objects actually observed. Sometimes we wish to draw conclusions about large groups (populations), even though we can only make observations on some of the group members (a sample); drawing such conclusions requires us to use inferential statistics.

NEW IMPROVED FORMULA 84

shown to be effective against cholera!

RESULTS OF MEDICAL STUDY ANNOUNCED

★ **An independent research laboratory** has released its preliminary report of a three-week study conducted by a leading medical authority and staff concerning the effects of new improved FORMULA 84 mouthwash.

★ **A number of people from different backgrounds** used new improved FORMULA 84 an average of 3.75 times daily during the period of the study. Their median age was 29.7.

★ **At the conclusion of the study** the available surviving participants were examined by the leading medical authority and his staff. Fifty percent of the examinations took place at a major midwestern hospital under controlled atmospheric conditions.

★ **The leading medical authority,** in his preliminary report, announced: "Not one of the examined subjects now displays symptoms of cholera. In my opinion, the daily use of new improved FORMULA 84 mouthwash is indispensable for good health."

KEY CONCEPTS

statistic   axis   sample

arithmetic average   descriptive statistics   population

median   inferential statistics

PROBLEMS

1. Even though you have not yet been formally introduced to specific statistics, you should be able to use the ideas discussed in this introductory chapter to evaluate some of the everyday statistics that turn up in newspapers, magazines, and radio and television broadcasts. Find an example of a statistic that is

   a. Used in an advertisement to show that a product is popular.
   b. Used in an advertisement to show that a product is effective.
   c. Presented by a professional polling organization.
   d. Used by a scientist to report the results of research studies.
   e. Used by a political or governmental figure in an attempt to gather public support or change public opinion.

   To the best of your ability, state whether each of the above includes enough information to reveal the original source of the data, whether it is the proper statistic for the intended message, and whether it is presented clearly or not.

2. You may have seen advertisements like the one on page 12. It describes the results of a medical "study" and includes some statistical information. Indicate whether or not you find the ad persuasive, and identify the specific items of information that made you think so. Even though you have not yet learned how to evaluate experimental or clinical studies with statistics, you should be able to comment on the information this ad gives about the source of data, about the appropriateness of the statistics cited, and about the soundness of the medical specialist's conclusion.

1 Descriptive Statistics

**Methods for describing or
summarizing information in a
manner that is clear and precise**

1. Data: The Raw Materials of Statistical Analysis

After reading Chapter 1 you should be able to do the following:

1. *Define the terms variable, observation, and measurement as they apply to the statistical process.*
2. *Specify some possible sources of nonobjectivity in data.*
3. *Determine the level of measurement involved,*
given a description of the variable under consideration.
4. *Determine whether a variable is continuous or discrete.*

Observations and Variables

Statistical methods begin with numbers and end with numbers. A group of numbers is manipulated to produce a single statistic, itself a number. Our primary goal is to use and interpret the numerical end products of statistical analyses and to understand some ways in which numbers convey information about groups and individuals. But first we must carefully consider the numerical raw material that goes into these analyses. This may seem obvious, but there is a great temptation when doing research or reading reports to pay more attention to the statistics themselves than to the nature of the data that produced them. Sometimes we are so anxious to obtain the statistical results of a study or experiment that we forget that statistics cannot be interpreted properly if derived from inappropriate data. So this chapter will focus on a subject you should consider every time you use or respond to statistics: the original source of the data.[1]

The term "statistic," in fact, originally meant little more than a collection of data. John Graunt (1620–1674) is often recognized as the first statistician, because he compiled a thorough count of births, deaths,

---

[1]*Data* is the plural of *datum*; a datum is a single fact or item of information.

diseases, and other characteristics of London residents in 1661–1662. This type of work began to appear with regularity in England and France during the late seventeenth century. "Statistics" was derived from the word "state," in that such compilations were of primary interest to governments. Today we use the term in a similar manner when we refer to data as statistics — such as the "vital statistics" we give the census taker, or the statistics from last night's baseball game. More often, however, statistics refers to analytical methods applied to those data; thus, we begin our study of statistics by distinguishing clearly between the raw material and the end products of the science.

Any collection of information about a particular subject may be called data, but for a group of data to be usable in statistical analysis, it must meet two conditions. First, it must represent systematically made observations on one or more variables, and second, it must be reducible to numerical form. The process of assigning numbers to observations is called **measurement,** and we will cover that topic later in this chapter.

To the scientist, observation means more than casually watching someone or something. A scientific **observation** is the act of carefully and objectively recording an item of information according to a specified rule or procedure. A chemist who measures and records the specific gravity of a substance is making an observation; so is a psychiatrist who records the verbal responses of a patient to an inkblot pattern. The important point is that observations are made according to a specified procedure and that they are made in the same way each time.

In the behavioral sciences there are many characteristics of people or groups that could be chosen for observation. If the characteristic can *change* in value or kind from observation to observation, it is called a **variable.** Variables associated with people include physical characteristics, such as height, weight, hat size, and so on. If these characteristics are measured for a number of people, they can change in value from observation to observation and are thus variables. Sometimes the value of the variable will not change from observation to observation; we might find two people with the same measured weight, for instance. Nonetheless, weight is still a variable because it is a characteristic that *can* change from observation to observation. Performances on various tasks can also be variables, such as scores on IQ tests or the amount of time required for different people to run a mile. These, too, can change in value from observation to observation.

Amplifying a statement made in the Introduction, we can say that statistical methods are used to summarize, analyze, and evaluate systematically made observations. Each observation is the recorded value of some variable, and collectively these observations form the raw material for statistical analysis.

Every statistical user and consumer must be aware of the fact that all data are not created equal. Data may differ drastically in quality and thus in the quality of statistics they produce. Statistical methods themselves do not automatically reject low-quality or inappropriate data; that is the responsibility of the statistician. Computations can be performed on any group of numbers. The resulting statistic will not, by itself, indicate whether the data it was based on is of high enough quality to allow meaningful interpretation. Unfortunately, statisticians are often negligent in this regard, and statistical consumers are frequently unaware of the importance of using data of the appropriate quality for the statistic at hand. Understanding the basic factors that determine data quality will help both statistician and consumer avoid misinterpretation and misuse of statistical methods. The quality factors we will consider are objectivity and level of measurement.

Objectivity in Measurement

Scientific observations should be made as objectively as possible. That is, they should be recorded impersonally, accurately, and in the same way every time. This is an easily stated goal that is sometimes quite difficult to achieve.

Take the matter of observing a car's speedometer reading, for example. In most cars speed is indicated by the location of a needle that is free to swing across a numbered scale; speed is taken as the scale number directly behind the needle at a given moment. Speedometer readings taken from the driver's seat can differ by 5 or 10 mph from speedometer readings taken from the front passenger's seat, since a person's view of which number is behind the needle depends on the angle from which he sees the speedometer.[2] Thus, in order to obtain objective observations, a series of speedometer readings must all be taken from the same location.

The speedometer problem is analogous to a problem that occurs frequently in the behavioral sciences, in which the observations consist of people's scores on paper-and-pencil tests. So-called objective tests are those which, when completed, yield exactly the same score regardless of who grades them. Multiple-choice, true-false, and (sometimes) fill-in-the-blank exams are objective, in the sense that there are specified correct answers for every question. Correctness of the responses can be determined by anyone who has the answer key. Essay exams, however, are considered subjective exams, because a person's score usually depends somewhat on the individual grader's opinion of both the question and the written response to it. Different graders are likely to have different criteria for determining correctness and will usually differ in the scores they assign to any given answer. For observations consisting of subjective test scores to be as consistent as possible, they should all be made by the same

---

[2]Errors in meter readings produced by viewing meters from different angles are known as parallax errors. Manufacturers of high-quality precision electrical meters try to eliminate parallax errors by placing a mirror directly under the needle. The viewer can then ensure that his head is in exactly the same location for all readings by maneuvering his head so that the mirror image of the needle is directly under the needle.

grader, for the same reason that all speedometer readings should be taken from the same position in the car.[3]

Nonobjectivity can also enter the observation process when the act of measurement itself influences the nature of the observation. Electrical engineers know, for example, that when a meter is used to measure voltage or amperage in a circuit, the meter itself becomes part of the circuit and thus alters the very characteristics that are being measured. Similarly, anthropologists are well aware that their presence in a different culture can alter the behavior of the people they are trying to observe. Developmental psychologists who observe play behavior of three-year-olds often try to do so from behind one-way mirrors, in order to remove the influence of the measuring device (the psychologists) from the observation. Pollsters and interviewers try to approach all of their human subjects in exactly the same way, since even the facial expressions and tones of voice displayed by observers can influence the responses they obtain.

Since truly objective observation is so difficult to attain, particularly in the behavioral sciences, a conscientious researcher will report the conditions under which his observations were made. That way, the consumer can decide for himself whether the data were obtained objectively or not. In fact, if the report is to appear in a professional research journal, it *must* carefully describe these conditions, or else the article will not be accepted for publication.

Level of Measurement

Many people are surprised to find that a study of basic statistics usually includes a brief survey of measurement and measurement theory. After all, most of us have been measuring things all our lives and are already familiar with measuring devices such as rulers, stopwatches, and speedometers. Is it really necessary for a student in the behavioral sciences to study measurement in order to use or understand statistics?

In the natural sciences one rarely needs to give much thought to the *theory* of measurement. Physicists, for example, measure variables such as time, displacement, and electrical potential. Although it may be technically difficult in some cases to obtain these measurements, the principle involved is simple: The measured object is compared directly with a standard scale or calibrated meter. The physicist can analyze his observations with little consideration of measurement theory because numerical values reported as measurements have magnitudes and properties that correspond closely to magnitudes and properties of the real variable. The relation between lengths of a two-centimeter rod and a one-

---

[3]Teachers who must use several assistants to grade large numbers of essay exams can still ensure that all tests are graded according to the same criteria — if the test consists of more than one question. Each grader evaluates the answer to one (or more) questions on all exams. This procedure is more objective than giving each grader a certain number of complete exams to score.

centimeter rod is described very accurately by the relation between numerical values one and two.

Measurement in the behavioral sciences, however, is often more uncertain. Psychologists, for example, sometimes assign numerical measurement values to observations on such variables as attitude strength, personality traits, and general intelligence. Quantities of these variables cannot be determined with the same certainty that is associated with measurement of physical variables. In such cases, we must distinguish between the underlying variable itself and the performance (or behavioral) variable that actually enters our statistical analysis.

As a psychologist, for instance, I might assume that you have an attitude with a particular strength toward the following statement: "Nuclear power is a safe, economical source of energy." An attitude is something that we can never observe directly, however. We infer its presence from observable behavior; we cannot put a ruler inside the head and measure it, or weigh it, or in any other way directly examine it. Thus, an attitude is an **underlying variable;** we assume its existence because doing so helps explain some observable behavior, but we will never be able to measure it directly. In order to assess your attitude, I could read you the above statement about nuclear power and then ask you to indicate your own degree of agreement or disagreement by placing a check mark on a scale such as the following:

STRONGLY AGREE	MODERATELY AGREE	NO OPINION	MODERATELY DISAGREE	STRONGLY DISAGREE
1	2	3	4	5

(This is known as a *Likert scale*.) Suppose you check category 2, moderately agree, and a friend checks category 4, moderately disagree. Responses on such a scale represent observations on a **performance variable** (or, equivalently, a **behavioral variable**); here, we can observe directly which category you and your friend each choose.

What can we conclude about attitude strength on the basis of these check marks? We know that the number 4 has twice the value of the number 2, but we do not know whether your friend actually feels twice as strongly as you do, because we do not know very much about the relationship between people's attitude strengths (the underlying variable) and the check marks they make on such a scale (the performance variable). We do not know how much attitude strength it takes to produce a check mark at a certain point on the scale. As we will see in this and the chapters that follow, there are some things we can say about the underlying variable and some things we cannot say about it from such data. In order to understand which statements about the underlying variable are appropriate, we must understand the basic concepts involved in assessing level of measurement.

Essentially, **level of measurement** indicates the degree to which numbers assigned to observations correspond with characteristics of the variable we are really interested in. With length measurements of 4 feet and 2 feet, the numbers 4 and 2 are related to each other in the same way that the properties they represent are related: one measurement (or length) is twice the other. With the Likert-scale example above, however, the number 4 is still twice the number 2, but we cannot assume that the respective values they represent on the underlying variable stand in a similar two-to-one relationship to each other. For these reasons, measurements of length and Likert-scale measurements of underlying attitude strength represent different levels of measurement. It is important to determine the level of measurement whenever statistics are used because statistics appropriate for some levels are not appropriate for others. After reading this chapter, you, too, should be able to make this evaluation and to classify any system of observation at one of the five following levels: nominal-, ordinal-, interval-, ratio-, or absolute-level measurement.[4]

Before looking at the differences among levels of measurement, however, we should note that data from all levels of measurement can be represented by numbers, and that it is not possible to determine the level of measurement simply by looking at the data. During a football game, for example, we can measure the number of yards a certain running back gains each time he carries the ball; these measurements will form a list of numbers. We can "measure" the telephone number of everyone who happens to attend a certain party, and these observations will also produce a list of numbers. A doctor can measure the drug dosage given to a patient during a number of treatments, or a teacher can record the IQ scores of class members. As we have just seen, a pollster can record Likert-scale responses to attitude statements. All this information represents measurement, and all can be presented in numerical form; but the systems of measurement used in these examples differ in level, and statistics that can be applied to some of these data are not appropriate for other data. An average[5] dosage level has meaning, an average telephone number has no meaning, and an average Likert response may or may not have meaning, depending on what the statistician chooses to say about it.

It should be mentioned that there is some debate among behavioral scientists on the matter of how much attention should be paid to the level-of-measurement issue.[6] Among psychologists, for instance, those

---

[4]This is an approach to measurement theory developed by S. S. Stevens. See, for example, Stevens, "Mathematics, Measurement and Psychophysics," in *Handbook of Experimental Psychology* (New York: Wiley, 1951).

[5]Average, as we will see in Chapter 3, can have several meanings. Here, consider it to mean the arithmetic average that we used in the Introduction examples (add up the measurements and divide by the number of measurements).

[6]S. S. Stevens has argued that statistics cannot have meaning for real variables unless the data represent a level of measurement appropriate for the statistics used. See, for example, his article "Measurement, Statistics, and the Schemapiric View," *Science* 161 (1968): 849–856.

Other (but by no means all) behavioral scientists contend that most statistics can be

who focus only on observable behavior (i.e., behaviorists) have little need to consider this subject because they restrict their analyses to behavioral variables; in such cases, behavioral variables represent the highest level of measurement because properties of numerical observations correspond to properties of the variables under study. In contrast, cognitive psychologists who make observations on behavioral variables and then make statements about underlying variables such as intelligence, attitude strength, attention, interpersonal attraction, and so on, should, in my opinion, always consider this subject before drawing conclusions about underlying variables.

In summary, then, level of measurement indicates the degree to which numbers (or names) assigned to observations correspond to real (but sometimes unobservable) characteristics of the variable under study. These characteristics will be considered for each level of measurement.

Nominal-Level Measurement

Nominal-level measurement[7] is the lowest level of measurement and can be defined quite simply: A nominal-level measurement scale is a system of assigning labels to observations so that each observation receives one label or another, but never more than one.

Walking through a parking lot and writing down the make of each car is nominal-level measurement, for example. The variable of interest, car make, can be measured as "Buick," "Ford," "Mercedes," and so on. Similarly, the numbers on football players' jerseys represent nominal measurement, since the numbers serve (only) to identify different players, and each player can have only one number, or label.

Nominal-level measurements may be turned into numbers suitable for statistical analysis by making frequency counts of the number of times each observation category appears in a group of observations. We might record, for instance, that the parking lot holds 13 Buicks, 12 Fords, and 2 Mercedes. More on this process is covered in the section on Absolute-Scale Measurement.

Nominal-level measurement is so simple you may wonder why we even identify it as measurement. One reason is that many kinds of observations in the behavioral sciences can be classified as being at no higher a level than nominal. Identifying users of the library as either "male" or "fe-

---

applied to numerical data representing all levels of measurement, if one is careful to interpret them properly. Proponents of this view include E. W. Adams, R. F. Fagot, and R. E. Robinson ("A Theory of Appropriate Statistics," *Psychometrika* 30 [1965]: 99–127) and B. O. Baker, C. D. Hardyck, and L. F. Petrinovich ("Weak Measurements vs. Strong Statistics: An Empirical Critique of S. S. Stevens' Proscriptions on Statistics," *Educational and Psychological Measurement* 26 [1966]: 291–309).

The debate on this issue is a subtle one, but interested students should be able to follow the essence of the argument through the articles cited above. Both sides of this issue are also examined in P. A. Games and G. R. Klare, *Elementary Statistics: Data Analysis for the Behavioral Sciences* (New York: McGraw-Hill, 1967).

[7]Sometimes called categorical measurement.

male'' is nominal-level measurement, for instance; classifying a patient as either schizophrenic, neurotic, or retarded is also nominal-level measurement. There are many useful methods for dealing with nominal data.

Ordinal-Level Measurement

Nominal scales place observations into distinct classes, but there is no rank or order among classes in this lowest level of measurement. The next higher level of measurement, **ordinal-level measurement,** assigns labels that can be placed in an order. If we observe three men and describe them as "tallest," "shortest," and "in between," we have performed ordinal-level measurement on the height variable, since the three classes (labels) can be rank-ordered. Similarly, we could visit a military base and "measure" the rank of every soldier we meet. Observations like "private," "colonel," and "captain" are also ordinal measurement, since they can be placed in order ranging from high to low. Thus, ordinal scales have all the properties of the nominal level, plus a specified order among measurement classes.[8]

Although rare in the natural sciences, ordinal-level measurement is quite common in the behavioral sciences. Social psychologists cannot quantitatively measure the absolute strength of attitudes, but they can ask people to arrange a list of attitude statements in order, ranging from the statement with which they least agree to the statement with which they most agree; here the categories (most agree, least agree, and intermediate ratings) can be placed in order. Similarly, classifying children's intellectual developments according to Piaget's well-known developmental stages is also ordinal-level measurement, since Piaget's stages have a prescribed order.

Groups of such rank-ordered lists are subject to many forms of statistical analysis that cannot be applied to nominal-scale data. In later chapters, for example, we will discuss the median and the interquartile range, two statistics that have meaning for ordinal-scale data. Also, many of the nonparametric statistics covered in Chapter 13 were developed specifically to deal with ordinal-level data.

When numbers are used as nominal- or ordinal-level *category names*, those numerical values should not enter the calculations as statements of quantity; with ordinal-level data, numbers simply represent rank order

---

[8]Strictly speaking, a group of category labels qualifies as ordinal-level measurement if the categories can be placed in rank order *and* if the categories exhibit the property of *transitivity*.

Suppose that all observations are described as being X, Y, or Z. If X is greater than Y, and if Y is greater than Z, this group exhibits transitivity if X is greater than Z.

A situation in which rank order exists, but where transitivity may not exist, occurs when we informally predict outcomes of basketball games. In a given season, suppose we know that Purdue will always beat Notre Dame, and suppose also that Notre Dame will always beat UCLA. Does it necessarily follow that Purdue will always beat UCLA? If the answer is no, we cannot assign ordinal-level designations like first-, second-, and third-best team on the basis of individual rank orders, since transitivity does not hold.

among observation categories and not "amounts" of the variable. Thus, statistics computed from ordinal-level data also refer to rank order among measurement categories, and not to amounts of the variable.

Ordinal-level measurement is still not appropriate for many higher statistics for a very important reason: Differences between adjacent classes in the rank order are not necessarily equal. We cannot assume, for example, that the difference in height between "tallest" and "in between" is the same as the difference between "in between" and "shortest." This deficiency precludes the use of statistics like the arithmetic average with ordinal data, even though rank-ordered observations may be labeled with numbers.

Interval-Level Measurement

If the measurement scale meets the requirements of ordinal measurement (distinct classes, order among classes) and has equal differences between adjacent measurement classes, it is at least **interval-level measurement.**

A classic example of interval-level measurement is the Fahrenheit temperature scale. It clearly meets the requirements for the nominal and ordinal levels and, in addition, has equal intervals between measurement units on the variable "temperature." This means that the difference on the underlying variable (degree of heat) between 30° and 31° is the same as the heat difference between 90° and 91°, and the difference between 25° and 30° is the same as the difference between 115° and 120°.

With interval-scale data, differences between measurement classes have meaning, and these differences can enter into calculations of statistics like the arithmetic average. Suppose, for example, that the noon temperatures on three successive days in Chicago were 50°, 70°, and 45°. The average noon temperature for these days is 55°. If, in Miami, the noon temperatures for the same three days were 75°, 80°, and 55°, we can compute an average for these observations (70°) and compare it with the Chicago average. If, however, we had measured each city's temperatures with an ordinal scale composed of "hottest," "medium," and "coolest" measurement classes, such a comparison between averages would tell us nothing — since "medium" was the average in both cities.

Interval scales like the Fahrenheit temperature scale do not, however, represent the highest level of measurement. They are limited by the absence of an absolute zero point. True, the Fahrenheit scale has a point that is labeled 0°, but that point does not represent the complete absence of temperature. It is possible to have temperature readings colder than 0°F. On an interval scale, a measurement of zero does not indicate zero quantity of the variable.[9]

---

[9]The Celsius temperature scale (sometimes called the centigrade scale) also represents interval-level measurement only, since 0.0°C represents the temperature at which water freezes, not absolute zero.

The Kelvin temperature scale has a zero point that means absolute zero, and it thus represents ratio-level measurement. For this reason, many calculations in chemistry and

Interval-scale data are suitable for many forms of analysis, but the absence of a zero point eliminates the possibility of using interval data in computations involving *ratios* of measurements. Consider, for instance, the error we would be making if we said that a temperature of 100.0°F is twice as hot as a temperature of 50.0°F. Here we would have inappropriately made a ratio of the measurements by using the word "twice." That is, we would have said that because $100.0/50.0 = 2$, 100.0° is twice as hot as 50.0°. The ratio is invalid because 0°F does not represent zero temperature — the complete absence of temperature occurs at 459.6°F below 0°F. So a reading of 50.0°F is really 509.6°F of temperature (50 + 459.6), and a reading of 100°F really represents 559.6°F above absolute zero. 559.6/509.6 does not represent a ratio of 2:1, or twice. In order to make ratio statements about measurements, we must use ratio-level measurement.

Ratio-Level Measurement

The highest level of measurement is called **ratio-level measurement,** because ratios of these measurements have meaning for the underlying variables they represent. In order to qualify as ratio-scale measurement, a measurement system must meet all the requirements of interval scales (distinct classes, order among classes, equal intervals between classes), and in addition have a zero point that represents absolute zero on the underlying variable.

Measurements of most physical variables qualify as ratio-scale data: length, weight, time, voltage, pressure, and velocity are some variables measured on ratio scales. A zero measurement of any of these truly means zero quantity of the variable, and we can therefore compare the magnitudes of these quantities in ratio statements: 2,000 lb is twice as heavy as 1,000 lb. The time interval 2 hr is twice as long as 1 hr.

Ratio-level measurement does appear in behavioral science research, although it is not so common there as in the physical sciences. Human reaction times, the amount of time required to finish a problem-solving task, the distance two people place between themselves when conversing, and changes in galvanic skin response (GSR) that occur when telling lies are just a few of the many ratio-level measurement scales employed in psychological and social sciences.

Notice that it may not actually be possible in some cases to observe zero values of the variable. The condition of absolute zero air pressure, for example, can be approached but not reached in the laboratory. Nonetheless, the measurement scale (air pressure) has a zero point and is therefore ratio-level measurement. The level of measurement is determined by the nature of the measurement scale, not by the values of observations actually obtained.

---

physics require temperatures to be specified in degrees Kelvin rather than degrees Celsius or degrees Fahrenheit.

Absolute-Scale Measurement

Nominal-, ordinal-, interval-, and ratio-level scales are commonly considered as the only levels of measurement. Data from any of these scales, however, may be converted into measurements on another very useful kind of measurement scale, the **absolute scale.**

Under the first four systems of measurement mentioned, we record either the name of an event or the value of some measurable quantity. If, instead, we record the number of times some event occurs, we have performed absolute-scale measurement. The variable of interest might be the number of times a rat presses a lever in a Skinner box during a given time period, the number of people who vote for Senator Longspeech in a given election, or the number of times the New York Mets win the World Series in a decade. We are also using absolute-scale measurement when the variable observed is the number of objects (or people) counted. Observing that there are twenty-seven Fords in the parking lot, or that there are seven men and five women on a jury, or that a family has four children, is absolute-scale measurement.

It is very important to recognize the way in which observations from the other measurement scales are turned into absolute-scale measurements, since absolute-scale data are appropriate for many useful kinds of statistical analysis that cannot be used with some other levels of measurement. If you record the fact that a person is female, you have performed nominal-level measurement. If you report that a group includes ninety-four females, you have performed absolute-scale measurement.

The reasons that military ranks such as private and captain are ordinal-level measurement categories are discussed on page 24. The armed forces also designate each rank with an ordinal number, so that an army buck private is an E-1, a private first-class is an E-2, and a corporal is an E-3 (E stands for enlisted). Now suppose that a squad consists of one E-3, three E-2s, and four E-1s. Would it be permissible to compute the arithmetic average of the ranks in the following way?

$$\frac{3 + 2 + 2 + 2 + 1 + 1 + 1 + 1}{8} = \frac{13}{8} = 1.625$$

Could we say that the average rank is E-1.625? The answer is no, because the rank designations are ordinal-level measurement, and ordinal data are not appropriate raw material for computing the arithmetic average.[10] It is a different matter, however, when we compute the arithmetic average of the number of people *in* each rank in the following way:

---

[10]Strictly speaking, the average would not be appropriate for the purpose specified here: determining the average rank. However, averages of nominal- and ordinal-scale numbers may be computed to ask certain kinds of questions such as, What is the probability that the observed distribution of numbers occurred by chance? While such uses are beyond the scope of this chapter, you may wish to read more on the matter in F. M. Lord's entertaining article "On the Statistical Treatment of Football Numbers," *American Psychologist* 8 (1953): 750–751.

$$\frac{1 + 3 + 4}{3} = \frac{8}{3} = \text{an average 2.667 soldiers per rank}$$

This average is appropriate because the absolute-scale measurements *can* be used to compute an arithmetic average. In fact, absolute-level measurements can be used wherever ratio-level measurements can.

Consider another situation where a psychiatric hospital classifies patients on one ward as being either "catatonic schizophrenic," "hebephrenic schizophrenic," or "paranoid schizophrenic." If the classification system assigns one and only one of these labels to each patient, the classification represents nominal-level measurement "type of schizophrenia." However, suppose that we learn there are 15 people classified as "catatonic," 12 classified as "hebephrenic," and 6 classified as "paranoid." We can now refer to a different variable that represents absolute-level measurement: the number of people in each category. For the first variable, the measurement categories observed were "catatonic," "hebephrenic," and "paranoid"; for the second variable, the measurement categories observed were "15," "12," and "6."

Determining the Level of Measurement

Perhaps the best way to study differences among measurement levels is to practice evaluating several scales. Above all, remember that the level of measurement depends on the variable you intend to talk about when you analyze the data and what your measurement system tells you about that variable. The procedure that follows is straightforward. First, determine whether or not the scale (your measurement system) meets the criteria for nominal-level measurement. If it does, then evaluate the scale with respect to ordinal-level criteria. Continue this process as far as possible. If, for example, a measurement scale meets the requirements for the interval level but not for the ratio level, it is an interval scale. The criteria to be used at each level are summarized in Table 1.1.

Consider first a simple example in which the measurement concepts are easy to follow. A men's track coach must select athletes to form a relay team. Suppose there are six runners available, only four of whom can make the team. The coach therefore decides to measure their individual abilities by holding a try-out race in which all six men run. At the end of

Table 1.1 *Properties of Measurement Scales*

PROPERTY OF SCALE	LEVEL OF MEASUREMENT			
	NOMINAL	ORDINAL	INTERVAL	RATIO
Discrete classes	X	X	X	X
Order among classes		X	X	X
Equal intervals			X	X
An absolute zero				X

the race each man's position is recorded. Assuming there are no ties, one man will be measured as first, another as fourth, and so on. What level of measurement is involved here?

The answer depends on what variable the coach is really interested in and what his data tell him about that variable. Suppose the crucial variable is "running speed"; the performance variable that he has measured in this case is "position at finish." This scale clearly meets the requirements for nominal-scale measurement: distinct classes. Each man receives a measurement-class label (his position) and can receive only one such label. Therefore, we can check the scale to see if it meets the ordinal-level requirement of order among classes. If we are interested in running speed, we now have to ask what observations on the performance variable position tell us about speed. Can we rank order the runners on the speed variable if we know their positions? Again, the answer is a clear yes. The first-place runner is faster than the second-place runner, and so on. The next step is to check the interval-level criterion: equal intervals between classes. As long as we are interested in running speed, the positional data do not pass the interval-level test since we cannot assume that there are equal intervals between the position categories. The first- and second-place runners may be of almost exactly the same speed, or they may differ drastically in speed. Thus, for the purposes indicated, the scale represents ordinal-level data. If it fails the test at the interval level, there is no need to check at the ratio level. (See Figure 1.1.)

Now, suppose that the coach records each man's time instead of his position. Each observation indicates the number of seconds one man takes to run the quarter mile. Two possible observations might be "man 1,

Figure 1.1   Ordinal-level measurement is involved in a race, where only the *order* of finish is recorded. If we are interested in running speed, then recording each runner's position at the finish will tell us something only about *order* among running speeds, not about the *degree* of differences among runners.

58 sec," "man 2, 56.4 sec." What level of measurement is involved in this case?

The variable of interest in this case is "running time," and we can tell a great deal about each runner's time from the stopwatch readings. The measurement system easily meets the requirements of nominal-scale measurement because each man receives one and only one time reading; equal times are considered equal, different times are considered different. Next we check the ordinal-level criterion and determine if the measurements can be rank-ordered. The time measurements can, of course, be placed in order from high to low, so we advance to the interval-level criterion. Since all seconds are considered equal in length, the measurement scale also qualifies as interval-scale measurement. Finally, we look for a zero point on the measurement scale that would represent an absolute zero quantity of the variable. The time scale has such a zero point, and we thus know that the measurement of time intervals is ratio-scale measurement. (See Figure 1.2.)

Now consider a final example more typical of the problems encountered in the behavioral sciences. Many school children take annual intelligence tests from which they receive IQ scores such as 95 or 115. These scores are currently at the center of several controversies regarding the relation between intelligence and race, intelligence and social class, and other issues. Much of the debate in these areas involves statistics computed from IQ scores. The validity of arguments made with these statistics depends on many factors, including the question of whether or not the original data (IQ scores) represent interval- or ordinal-level data. Determining the level of measurement involved with IQ scores is not, however, an easy matter.

IQ test scores meet the requirements for nominal-scale data; most tests yield a single IQ score for each child. The scores, furthermore, can be ranked from high to low, so the IQ-score scale is at least ordinal-level measurement. We run into a problem, however, when we apply the check for interval-scale measurement. Do we know that equal differences on the performance variable (test-score scale) are associated with equal differences on the underlying variable (intelligence)? Can we assume that the difference in intelligence between two children who score 100 and 101 is the same as the difference in intelligence between two children who score 140 and 141? If the answer to these questions is yes, then it is appropriate to refer to the mean (average)[11] intelligence (underlying variable) of one group being, perhaps, different from the mean intelligence of another group. However, if the answer to these questions is no, then we must speak instead of the mean *score* (performance variable) of one group as being different from the mean *score* of another group. The difference between these two kinds of statements is subtle, but very important.

---

[11]This discussion applies only to the use of the mean, or arithmetic average; if the median is used as the measure of "average," then it *is* appropriate to make statements about the underlying variable because the median can be properly used with ordinal-level data. See Chapter 3 for the differences between mean and median.

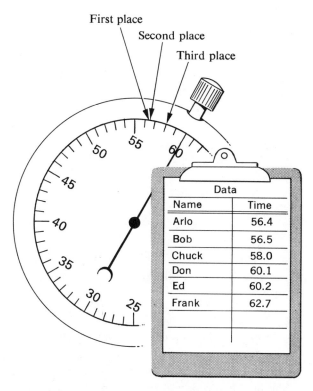

Data	
Name	Time
Arlo	56.4
Bob	56.5
Chuck	58.0
Don	60.1
Ed	60.2
Frank	62.7

Figure 1.2   When time is the variable measured, ratio-level measurement is involved. Here we can tell not only the order of finish but also distances (in time) between those who finished. The difference between first and second place is smaller than the difference between second and third place. Because the time scale has a zero point, it is also proper to say that a run of 90 sec is twice as long (in time) as a run of 45 sec.

Which kind of statement is correct is still a matter of serious disagreement.

At this point we have not yet discussed specific statistics, so it may be difficult for you to appreciate the need for determining level of measurement. The data requirements for each statistic will be discussed when each is introduced, together with some reasons for those requirements. For the present, we can perhaps suggest the importance of measurement theory by simply listing the data requirements for some familiar statistics. Table 1.2 shows that when data represent higher-level measurements, the statistician can make more use of statistics than when lower-level measurement is involved. This table should be interpreted with caution, however. Where an X appears, the statistic may be used in statements both about the underlying variable and about the observed variable. Where no X appears, use of the statistic is limited to statements about the observed performance variable. For instance, assume that the Likert-scale data mentioned earlier (pp. 21–22) represent ordinal-level measurement of attitude strength. If we compute a mean (arithmetic average)

Table 1.2 *Permissible Operations on Data from Various Levels of Measurement. For Interpretation, See the Text.*

	LEVEL OF MEASUREMENT				
STATISTIC	NOMINAL	ORDINAL	INTERVAL	RATIO	ABSOLUTE
Mode	X	X	X	X	X
Median		X	X	X	X
Mean			X	X	X
Range		X	X	X	X
Interquartile range		X	X	X	X
Standard deviation			X	X	X
Pearson *r*			X	X	X
Spearman rho		X	X	X	X

of such responses, we can talk about the mean response (performance variable), but we cannot talk about the mean attitude strength (underlying variable).

As Table 1.2 suggests, a crucial difference between levels of measurement involves the difference between ordinal and interval levels; all statistics appropriate for interval-level data are also appropriate for ratio- and absolute-level data. (This is true for all statistics covered in this book.) Thus, the important question in many cases is simply deciding whether or not data represent interval-level measurement.

Continuity

There is a final factor involved in data collection you should be aware of. It is sometimes useful to consider the **continuity** of the underlying variable, particularly when preparing data for graphical presentation, or in choosing among several possible statistics for a particular task.

When we speak of continuity, we refer to the fact that variables are either *continuous* or *discrete*. A variable is continuous if it does not have a minimal size unit. Length is a good example of a continuous variable, because there is no minimal unit of length. An object may be 12 in. long, it may be 12.5 in. long, or it may be 12.541379523 . . . in. long. Theoretically we can speak of tenths of an inch, ten-thousandths of an inch, or ten-trillionths of an inch. Although we cannot measure all possible different length values with absolute accuracy (some values will be too small for any measuring instrument to register), it *is* possible for objects to exist at an infinite number of different lengths. Length is a continuous variable because possible values of this variable exist in an unbroken continuum between every pair of points on the measurement scale. Select any two points on the measurement scale of length, and it is still theoretically possible to identify an infinite number of other points between them.

Discrete variables, on the other hand, *do* have a minimal size unit. The amount of money in your pocket at this moment is a good example of a discrete variable, because in United States currency the variable (amount of money) has a minimal unit. You can have $1.35 or $1.36, but you cannot have $1.3572.[12] Different amounts of money cannot differ by less than the minimal size unit; all quantities of the variable are multiples of the minimal unit size. Since absolute-scale measurements are frequency counts, all measurements at that level are observations of discrete variables. The number of children per family, for instance, is a discrete variable since the minimal unit is one child. Families may have two children or three children, but not 2.5 children.

Discrete variables may, however, produce statistics representing values that the variable cannot have. We have all heard, for example, that the average American family has 2.3 children, when, in fact, no American family has 2.3 children.

When trying to decide if a variable is discrete or continuous, look for the existence of the minimal unit. If some quantity of the variable cannot be subdivided, the variable is discrete.

It is very important that you consider the underlying variable and not simply how the data look when making the discrete/continuous determination. When measuring devices of any kind — voltmeters, Likert scales, rulers, true/false exams — are used to measure presumably continuous underlying variables, they produce data that look like discrete-variable data, even though the variable is continuous. The measuring device may produce crude data or it may produce very precise data, but it will *look* discrete because every measuring device has a minimal scale unit. This does not mean, however, that the measured variable is necessarily discrete.

Consider, for instance, weighing yourself on the bathroom scales. Readings from those scales are usually reported to the nearest whole pound, even though people's weights do not always come in whole pounds. You would say that you weighed 152 lb on such a scale, even if you actually weighed anywhere between 151.5 and 152.5 lb. In this case, 151.5 and 152.5 are the **true limits** of the number 152, and with a continuous variable such as weight, the number 152 would mean all values between 151.5 and 152.5. In contrast, if a student reported that he had 152 books in his library, the variable (number of books) is discrete, and he would mean only that he has 152 books, not somewhere between 151.5 and 152.5 books. Determining true limits of numbers becomes important when preparing frequency distribution histograms and cumulative graphs (Chapter 2) and when computing values of percentiles and percentile ranks (Appendix 3).

If the measuring instrument is more precise than the bathroom scales, the true limits of a number may represent a smaller range of possible true values. A chemist's precision balance, for instance, may report weights to

---

[12] Sticklers for detail may comment that the *mill* (one-tenth of a cent) is the minimal monetary unit. However, if we consider only the amount you can actually carry with you in cash, the cent is the minimal unit.

the nearest hundredth of a gram, so that individual readings would be 14.57 g, 19.21 g, and so on. In this case, the true limits of each number are found by subtracting or adding half of the smallest unit measured. For the chemist's balance, that would be ±0.005 g, so that the true limits of 14.57 would be 14.565 and 14.575.

C Rounding Off

A final topic that we should consider is the process of rounding off. Rounding off is likely to be necessary in any statistical work, so we should establish some conventions that we will use in all the statistical exercises that follow.

Rounding off is the process of leaving out some of the digits that appear in statistical computations; when reporting the results of a statistical calculation, we leave off digits that interfere with clear communication and add nothing to the interpretation of the statistic. The arithmetic mean of 4, 2, and 2 is 2.66666666 . . . , for instance. At some point in the process, it is usually advisable to *round off* some of the extra digits.

Whenever possible, only the answer should be rounded off. The convention is to include one more decimal place in the answer than appeared in the original data. With data like 10, 19, 13, or 4, for instance, and a statistical answer of 3.68235, the final answer would appear as 3.7. If the data were numbers such as 10.1, 19.5, 13.9, and so on, the answer would be rounded to two decimal places, 3.68. Carrying any additional digits in the answer would be meaningless and misleading, since the original measurements themselves did not conform to that level of accuracy.

When rounding off, look one place to the right of the last digit you'll keep. If the digit beyond the last one to be kept is greater than 5, then raise your last digit by 1; if the digit beyond is less than 5, then leave the last one as is. Thus, rounding to two decimal places:

3.5682 is rounded to 3.57, but 3.5642 is rounded to 3.56
↑——digit beyond last digit——↑
to be kept

What happens when the "digit beyond" is 5? Mathematicians have adopted a convention for that situation, too. When a 5 occupies the critical spot, return to the last digit you're keeping. If it is an even number, leave it as is. If it is an odd number, raise it by 1. Thus:

3.645 is rounded to 3.64, but 3.675 is rounded to 3.68

Using this convention for 5's, about half the time you'll be rounding *up* and half the time simply throwing away the 5. In a series of many such operations the net results of these two processes should cancel each other.

Before reaching the answer, however, you should always carry as many decimal places as you can conveniently handle; rounding errors introduced in the middle of problems multiply their effects on the answer. For instance, a problem might ask you to find the product of 5.16666666 . . . and 8.666666666. . . . If you round these to two decimal places before multiplying, you'll find that:

$$(5.17)(8.67) = 44.8239 = 44.82$$

If the rounding is completed after multiplying:

$$(5.16666666 . . .)(8.666666666 . . .) = 44.777777 . . . = 44.78$$

The magnitude of the error only increases as more steps are included.

Consider the Source

In a well-written scholarly report the author indicates very clearly which parts of the work he considers to be facts and which are his interpretations of those facts. He realizes that other reasonable people are likely to come to different conclusions regarding the validity of the facts and his interpretation of them, so he specifies the sources of his information and the reasons for his interpretations.

When the report contains statistics, the "facts" are the original data, and interpretation of those facts includes both the choice of statistics and the conclusions drawn from them. In this chapter we have considered some means of evaluating the quality of those facts: criteria of objective data and some characteristics of different levels of measurement. When computing statistics from your own data, you should perform these evaluations before choosing statistics to compute; the importance of this step in the statistical process should become clearer as we consider individual statistics and methods. Whether or not you can perform these evaluations on someone else's data depends on whether or not he or she tells you very clearly *what* was observed and *how* the observations were made.

Unfortunately, advertisements, news reports, and friends' arguments don't always qualify as scholarly reports. Quite often we cannot answer the question, Where did the original data come from?

SUMMARY

Data, the raw materials of statistical analysis, consist of a group of systematically made observations on a variable. A scientific observation is the act of carefully and objectively recording an item of information according to a specified rule or procedure. A variable is a characteristic of a person (or object, or group) that can change in value or kind from observation to observation.

Sometimes we become so concerned with the statistical results of a study that we forget that statistics cannot be interpreted properly if derived from inappropriate data. Data differ in quality, and all data are not appropriate for all statistics. The producer and consumer of statistics must be able to evaluate data with respect to two quality factors: objectivity and level of measurement.

Objective data are those that are recorded impersonally, accurately, and in the same way every time an observation is made. In the behavioral sciences it is often difficult to obtain all measurements in the same way every time, or in such a way that the act of making an observation does not affect the nature of the observation itself. This is one reason why experimental reports should include a description of methods used in data collection, so that the reader can evaluate the objectivity of the data. Another reason is so that other researchers trying to replicate the study can use the same methods.

Systems of observation (measurement scales) differ in another quality factor: level. Level of measurement refers to the degree to which numbers assigned to observations correspond to properties of the object or performance observed. The lowest level of measurement, nominal measurement, is obtained when labels are assigned to observations so that each receives one, and only one, label. Ordinal-level measurement, the next higher level, occurs when measurement classes can be placed in a rank order. Interval measurement represents a higher level than ordinal measurement and occurs when the measurement scale has equal distances between its adjacent classes. The highest level of measurement, ratio measurement, is obtained when the measurement system meets all the requirements for interval-level measurement *and* uses a scale with an absolute zero point.

Under the nominal, ordinal, interval, and ratio systems of measurement, the names of events or quantities of some measurable characteristic are recorded. If, instead, the number of times some event occurs, or frequency counts of people or objects, are recorded, the absolute scale of measurement is involved.

It is important to determine the level of measurement before performing statistical analyses, because data from some levels are not appropriate for all statistics.

Another characteristic of data that is sometimes important is continuity. Continuity refers to the fact that variables are either continuous (can have an infinite number of values between any two points on the measurement scale and do not have a minimal size measurement unit) or discrete (have a minimal size measurement unit and can exist only at a finite number of values between specified points on the scale).

KEY CONCEPTS

data	performance (or behavioral) variable	ratio-level measurement
measurement		absolute scale
observation	level of measurement	continuity
variable	ordinal-level measurement	true limits
underlying variable	interval-level measurement	rounding off

SPOTLIGHT 1   **Garbage In, Garbage Out**

Several years ago *Time* magazine took a hard look at the American fascination with statistics — and the doubtful quality of some of those statistics. In an editorial essay entitled "The Science and Snares of Statistics," *Time* suggested:

This dedication to numbers has created its own pitfalls for the innocent — and opportunities for the purveyors. There is an air of certainty about the decimal point or the fractionalized percentage — even in areas where the measurement is statistically absurd or the data basically unknowable. A classic example is a survey made some years ago, which solemnly reported that 33⅓% of all the coeds at Johns Hopkins University had married faculty members. True enough. Johns Hopkins had only three women students at the time, and one of them married a faculty member. The American Medical Association announces not that very few people dream in color, but that "only 5% of Americans" dream in color. New York City has 8,000,000 rats. How does anybody know? Statisticians have a phrase for this, borrowed from the computer industry on which they now rely. The phrase is "garbage in, garbage out" — meaning that the result that comes out is only as good as the material that is fed in.

For the sake of drama or publicity, numbers are slapped on nearly everything — and the bigger the number the better. . . . Newsmen during the Detroit race riots pressed a harried fire chief for damage estimates. His guess: $500 million. So far, in the cooling aftermath of riot, insurance companies are processing only $84 million worth of damage claims, and the overall loss is now put at $144 million. For newsmen, the National Safety Council issues forecasts of expected highway deaths over holiday weekends usually with a prediction tacked on of "record fatalities." What the forecast never says is that the record is due to population increases and wider use of automobiles, and that the fatality rate is usually just as high proportionately on other weekends — holiday weekends are just a bit longer. . . .

Since it is obviously impractical to poll the nation on anything less important than the selection of a President, one cherished statistical tool is the sample. Not even statisticians can agree on how big or good a sample can be relied upon as representing the whole. Dr. Alfred C. Kinsey's celebrated reports were criticized by statisticians not so much for their moral implications but because they made sweeping presumptions on the basis of too small a sample (in the male study, only 5,300 men provided data). The Nielsen ratings, by which television programs live or die, have been justly attacked because Nielsen recorders are necessarily hooked to the sets of those viewers willing to have a recorder — a special class by definition, whose tastes may or may not correspond with those of the unpolled millions of the total TV audience.

Still the state of the art of statistics has come a long way since 1661, when its founding father, London Haberdasher John Graunt, began a careful count and found that more boy babies died in infancy than girls, and concluded that therefore there must be more women than men in Britain. Today's scientists, who no longer believe that anything is absolutely certain, also believe that many things are predictably probable. And it is the computer, fed with vast amounts of past data, that can project or at least outline the alternatives of several possible futures. "The computer has enshrined statistics," says M.I.T.'s Professor Harold A. Freeman. "Without it, statistics would still be a grubby business." Where once all they had to do was count, and perhaps draw graphs, statisticians are now "programmers," with a mystique all their own. Unquestionably, for the moment, numbers are king. But perhaps the time has come for society to be less numerically conscious and therefore less willing to be ruled by statistics.[13]

Since it doesn't seem likely that society will become "less numerically conscious," at least in the near future, a better suggestion might be that we try to become more statistically sophisticated, more familiar with proper usage of statistics and more aware of their limitations. We might move in this direction by including statistics in the standard high school — or even elementary school — curriculum. Maybe the time has come for Reading, 'Riting, and Regression analysis.

PROBLEMS

1. State the level of measurement most likely involved when one makes observations on the following variables:

   a. Human reaction times, measured before and after drinking an alcoholic cocktail.

   b. A student's first, second, and third choice of living quarters among campus residence halls.

   c. The number of shares of stock sold on the New York Stock Exchange on a certain day.

   d. Classification of psychiatric-hospital patients as either "neurotic" or "psychotic."

   e. Recording the amount of light necessary to just detect a small visual stimulus in a dark room (luminance, the variable measured, may be thought of here simply as physical energy).

   f. A hospital patient's temperature.

   g. A racing-car driver's position at finish in each of eight Grand Prix races.

   h. A racing-car driver's maximum speed in each of eight Grand Prix races.

---

[13]Condensed from "The Science and Snares of Statistics," *Time* (September 8, 1967): 29. Reprinted by permission from *Time*, The Weekly Newsmagazine. Copyright Time Inc., 1967.

2. What level of measurement is involved if
   a. A researcher records the number of times a laboratory rat depresses a bar in its experimental chamber during a one-hour test session.
   b. A psychiatrist has patients give numerical ratings of their own anxiety levels, on a scale where 0 represents "no anxiety" and 100 represents "extreme anxiety."
   c. A medical technician estimates the number of red blood cells in a laboratory sample.

3. Indicate whether each of the following variables is discrete or continuous. If discrete, give the smallest unit by which the variable may change value.
   a. Human reaction times.
   b. The weekly incomes of individual families in Chicago.
   c. The number of times a laboratory rat presses a lever in an experimental chamber (Skinner box) during a one-hour test session.
   d. The amount of water consumed daily by a laboratory pigeon.
   e. A hospital patient's temperature.

4. Indicate whether the following measurements represent discrete or continuous measurement. If discrete, give the smallest unit by which the variable may change value. If continuous, give the true limits of the reported observation.
   a. It is determined that a small section of a human retina (light sensitive tissue at the back of the eyeball) contains 12,142 receptor cells.
   b. The noon temperature today was 10° Celsius.
   c. A runner finishes third in a race having ten entrants.
   d. A shade tree is measured as 52 feet, 1 inch tall.

5. What variable (or variables) might you use as a basis for your observations if you are interested in studying
   a. The relationship between smoking and health.
   b. Overcrowding in mental hospitals.
   c. The effect of rehearsal on memorization of nonsense syllables.
   d. Attitudes of U.S. citizens toward government-supported health care.
   e. The crime rate in a certain city.

6. Conflicting evidence has been published regarding all of the issues listed in problem 5 above. For each part of problem 5, describe at least one possible source of bias that may explain why different researchers could obtain different kinds of evidence and come to different conclusions about each issue.

7. For each variable that you listed in problem 5, indicate the level of

measurement involved. Also indicate whether your variables are continuous or discrete.

8. What are some possible sources of bias that might be a problem if you attempt to survey or measure
   a. The number of incidents of physical violence within the family that occur in a certain neighborhood.
   b. The number of drivers who actually come to a complete stop at a stop sign by a remote rural intersection.
   c. Family incomes of people who choose to play the state lottery compared with family incomes of people who do not play.

9. For each of the variables given in problem 8, indicate the level of measurement, and whether each variable is continuous or discrete.

10. Explain why each of the following may represent biased observation:
    a. Televised interviews with supermarket shoppers, asking for their opinions about one brand of margarine.
    b. Results of an opinion survey conducted by Congressman A. The congressman sent questionnaires to all constituents and asked them to mail in replies.
    c. A sociological study conducted to determine the degree of physical violence that occurs within the family. Surveyors selected 1,000 families across the country to interview by telephone.

11. For each example listed, indicate the true limits of the given measurement:
    a. A kitchen "diet" scale indicates that a food sample weighs 53 grams;
    b. A postal scale indicates that a letter weighs 3.7 ounces;
    c. On a test of manual dexterity, Susan completed an object-sorting task in 49.2 seconds, according to the stopwatch;
    d. A micrometer reading shows that a pencil is 0.256 inches in diameter;
    e. A Likert scale used in attitude surveys has seven categories for making responses, numbered "1" through "7"; Bill says that his attitude strength on a particular issue is best represented by a "5."

12. Round off the following numbers to one decimal place:
    a. 47.9421   c. 968.75   e. 0.1620   g. 2.4583
    b. 132.79814 d. 0.04   f. 98.009   h. 77.492

13. Round off the following numbers to three decimal places:
    a. 98.905831   c. 175.001009   e. 232.050509
    b. 14.9904   d. 1078.55555   f. 984.656558

14. Give the true limits of each of the following measurements:

    a. 160 lbs

    b. 421.17 grams

    c. A machine part 1.003 inches in diameter

    d. 2 hours, 27 minutes

15. What should you assume are the true limits of the following "rounded" numbers?

    a. $110,000   c. A crowd of 66,000 people

    b. 7,600 barrels of oil   d. 12 million light years

2. The Frequency Distribution: Picture of the Group

After reading this chapter you should be able to do the following:

1. *Prepare a frequency distribution table or graph for a group of observations.*

2. *Use the table or graph to estimate the range of score values, the range of frequencies, the typical group member, and variability among group members.*

3. *Detect skew from a frequency distribution graph.*

4. *Construct and interpret a relative frequency distribution.*

5. *Estimate percentile ranks and percentiles from a cumulative proportion graph.*

6. *Discuss several ways in which graphs can be misleading.*

Statistics and the Real World

Statistical procedures are sometimes defined as methods for making decisions in the face of uncertainty. They were defined in the Introduction as methods for summarizing, analyzing, and evaluating information. These two definitions do not really conflict. One states the general purpose underlying most statistical applications; the other describes the general approach to that goal. Taken together, they express the value of, and the need for, statistics.

Most people using statistics are really interested in learning more about important variables. Does loud factory noise cause permanent hearing loss among workers? Is the "new math" a better method of teaching arithmetic than traditional methods? Our informal observations on such matters may not produce clear answers to these questions. Some children seem to learn arithmetical concepts easily regardless of the method used; others do poorly with either method. Similarly, there are many factors that might contribute to hearing loss, and we cannot assign the blame to factory noise simply because some people exposed to the noise become

42

hard of hearing. In short, these are complex situations in which there is some uncertainty; there is a great deal of variability among individual observations that cannot be immediately explained by casual observation. Such questions must be approached statistically. We take many observations and subject them to statistical analysis in the hope that some order will appear out of what otherwise seems to be chaos. Statistical analysis may reveal characteristics of a group of observations that are not apparent from a single observation.

Frequency Distributions

One concept that is central to all statistical operations is the **distribution.** Distributions are involved every time we deal with variables and observations or potential observations. We will begin our discussion of specific statistical methods by introducing one kind of distribution, the empirical frequency distribution, and in later chapters we will discuss the *probability distribution* and theoretical distributions known as *sampling distributions*.

Empirical frequency distribution is just another name for a group of observations (on a variable) that have been counted and sorted. It is a tally of the number of times (frequency) with which different events or values of a variable occur in the group of observations. The tally thus shows the way values or events are distributed in the total group of observations, hence the term "frequency distribution." (The word "empirical" signifies that the values and frequencies were obtained from actual observations and not from a theory.) In a sense, the empirical frequency distribution is a group picture of the data.

In its simplest form, the frequency distribution[1] is just a two-column table, with one column listing event categories or values of the variable and the other column listing their respective frequencies in the group.

Table 2.1 represents a typical frequency distribution for a nominal-level variable. (The symbol f stands for frequency.) Suppose we are attempting to describe the student body of a small college in Indiana, and in so doing we prepare a list naming each student and his or her home state. Such a list might read: Elizabeth Williams (Indiana), Roger Bates (Illinois), John Wilski (Indiana), and so on. A nominal-level observation is made when we classify each student by home state. To construct the frequency distribution, we simply count the number of observations in each category. The concepts involved in preparing and interpreting such a table are simple and straightforward. Visual inspection of the table indicates, for instance, that by far the largest number of students come from Indiana.

Construction of even such a simple table, however, requires some judgment. For instance, in what order should the measurement categories be listed? How many categories should be listed? With nominal-level

---

[1]The word "empirical" is implied when it is clear that we are talking about data that have been collected.

Table 2.1 *The Number of Students from Various States Attending One Small College*

HOME STATE	NUMBER OF STUDENTS f
Indiana	250
Illinois	110
Ohio	55
Kentucky	30
Michigan	20
Other U.S.	37
Foreign	12
Total	514

measurements, the observation categories may actually be listed in any order, since, by definition, nominal categories do not stand in specific ordinal relation to each other. In this case, the statistician chose to list the specific states beginning with the highest frequency observed, and proceeding down through lower frequencies. The statistician also decided, arbitrarily, not to list specific states sending fewer than 20 students to this school. Twelve other states are represented, but they are grouped together under the "Other U.S." category because the statistician felt that listing them would add little to the information value of the table.

When observation categories represent ordinal- or higher-level measurement, however, they must be listed in their proper order.

Consider a frequency distribution table showing the results of a hypothetical public-opinion survey. Suppose that 100 residents of Ficticia were selected at random[2] and asked to respond to the following statement: "Too much money is being spent on national defense." Each person was asked to rate his degree of agreement or disagreement with that statement by placing his opinion in one of the following categories: "strongly agree," "agree," "neither agree nor disagree," "disagree," or "strongly disagree."

Before compiling the frequency table we must determine the level of measurement involved. Each observation on the variable "opinion category" fits in one, and only one, category, and the categories can be placed in rank order, ranging from "strongly agree" to "strongly disagree." However, we have no basis for assuming that the difference between "strongly agree" and "agree" is the same as the difference between "disagree" and "strongly disagree" or as that between any other pair of adjacent categories. Thus, the data must represent ordinal- but not interval-level measurement.

If the data represent nominal-level measurement, we can list observation categories in any order in the table. If they reach ordinal-level

---

[2]Methods of selecting people to participate in surveys and the meaning of random are covered in Chapter 9.

Table 2.2 *Responses of Ficticia Residents to the Statement: "Too Much Money Is Being Spent on National Defense"*

RESPONSE CATEGORY	*f*
Strongly agree	10
Agree	16
Neither agree nor disagree	20
Disagree	36
Strongly disagree	18
Total	100

measurement or higher, however, the categories must be listed *in order* in the table. The obtained frequency table is shown in Table 2.2. Again, simple visual inspection of the frequency distribution table reveals characteristics of the entire group of data. One can get a "feel" for the distribution of the variable (agreement with the statement) simply by looking at the frequency numbers. More important, such a table can serve as the starting point for further statistical computations. While the basic observations represent ordinal-level data, the frequencies in the table may be treated as absolute-scale values and may be used in the computation of many statistics.[3]

Grouped Frequency Distributions

Construction of frequency distribution tables like Tables 2.1 and 2.2 is a simple and direct undertaking. Observations in each category are simply counted. Interpretation of such tables is also uncomplicated, since the tables specify exactly the nature of each observation in the group. Unfortunately, not all data lend themselves to such direct tabulation.

A problem occurs when there are simply too many values of the variable or too many categories of events to list in a table. This happens most often when the variable is continuous and can (theoretically) exist at an infinite number of values, but it can also occur with a discrete variable that can assume many values.

Suppose an intelligence test were given to, say, 10,000 people. Each person would receive a score such as 121, 104, 87, 112, and so on. A list of the 10,000 scores obtained would constitute 10,000 observations on the variable "IQ score." Summarizing these scores in a frequency distribution table might be difficult because we would probably find that at least one person scored 30, another scored 31, another 32, and so on up through a score of 169. Our frequency distribution table would thus have to list 140 different values (one for each score from 30 through 169) and the frequency obtained for each. Reading such a table to ascertain group char-

---

[3] See Chapter 1 if this is not clear.

acteristics would be almost as difficult as reading the original list of scores. This is a situation that calls for a **grouped frequency distribution.**

In the grouped data table, instead of listing each possible score, we list a small number of **score intervals** and count the number of observations that fall into each. Let's look at one such table and then examine the steps that go into constructing and reading the grouped frequency distribution.

Table 2.3 is a grouped frequency distribution for 10,000 hypothetical IQ scores. Like the tables discussed earlier, it has two columns, one listing values of the variables, the other listing the obtained frequencies. Furthermore, since IQ scores represent at least ordinal-level measurement, the values are listed *in order* in Table 2.3. Usually, the high values are placed at the top and low values at the bottom, although this is a convention that is sometimes reversed.

In this distribution the lowest score obtained was 30 and the highest was 169. Notice that the frequency table includes these scores and all scores in between. This is accomplished by determining the **range** of obtained scores and dividing it into a relatively small number of equal-size score intervals. Here, the range of scores is determined as follows:

$$\text{Range} = \text{High score} - \text{low score} + 1$$

$$\text{Range} = 169 - 30 + 1 = 140$$

The scores registered by these people thus cover 140 points on the IQ scale, that is, the distance on the scale from 30 through 169.[4] To divide this range into a manageable number of intervals, we must first select the number of intervals desired. This is a crucial step in the grouping process, since choosing too many or too few intervals will produce a table that is unclear and confusing. For most grouping purposes, ten to twenty intervals are appropriate; Table 2.3 lists fourteen intervals.

After determining the range and the number of intervals, the next step in the grouping process is to determine the **interval size.** To do this, divide the range by the number of intervals. For Table 2.3, this is:

$$\frac{\text{Range}}{\text{Number of intervals}} = \frac{140}{14} = 10$$

Thus, each interval will be ten IQ points wide. In the real world (as opposed to the artificial world of textbook examples), dividing the range by the number of intervals usually produces a quotient like 10.58934 or 7.39021. In these cases, choose a value for interval size that is a whole number, such as 10 or 7. This may result in using one interval more or less than you had originally intended, but it will greatly simplify construc-

---

[4]Why is the 1 added in to obtain the range? Consider a much simpler distribution in which the highest score obtained is 100 and the lowest is 99. Subtracting the low score from the high to obtain the range would give us $100 - 99 = 1$, the distance *between* scores. However, we want to determine the distance covered by the high and low scores *and* the distance between them. Hence, we add 1 to make sure that a high score of 100 and a low score of 99 cover a range of two scores. This process will be covered in more detail when we compute the range in Chapter 4 (pp. 94–97).

Table 2.3 *Grouped Frequency Distribution for 10,000 Intelligence Test Scores*

IQ	f
160–169	3
150–159	20
140–149	110
130–139	310
120–129	820
110–119	1,812
100–109	2,352
90–99	2,300
80–89	1,450
70–79	560
60–69	200
50–59	40
40–49	20
30–39	3
Total	10,000

tion of the table, and make it much easier to read.[5] In fact, many statisticians prefer to use only round values such as 5, 10, 100, and so on, as interval sizes; this may give you several more or several fewer intervals than you would like to have, but the practice does produce a clearer table. In brief, choosing an interval size is another place where there are no absolute rules. There are only guidelines, within which you must use your own judgment.

At this point, table construction can begin. Each interval is now identified by its **limits.** The lowest interval in the table has a **lower limit** of 30 and an **upper limit** of 39 and thus covers an interval of ten IQ points. (Not nine, but ten. If you doubt this, count the number of score values involved, beginning with 30 and ending with 39.) The score of 30 was chosen as the lower limit because it is the lowest score in the group. The next interval begins with a lower limit of 40, the interval above that with a lower limit of 50, and so on. Intervals are listed until an interval is specified that includes the highest value obtained, 169.

All that remains to be done is the tabulation of the frequencies. A separate tally sheet may be kept, on which each observation is tallied in the appropriate interval. A score of 46 would produce a tally in the interval 40–49; so would a score of 49. A score of 132 would fall into the interval 130–139. After all observations have been tallied, the tallies are

---

[5]Many statisticians prefer to use intervals that are an odd number of units wide. This is because computations made from frequency tables and some graphing procedures use the interval's midpoint, the value in the middle of the interval, and all intervals can have a whole number for a midpoint if the interval size is odd rather than even.

counted and entered in the frequency table as numbers. When many observations are involved, such as the 10,000 in this example, the required bookkeeping task is extensive, and many statisticians understandably use computer programs to determine the interval into which each score goes and to perform the frequency counts.

Using Frequency Distributions

Presentation of a group of observations in a frequency table is usually an initial step in any form of data analysis; sometimes it can be the last step. Visual inspection of a frequency table may be all that is necessary to learn interesting characteristics of the group. Among other things, the table may reveal the following:

1. *The range of frequencies observed and the range of categories observed.* Michigan sent only 20 students to the college, whereas Indiana sent 250. These values indicate a *range of frequencies* observed, 20 to 250. When the variable represents ordinal, interval, or ratio measurement, the table also indicates the *range of values* in the group. In Table 2.3, for instance, the IQ scores covered a wide range of values, from about 30 to 169.

2. *The typical observation.* A look at the frequency distribution can tell you approximately where most of the observations fall or what is the most frequently occurring observation. Most of the scores recorded in Table 2.3 seem to fall between 80 and 119, so we might call scores in this part of the scale *typical* IQ scores. As we will see in the next chapter, there are several specific statistics designed to indicate the typical group member; the frequency distribution can serve as a starting point for computing these.

3. *Variability among frequencies.* Do all the events or values listed in the table occur with about equal frequency, or are some much more prevalent than others? From Table 2.2 we can tell that while all shades of opinion were observed, some opinion categories were reported much more often than others. In Chapter 4 we will study some statistics designed to measure variability. However, if all we want is a crude estimate of differences among frequencies for different values of the variable, a quick inspection of the frequency table may suffice.

Frequency Distribution Graphs

Frequency distribution tables have another important use — they serve as a starting point for making frequency distribution graphs.

Graphs are pictorial devices that use distances and areas on paper to represent quantities. Frequency distributions are often presented graphically because some group characteristics are easier to appreciate this way. Although graphs add little information that could not be gleaned from a table, they are visually more striking and sometimes easier to interpret. *The Wall Street Journal*, for example, typically prints a frequency distribution graph on its front page to summarize one or another aspect of business activity. Business trends that might be discernible only with difficulty from a table can be seen at a glance from the graph.

In its simplest form, the frequency distribution graph is built on two perpendicular lines, or **axes,** with each axis representing one column of the simple frequency distribution table. These are called, logically, the **horizontal axis** and the **vertical axis.**[6] News and advertising media sometimes employ very imaginative methods of picturing frequency distributions in which the axes are not actually drawn in, but they are always there, either explicitly or implicitly.

By convention, the vertical axis usually represents frequencies and the horizontal axis represents event categories or values of the variable. In this framework, the data may be represented in a **bar graph,** a special kind of bar graph called a **histogram,** or a **frequency polygon.**

Figure 2.1 shows a bar graph for the student data summarized in Table 2.1. Although the table does show you that most of these students come from Indiana, the graph gives the same information more visual impact.

Notice that the vertical axis is scaled so that different points on the axis represent different frequency (f) values, including a point representing zero frequency and a point representing 250, the highest frequency observed. Except in special cases[7] the frequency axis should begin with a value of zero at the bottom. Since distances on this axis must accurately represent frequencies, the axis is scaled *linearly*; that is, each small mark represents an f of 10, and all the marks are equally spaced on the line. An axis is linearly scaled if a given distance on the line represents the same range of values at all parts of the line.

Observation categories appear at regular intervals along the horizontal axis. When observations are made at the ordinal level, of course, categories must appear in their natural order. Since these data represent nominal-level measurement, the geographical areas may be arranged in any convenient order.

Information about observed frequencies is pictured with a bar over each category (geographical area), rising to the appropriate height in the frequency direction.

All the characteristics of the group of observations that can be extracted from the frequency table can also be extracted from the graph. The graph shows at a glance the highest and lowest frequencies observed, and communicates a feeling for the relative magnitudes of the various frequencies.

The bar graph has a special name when the observations represent data

---

[6]The horizontal axis is sometimes called the *abscissa* and sometimes the *x axis*. If the term abscissa is used for the horizontal axis, then the vertical axis is called the *ordinate*. When the horizontal axis is referred to as the *x axis,* the vertical axis is called the *y axis*. In most cases, it is permissible to use whichever pair of terms you prefer.

[7]See pp. 61–64, "Misleading Graphs," for a discussion of what happens when the frequency axis does not begin at zero.

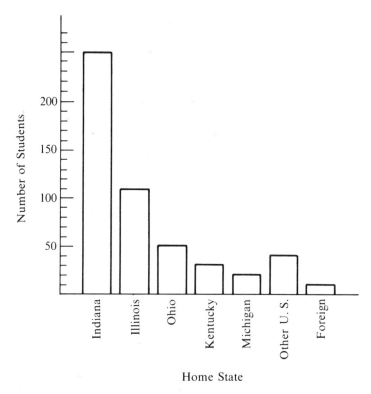

Figure 2.1   Bar graph for student data tabulated in Table 2.1.

above the nominal level: **histogram.** For instance, the distribution of scores in Table 2.3 can be pictured in histogram form, as in Figure 2.2. Although this graph looks quite similar to the bar graph in Figure 2.1, there are some subtle differences.

First, since this is a graph prepared from a grouped-data frequency table, there is one bar for each score interval rather than one for each possible score value (in Figure 2.1, there is one bar for each observation category). The height of each bar thus shows the number of observations that appeared in that interval.

Second, the IQ-score axis is scaled linearly, just like the frequency axis. Each mark on this axis represents ten IQ points, and the marks are evenly spaced. With interval- or ratio-level data, distances on this axis represent *quantities* of the variable (measured IQ) and thus must be scaled linearly.[8]

Third, notice that the bars are in contact with each other. With data above the nominal level, one usually brings the bars into contact with each other because each bar is supposed to span the *entire* distance in the interval. Technically, this is not the distance between the interval's upper and lower limits, but, rather, the distance between the interval's **upper**

---

[8]See the discussion on pp. 62–63 on improper deviations from linearity.

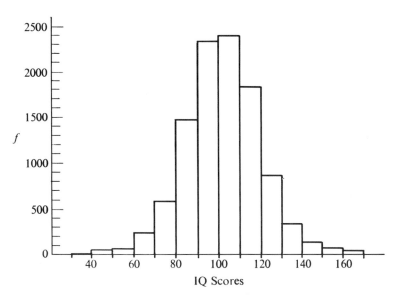

Figure 2.2 Frequency histogram for the 10,000 IQ scores in Table 2.3. The histogram bars are equal in width to ten IQ points, the size of the intervals in the frequency table. Each bar is centered on the interval it represents.

true limit and its **lower true limit.** If you mastered the concept of a *number's* true limits in Chapter 1 (pp. 33–34), you will have no problem understanding the concept of an interval's true limits. *Read the following sentence slowly*: The upper true limit of an interval is the upper true limit of the number that is the interval's upper limit. For the interval 37–44, for instance, 44 is the upper limit, and 44.5 is the upper true limit. Similarly, the lower true limit of that interval is 36.5. Upper and lower true limits of the interval 29–36 are, respectively, 28.5–36.5. The distinction between limits and true limits may seem trivial at this point, but it becomes especially important later in this chapter when we estimate and compute percentiles.[9]

Suppose, instead, that we were graphing a distribution of human reaction times, and that one interval in the grouped-data distribution included reaction times between 0.300 and 0.319 seconds. The true limits of this interval would be 0.2995 and 0.3195 seconds, and interval size would be 0.0200 seconds.

When the data consist of nominal-level observations, the bar graph is an effective means of picturing the group. When the data are at the ordinal level or above, we call the graph a histogram. Also, when the data represent ordinal- or higher-level data, we may use a **frequency polygon** graph instead of histogram.

Figure 2.3 is a frequency polygon made from the scores summarized in

---

[9]Some statisticians refer to our concept of "limit" as "apparent limit" in order to distinguish these limits from true limits.

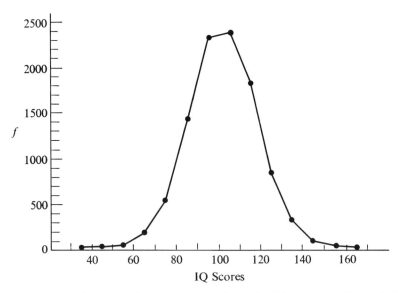

Figure 2.3 Frequency polygon for the same scores shown in Table 2.3 and Figure 2.2. Each interval in the frequency table is represented by a dot over the interval's midpoint.

Table 2.3. It shows exactly the same information as Table 2.3 and Figure 2.2. Instead of a bar for each interval, however, the polygon has a dot. A dot representing each interval in the table is placed *over the middle* of the interval at a height corresponding to the appropriate point on the frequency axis, and the dots are then connected with a line.

Frequency polygons are easier to draw than histograms, and they offer the added advantage of allowing several polygons to be placed on the same pair of axes. An example of this is shown in Figure 2.5.

The Shape of the Distribution

Since the frequency distribution graph contains no more and no less information than the table it was made from, you may still be wondering if it is really worthwhile to go to the trouble of making a graph when a table might do. While it is technically true that the information content of the graph and table for the same data is identical, it is usually easier for the reader to extract certain kinds of information from the graph.

Remember that you are probably using statistics to examine the way a variable is distributed among a group of observations. You may be interested in finding the typical group member, the range of observed values, or the extent to which different score values differ in frequency. These characteristics may be estimated from either the table or graph, although most people probably find graphs easier to use for these purposes. In addition, graphs are especially useful for evaluating one other characteristic of distributions: **skew.**

Skew is a term used to describe the extent to which a distribution deviates from symmetry. All distributions are either skewed or symmetric, and two varieties of each kind are shown in Figure 2.4. These are frequency polygons that might represent observations on any variable. Figures 2.4(a) and 2.4(b) are **symmetrical** distributions because the left half of the polygon is a mirror image of its right half. Figures 2.4(c) and 2.4(d) are **skewed** distributions because the right half of each figure is not a mirror image of its left half.

Suppose that Figure 2.4(c) is a frequency polygon showing test scores registered by a large class of students on an introductory statistics exam. Test-score values would be arranged along the horizontal axis, and the vertical axis would scale the observed frequencies. You can tell from the figure that a relatively large number of people got low scores on the test and few people scored high. When many people score low and only a few score high, the resulting distribution is said to be **positively skewed.** If just the opposite happens, that is, if many score high and a few score low, then the distribution is **negatively skewed.** To many, these designations of positive and negative seem just backwards from the way they should be, since a positive skew means a high number of low values. But remember that the *tail* of the distribution, the part with low frequencies, points in the direction of the skew. A negatively skewed distribution like that in Figure 2.4(d) has a tail pointing in the negative direction on the horizontal axis.

The presence of skew in a distribution can sometimes be an important

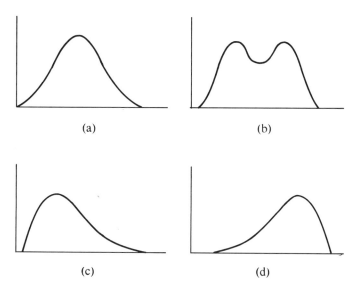

(a)   (b)

(c)   (d)

Figure 2.4 Symmetric and skewed frequency distributions. In each distribution the vertical axis represents frequencies obtained and the horizontal axis represents values of the variable or event categories observed. Examples (a) and (b) are symmetrical distributions because the left half of each is a mirror image of its right half. Distribution (c) is positively skewed, the kind of distribution that might be produced by scores from a difficult examination. An easy examination might produce a negatively skewed distribution, as in (d).

item of information. For the example described in the preceding para-graph, for instance, the appearance of positive skew in a frequency polygon might be a tip-off that the task was difficult for the group as a whole, and the appearance of negative skew might similarly indicate that the test was probably easy. Or, the presence of positive skew might indicate a *floor effect* — a value below which the variable cannot go; a negative skew might in the same way mean there is a *ceiling effect* — a value above which the variable cannot go. In no case does skew *prove* that the task was easy or hard, or that there is a floor or a ceiling, but it can suggest that you look for these characteristics.

Consider the frequency polygon that might appear if we graphed annual incomes of U.S. citizens. Different income values would be represented on the horizontal axis, and the vertical axis would represent the frequency with which these values occur in the population. Which graph — 2.4(c) or 2.4(d) — would more likely be the result? The income axis would cover a range of dollar values extending from zero through many millions. Most observations would tend to fall near the low end of the distribution, meaning that we could describe the distribution of annual incomes as positively skewed.

The frequency polygon for IQ scores in Figure 2.3, on the other hand, is nearly symmetrical. By indicating that the distribution is not skewed, we say that there are just about as many high IQs in the group as there are low IQs.

Skew can thus be a useful term for characterizing a distribution. It is a group characteristic that can usually be detected easily from a frequency distribution graph and with a little more difficulty from a frequency table. As we will see in the next chapter, skew can also be detected by examin-ing the relation between two statistics, the group mean and the median. The degree of skew can also be computed directly and exactly from the scores themselves, but that procedure is beyond the scope of this book.

The Relative Frequency Distribution

Tabled and graphed distributions are both group pictures of the data. They may show all you care to know about the way a variable is distrib-uted in a group; if not, the table will serve as a starting point for computing more statistics. One instance in which the simple frequency distribution is inadequate is when you want to *compare* one distribution with another and the two groups contain different numbers of observations. In this case, the frequency distribution must be modified to produce a **relative frequency distribution.**

Table 2.4(a) is a double frequency distribution showing the results of a chemistry examination given to students at two high schools. The scores were grouped into ten score intervals, each interval covering eight score points. Suppose you wish to compare the performances of these two classes by graphing both frequency distributions on the same pair of axes, as in Figure 2.5(a).

It is not easy to compare the two frequency distributions directly, because the two groups are unequal in size. The two frequency polygons are located at different parts of the vertical axis, and vertical distances on one polygon are not of the same relative importance as vertical distances on the other. Direct comparison of different-size groups can be accomplished if the table and graph are prepared as relative frequency distributions, however.

Relative frequency distributions for the same chemistry exam scores are shown in Table 2.4(b) and graphed in Figure 2.5(b). Each entry in the relative frequency table was determined by dividing the frequency for that interval by the size of the group involved. Twelve people from Adams H.S. scored in the interval 77–84. The corresponding relative frequency for that interval in Table 2.4(b) is found by dividing 12 by the total number of people from Adams H.S., 100. Thus:

$$\frac{12}{100} = 0.12$$

Forty-five people from Jefferson H.S. scored in the interval 77–84. Since there were 500 students from Jefferson, the corresponding relative frequency for this interval is:

$$\frac{45}{500} = 0.09$$

It should be clear that you cannot construct a relative-frequency table until you first construct a simple-frequency table. As a check on your computations, be sure that the relative frequencies for each group add up to 1.00.

Table 2.4 *Frequency Distribution (a) and Relative Frequency Distribution (b) for Chemistry Exam Scores Obtained by Students at Two High Schools*

CHEMISTRY EXAM SCORES			CHEMISTRY EXAM SCORES		
SCORE	ADAMS H.S. (*f*)	JEFFERSON H.S. (*f*)	SCORE	ADAMS H.S. (RELATIVE FREQ.)	JEFFERSON H.S. (RELATIVE FREQ.)
93–100	3	30	93–100	0.03	0.06
85–92	8	35	85–92	0.08	0.07
77–84	12	45	77–84	0.12	0.09
69–76	25	60	69–76	0.25	0.12
61–68	18	75	61–68	0.18	0.15
53–60	12	75	53–60	0.12	0.15
45–52	8	65	45–52	0.08	0.13
37–44	6	50	37–44	0.06	0.10
29–36	5	35	29–36	0.05	0.07
21–28	3	30	21–28	0.03	0.06
Total	100	500	Total	1.00	1.00
(a)			(b)		

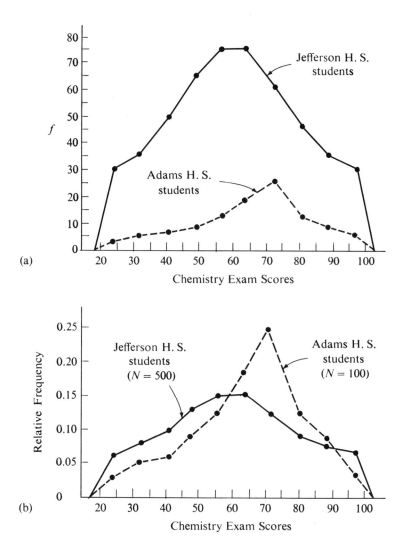

Figure 2.5 Comparison of chemistry exam scores for two groups of students, using (a) two frequency polygons and (b) two relative-frequency polygons. Direct comparison of group characteristics is more difficult in (a), because the two groups are unequal in size and the two polygons are widely separated on the frequency axis. In (b) it is easier to see that Adams H.S. students performed slightly better overall than did Jefferson H.S. students.

The advantages of turning frequencies into relative frequencies can be seen by comparing Figure 2.5(a) with Figure 2.5(b). The latter allows the two groups to be compared directly, even though different numbers of students are in each group. It is easy to see, for example, that the Adams H.S. scores are negatively skewed, while the distribution of Jefferson H.S. scores is almost perfectly symmetric. Furthermore, the Adams score distribution is more peaked than the Jefferson distribution. These observations suggest that the Adams students found the test somewhat easier than the Jefferson students, and that there is more variability among Jefferson students.

Notice that the vertical axis on the relative-frequency graph (Figure 2.5(b)) covers only the relative-frequency values from 0 to 0.25, the highest obtained.

◻ Estimating Percentiles — The Individual Within the Group

Up to now, we have been concerned with the group as a whole, with the characteristics of the entire collection of observations. But sometimes we need to use statistics to find out something about an individual within the group. If you have scored 80 on a chemistry test, how do you stand with respect to your classmates who also took the test? Jim ran the Boston Marathon in 2 hr, 30 min; how does his performance compare with the 3,900 other runners in the same race? The group picture, or frequency distribution, yields an answer to such questions.

In the chapters that follow we will use specific statistics derived from the frequency distribution to evaluate individual scores. Here, we will consider a means of comparing an individual to a group that makes use of the relative frequency distribution constructed in the preceding section.

The individual's standing in the group can be expressed by stating his **percentile rank**.[10] This is simply the percentage of the group that scored below him. If he scored 60 on a chemistry exam, and if 34 percent of the people taking the test scored below 60, his score would have a percentile rank of 34 *in that group*. If the same score beat 90 percent of the scores in another group, then its percentile rank would be 90 in *that* group.

Percentile ranks are widely used in educational circles, usually in connection with examination scores. If you take, say, the SAT (Scholastic Aptitude Test) or the GRE (Graduate Record Exam), you will receive your score in the mail together with a booklet showing how to find the percentile rank associated with that score. The booklet might say, for instance, that an aptitude test score of 680 has a percentile rank of 88 *among all students who have taken that test in the last four years* (a percentile rank is meaningless without reference to some group). The advantage of percentile ranks as descriptions of individual performance is that they are easily understood by students and teachers alike.

A concept inherent in the idea of percentile rank is the **percentile**. The 10th percentile, for instance, is the score *value*[11] (or value of the variable) with 10 percent of the observations falling below it. Saying that a certain

---

[10]Some statisticians prefer the term *centile rank* to percentile rank.

[11]Notice that a percentile is a point on the measurement scale with a certain percentage of observations falling below it — not necessarily a score value that was actually obtained.

score has a percentile rank of 50 is just another way of saying that the score is at the 50th percentile. If you want to find a *percentile*, find the score value with a given percentage of observations below it; if you want to find the *percentile rank* of a score, begin with a score and find the percentage of observations below it.

One way to compute percentile ranks would simply be to count the number of observations that fall below a given score and then calculate the appropriate percentage. If the list of scores is very long, however, this can be a tedious process. The more usual approach is to *estimate* percentile ranks — and percentiles — from a grouped-data table. The grouped-data table is first converted into a **cumulative proportion table.** Then the percentiles or percentile ranks are estimated from a **cumulative proportion graph** (made from the cumulative proportion table) or computed directly from the table. Since this book is not intended as a comprehensive computational guide, and since the latter method is best explained in many steps, it is covered in Appendix 3. Here we will consider the briefer (but less precise) method of estimating percentiles or percentile ranks from the graph.

The required graph is constructed from a cumulative proportion table, which, in turn, is constructed from a relative frequency table. Table 2.5 shows how this is done. The score values and relative frequencies are the chemistry exam data from Table 2.4, showing the Adams H.S. scores only. Entries in the cumulative proportion column specify the proportions of the distribution that fall *in or below* each interval; they are computed directly from the relative frequency values. Beginning at the lowest interval, the proportion of the scores that fall in or below that interval is, of course, the relative frequency of that interval, 0.03. The cumulative proportion of the next interval, 29–36, includes everything in that interval

Table 2.5 *Relative-Frequency   and   Cumulative Proportion Tables for Chemistry Exam Scores from Adams H.S. Students*

CHEMISTRY EXAM SCORES (ADAMS H.S.)

SCORE	RELATIVE FREQUENCY	CUMULATIVE PROPORTION	
93–100	0.03	1.00	→ 1.00 = 0.03 *Plus* ↘ 0.97
85–92	0.08	0.97	
77–84	0.12	0.89	
69–76	0.25	0.77	
61–68	0.18	0.52	
53–60	0.12	0.34	
45–52	0.08	0.22	
37–44	0.06	0.14	→ 0.14 = 0.06 *Plus*
29–36	0.05	0.08	→ 0.08 = 0.05 *Plus* ↘ 0.08
21–28	0.03	0.03	↘ 0.03

(0.05) plus everything below it (0.03), that is, 0.08 of the distribution. Similarly, the cumulative proportion for the interval 37–44 includes everything in that interval (0.06) plus everything below it (0.08), which is 0.14. This process is continued upward until the cumulative proportion has been calculated for every interval. This should turn out to be 1.00 for the uppermost interval, because every observation in the group either falls in or below that interval.

Percentiles and percentile ranks can then be estimated from a graph of the cumulative proportion table. This graph is similar in many ways to the frequency polygon and relative-frequency polygon, but there are some important differences.

On the cumulative graph the horizontal axis represents the range of test scores observed, just as on the frequency polygon. The vertical axis represents cumulative relative frequencies or cumulative proportions. Since the range of these in any group will always extend from 0 to 1, this axis is always scaled from 0 to 1.

Figure 2.6 is the completed cumulative proportion graph. On cumulative graphs it is very important to plot the points over the upper true limits of the intervals, because each point stands for all values below that upper limit. (On the frequency polygon points were drawn over the middle of the

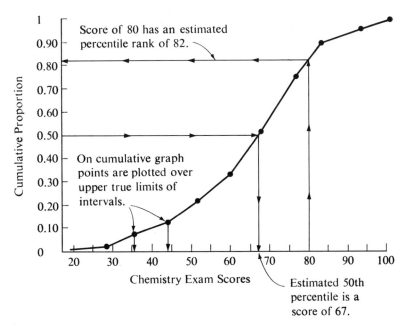

Figure 2.6   Cumulative proportion graph constructed from the data in Table 2.5. Each point on the curve represents a test-score value and the proportion of the observations that fall below that value (cumulative proportion). The text discusses a method for estimating percentiles and percentile ranks from such a graph.

intervals, since each point represented only the observations in that interval.) Estimates made from cumulative graphs will not turn out correctly if the points are not located above the upper true limits of each interval. Thus, the point representing the cumulative proportion for the interval 37–44 is placed over the point on the horizontal axis representing 44.5, not over the middle of the interval. When all points are plotted, they are connected with lines to complete the curve.

If, say, we want to find the test score corresponding to the 50th percentile, we find the score with a cumulative proportion of 0.50. We can estimate this score by drawing a line *horizontally* from the point on the vertical axis representing a cumulative proportion of 0.50 until it intersects the curve. (This is illustrated on Figure 2.6.) From this point of intersection, another line is drawn straight down until it intersects the horizontal axis. The point of intersection with the horizontal axis is taken as the 50th percentile, the score value below which 50 percent of the observations fall. In the chemistry exam example this estimate turns out to be a score of about 67.

On the same graph we can reverse the process and estimate the percentile rank of a given score. What percentage of the Adams H.S. students taking this test scored below 80? To estimate a percentile rank, find the score on the horizontal axis and proceed straight up until meeting the curve; then proceed horizontally until you reach the cumulative proportion axis. This process is also illustrated in Figure 2.6, showing that a score of 80 in this group is associated with a cumulative proportion of about 0.82. A cumulative proportion of 0.82 represents a percentile rank of 82.

Although percentiles and percentile ranks are easy to interpret, these statistics are not as widely used as one might expect. The reason is that they cannot be subjected to further statistical analysis except under special conditions. Usually, percentiles cannot be averaged, added, or multiplied in meaningful ways. Figure 2.7 illustrates the problem. The percentile values are not spaced evenly along the test-score scale. The 60th

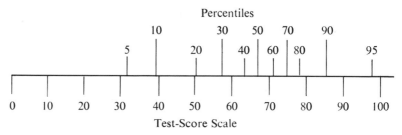

Figure 2.7   Various percentiles are shown for the cumulative-proportion graph in Figure 2.6. For this distribution of test scores, percentiles between 40 and 80 are relatively close together on the test-score scale, while other percentiles are more distant from each other. The difference in test scores between the 40th and 50th percentiles, for instance, is much less than the difference in test scores between the 10th and 20th percentiles.

Table 2.6 *Summary of Names of Percentiles*

PERCENTILE	NAME	SYMBOL
10th	First decile	D_1
20th	Second decile	D_2
25th	First quartile	Q_1
30th	Third decile	D_3
40th	Fourth decile	D_4
50th	Fifth decile	D_5
	Second quartile	Q_2
	Median	
60th	Sixth decile	D_6
70th	Seventh decile	D_7
75th	Third quartile	Q_3
80th	Eighth decile	D_8
90th	Ninth decile	D_9

percentile, for instance, is associated with a test score of about 71, and the 50th percentile is associated with a test score of about 67; at this part of the scale a difference of four test-score points covers ten percentile points. However, the 10th percentile falls at a score of about 39 and the 20th percentile at a score of about 50. At the lower end of the scale ten percentile points cover about eleven test-score points. Percentile points must therefore be treated very much like ordinal-scale data. This is the reason that statisticians turn to more sophisticated methods for describing individual performance, such as standard scores (discussed in Chapter 4).

Some percentiles have special names that are useful in certain applications. Percentiles that divide the distribution into quarters are called **quartiles.** The first quartile is the 25th percentile (symbolized Q_1), the second quartile (Q_2) is the 50th percentile, and the third quartile (Q_3) is the 75th percentile. Thus, if a college accepts only high-school students whose academic standing is above the third quartile in their class, the college accepts only those students in the top quarter of their class. Quartiles are useful for computing a variability statistic, the interquartile range (discussed in Chapter 4). The second quartile is often used as a measure of central tendency (see Chapter 3). In this case Q_2 is called the median. At other times the term **decile** is used to refer to the percentiles that divide the distribution into tenths. The value that has 40 percent of the observations below it, for instance, is the fourth decile. Table 2.6 summarizes the names assigned to the various percentiles.

Misleading Graphs

In the Introduction a graph of miles-per-gallon averages was used to show that graphs can be technically correct and yet misleading. Similarly, there are several characteristics of frequency-distribution graphs of which you should be aware in order to avoid misinterpretation. Even well-meaning statisticians sometimes produce the following mistakes: (1) improper deviations from linearity, (2) details magnified out of proportion, (3) distracting changes in areas of graphs.

1. *Improper deviations from linearity.* This is an error that can appear on either axis of the graph. Suppose, for example, that the data from Table 2.1 were graphed as in Figure 2.8. Although the information is accurate, the *impression* one gets from a first glance is that the number of students is about half the number from Illinois. The problem is that the vertical axis representing frequency is not linearly scaled. A linear scale is one on which equal distances represent equal numbers. Thus, Figure 2.1 has a **linearly scaled frequency axis** because each mark represents a frequency of 10, and the marks are evenly spaced.

When the observation categories represent interval-, ratio- or absolute-scale measurement, the horizontal axis must be linearly scaled as well. Improper deviations from linearity occur occasionally in scientific reports, but more often in advertisements and news items. When presented with a frequency distribution graph, check each axis for this possibility.

There *are* some occasions when it is desirable and proper to scale one

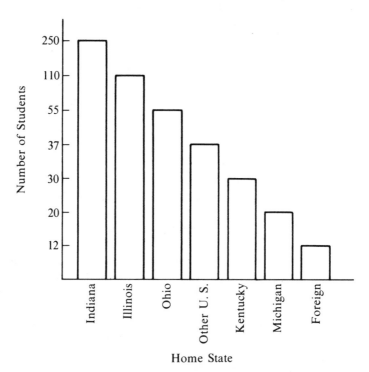

Figure 2.8   Bar graph with an improperly scaled vertical axis. The distances on the frequency axis do not have meaning, because the indicated frequencies are not scaled linearly. While the information conveyed by Figure 2.8 is essentially the same as that conveyed by Figure 2.1, Figure 2.8 gives a misleading impression.

or the other axis nonlinearly. Scientific graphs, for example, sometimes use axes that are scaled logarithmically. Again, however, distances on the graph have meaning with respect to the logarithmic scale, and the axes should be clearly labeled to indicate this.

2. *Details magnified out of proportion.* This is not necessarily an error by the graph's maker, but rather a misinterpretation by the graph's reader. It arises from the fact that the vertical axis on a frequency distribution does not have to begin at zero frequency. Remember the example in the Introduction (Figures I.1 through I.5) in which the vertical axis was cut and stretched to emphasize the trend toward lower fuel efficiency. The same procedure is often applied to frequency-distribution graphs. All that is required of the frequency axis is that it be properly scaled and that it include the highest and lowest frequencies observed (the range of frequencies observed). Sometimes it is hard to get a feeling for the magnitude of those frequencies when the vertical axis does not extend down to zero frequency, and sometimes the cutting and stretching make it easier to examine differences in frequencies among observation categories. However, in all such cases it is worth taking the trouble to imagine a spot near the bottom of the page (or lower) where the zero frequency would be located, in order to get an idea of the magnitude of those frequencies and the importance of those differences.

3. *Distracting changes in areas of graphs.* This is another problem that appears regularly in graphs presented by the news media. In an attempt to make the appearance of histograms more striking, illustrators sometimes replace the standard histogram bars with figures of various kinds. Figure 2.9 is an example of this practice with a bar graph. The graph shows the number of undergraduate students enrolled in various colleges at the University of New Hampshire during the school year 1976–1977. The illustrations were drawn in place of bars so that the height of each figure represents the appropriate frequency. However, the taller figures are also larger in *area* than the smaller figures, and the respective areas of the figures are not proportional to the frequencies. There were 3,417 students in Liberal Arts, and 945 students in Physical Sciences and Engineering during this year, and the respective heights on the drawings accurately represent these frequencies. However, the *area* of the Liberal Arts figure is about fourteen times the area of the Technology figure, and it is difficult to discount this factor during inspection of the graph.

This kind of problem rarely occurs, if ever, in scientific work, but it appears often in other statistical applications.

This is surprising because there are alternative methods of illustration that can be equally striking but less misleading. One of these is the so-called pictograph, in which small pictures are used instead of bars. All pictures are the same size and each might represent, say, 200 students. For the enrollment data in Figure 2.9, seventeen of these little figures could be placed over the Liberal Arts label to represent about 3,400 students, and four and three-quarters figures could be placed over the Technology label to stand for that college's 945 students.

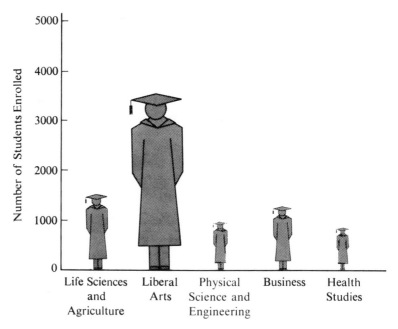

Figure 2.9   Number of students enrolled in various colleges at the University of New Hampshire during the 1976–1977 school year. Instead of equal width bars to represent each college, human figures are drawn to the appropriate heights. Unlike bars, however, human beings become wider as they become taller, and thus these sketches cover areas that do not reflect frequency values. The College of Liberal Arts enrolled about 3.7 times as many students as the College of Physical Sciences and Engineering, and the heights of the figures accurately reflect this relation. Unfortunately, the area covered by the Liberal Arts figure is about 14 times larger than the area covered by the Physical Sciences and Engineering figure. Such distracting differences between areas are difficult to ignore when reading graphs.

SUMMARY

In order to apply statistics to any problem, questions about actual situations must be translated into statements about observable variables. Statistical results appear as statements about variables, and choosing the appropriate variable to study is a critical part of the statistical process.

Sometimes we can learn everything we wish to know about a group of observations simply by inspecting a group picture of the data. Two such group pictures are the frequency-distribution table and the frequency-distribution graph. Each simply shows the number of times (the frequency) that each event or value of the variable occurred in a group of observations. Some group characteristics that can be gleaned from table or graph include the range of values observed, the range of frequencies, the typical group member, and the degree of variability among group members. If precise values of these characteristics are desired, the frequency table will serve as a starting point for further statistical analysis.

The most common types of frequency-distribution graphs are the bar graph, the histogram, and the frequency polygon. Graphs are particularly useful for detecting skew in a distribution.

Distributions may be either symmetric or skewed. Skew is a departure from symmetry and may be either positive or negative. In a positively skewed distribution, most observations appear at the low end of the measurement scale. Only a few have very high values. Such a distribution might be produced by a group of scores from a very difficult performance task. In a negatively skewed distribution most observations cluster at the high end of the scale. Only a few low values appear, suggesting that the task was rather easy for the group.

The relative frequency distribution is particularly useful when comparing two distributions from groups of unequal sizes. A relative frequency distribution shows the proportion of the entire group that falls in each event category.

There are several pitfalls to be avoided when making or reading graphs. An axis with improper deviations from linearity does not accurately represent quantities, and the graph constructed on that axis will have a distorted shape. Another graphing problem is magnifying relatively minor characteristics of a distribution out of proportion by cutting and stretching either axis. Also, single illustrations should not be used in place of graph bars, because their areas do not accurately represent quantities.

KEY CONCEPTS

empirical frequency distribution

frequency table

grouped frequency distribution

range

score intervals

lower limit (of an interval)

upper limit (of an interval)

horizontal axis

vertical axis

bar graph

histogram

frequency polygon

upper true limit (of an interval)

lower true limit (of an interval)

symmetrical distribution

positive skew

negative skew

relative frequency distribution

percentile

percentile rank

cumulative proportion table

cumulative proportion graph

percentile quartiles, deciles

linearly scaled axis

SPOTLIGHT 2   Clarity or Confusion?

Graphs appear in almost all forms of printed matter. They are an especially effective means of summarizing and clarifying what might otherwise be a confusing mass of information. They can also be visually striking and attractive, which no doubt appeals to authors and editors who wish to break up the monotony of the printed page. Graphs may clarify one aspect of the data or contribute a striking artistic touch; even "properly" made graphs, however, can easily be misinterpreted.

COMPARING THE DEFENSE BUDGETS

In Current Billions   In Constant (1958) Billions   As Percentage of G.N.P.

Figure 2.10   Comparison of the defense budgets.

Source: Time (October 8, 1973): 35. Reprinted by permission from *Time*, The Weekly Newsmagazine; Copyright Time Inc. 1973.

Figure 2.10 was used by *Time* magazine to compare the U.S. defense budgets for 1954 and 1974. The graph clearly shows that the two budgets can be compared on several bases, and the difference between them depends on the base used. The chart (like Figure 2.9) is a dressed-up bar graph, in which heights of the soldier figures represent the magnitudes. The outlined soldier in the third column is slightly more than two times as tall as his companion solid figure, because 12 percent is slightly more than twice as much as 5.9 percent. However, the *area* covered by the taller figure is more than four times the area covered by the smaller figure. When evaluating this graph, it's hard to pay attention to the heights while discounting the areas.

Figures 2.11(a) and (b) illustrate a more subtle problem associated with interpreting graphs. Figure 2.11(a) shows the results, as published, of a

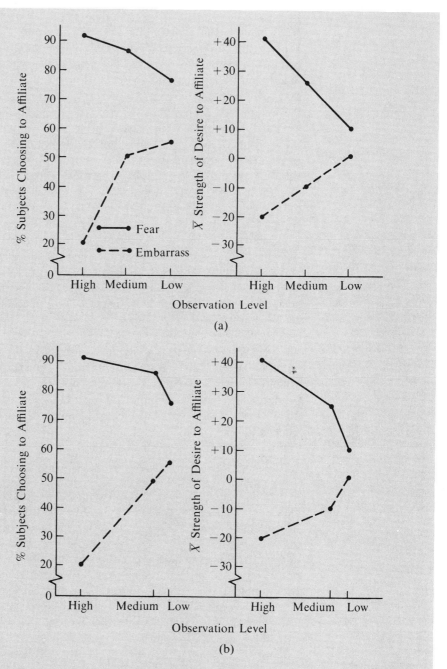

Figure 2.11 (a) Results of the Fish, Karabenick, and Heath (1978) study. (b) One possible way to regraph these results, as discussed in the text.

Source: B. Fish, S. Karabenick, and M. Heath, "The Effects of Observation on Emotional Arousal and Affiliation," *Journal of Experimental Social Psychology,* 14 (1978): 263. Reprinted by permission of the authors and Academic Press.

study by Drs. Barry Fish, Stuart Karabenick and Myron Heath.[12] In their experiment, three different groups of college students were told they were going to take part in a study of physiological psychology wherein "the reactions of the sense organs to various kinds of stimulation would be assessed." The researchers, however, were really interested in studying the extent to which different levels of fear or embarrassment influence affiliative desire, i.e., the desire to be alone or with other people. The "low," "medium" and "high" designations on the horizontal axes refer to the amounts of scrutiny (observation level) that each group would be subjected to during the supposed "experiment" in physiological psychology; more scrutiny was assumed to raise either fear or embarrassment in the subjects. Before participating in the "physiological study," subjects were asked to indicate on a rating scale the extent to which they wanted to wait (before starting the experiment) either alone or with other subjects. A positive rating (from 0 to 100) meant the subject wanted to be with other people; a negative rating (from 0 to −100) was given by people wanting to wait alone. Some subjects received further instructions that were designed to arouse fear; other subjects were given instructions designed to arouse embarrassment.

The results of the study, shown in Figure 2.11(a), are quite interesting. Subjects who were presumably most embarrassed or anticipated the most embarrassment (the "high" observation level) produced the lowest affiliation scores; subjects with, presumably, the highest level of fear produced the highest affiliation scores. Thus this experiment very neatly demonstrates that fear and embarrassment operate differently on self-rated affiliation.

Limits must be placed on the interpretation of these graphs, however. Notice that the "high," "medium" and "low" observation levels are evenly spaced on the horizontal axis. This is appropriate. What we do not know, though, is how much embarrassment or how much fear is associated with each observation level. Thus the observation levels are most likely related to each other in an ordinal fashion rather than in an interval fashion. Figure 2.11(b) may be a more accurate representation of the different amounts of fear and embarrassment or the "perceived degree of observation." Furthermore, in the right-hand graphs the vertical axes, which represent average $\bar{X}$ ratings, have linearly scaled rating points. It is likely that affiliative strength is measured at the ordinal level with these ratings; thus it is possible that the vertical axis of the average ratings should be scaled differently.

Figure 2.11(a) is a perfectly acceptable form because it clearly communicates the results of this study. It is important, however, not to let the linearly scaled axes of these graphs mislead us into thinking that they illustrate the precise *shape* of the relationship between these variables.

---

[12]B. Fish, S. Karabenick and M. Heath, "The Effects of Observation on Emotional Arousal and Affiliation," *Journal of Experimental Social Psychology*, 14 (1978): 256–265.

PROBLEMS

1. Thirty college students in a statistics class are asked to list their class standing. The responses given were:

Freshman	Sophomore	Junior	Sophomore	Freshman
Junior	Sophomore	Sophomore	Senior	Junior
Freshman	Sophomore	Junior	Sophomore	Freshman
Sophomore	Senior	Freshman	Senior	Junior
Senior	Sophomore	Sophomore	Junior	Sophomore
Sophomore	Senior	Sophomore	Junior	Junior

Summarize these observations with a frequency table.

2. Prepare a histogram for the frequency table derived from the data in problem 1.

3. Given below are limits for one interval from each of five different grouped-data frequency distributions. For each, give the interval's true limits, midpoint, and interval size.

a. 46–50 d. 1.01–1.41
b. 461–500 e. 201–300
c. 0.01–0.10

4. Listed below are limits for one interval from each of six different grouped data frequency distributions. For each, give the interval's midpoint and interval size.

a. 16–20 d. 591–620
b. 101–200 e. 7001–9500
c. 0.7–1.0 f. 0.17–0.18

5. Each pair of numbers represents the highest and lowest values from a series of observations. Find the range of each series as demonstrated in this chapter. Then, assume that you are preparing to group the data into approximately 15 intervals. For each case, give a reasonable choice of interval size, the limits of the lowest interval, and the true limits of the lowest interval. (There is no single correct answer for each of these; the answers given in Appendix 4 represent some good choices, but other values may also be appropriate.)

a. 98 points; 42 points
b. 55 miles per hour; 15 miles per hour
c. 1,079 meters; 273 meters
d. 221 milliseconds; 47 milliseconds
e. 1 hour 27 minutes; 45 minutes

6. In a frequency distribution of 420 observations, what relative frequency is associated with an interval's frequency (f) of

a. 137 c. 94 e. 150
b. 15 d. 29 f. 67

7. Complete the following summary table. (Complete the cumulative proportion column only if you covered the appropriate text section.)

LIMITS	MIDPOINT	FREQUENCY f	CUMULATIVE PROPORTION
41–50	———	4	———
31–40	———	11	———
21–30	———	26	———
11–20	———	15	0.2807
1–10	5.5	1	0.0175

8. Complete the following summary table (complete the cumulative proportion column only if you covered the appropriate section in the text).

LIMITS	TRUE LIMITS	MIDPOINT	FREQUENCY f	CUMULATIVE PROPORTION
160–166	——–——	——	1	——
——–——	——–——	——	10	——
——–——	——–——	——	32	——
——–——	——–——	——	50	——
——–——	——–——	——	33	——
——–——	——–——	——	28	——
——–——	——–——	121	20	——
111–——	——–——	114	15	——
104–110	103.5–110.5	107	8	——
97–103	96.5–103.5	100	3	——

9. Complete the following summary table (complete the cumulative proportion column only if you covered the appropriate section in the text).

LIMITS	MIDPOINT	f	RELATIVE FREQUENCY	CUMULATIVE PROPORTION
82–90	——	7	0.00875	———
——–——	——	31	0.03875	———
——–——	——	42	0.05250	———
——–——	——	65	———	0.90000
——–——	——	117	———	0.81875
——–——	——	132	———	0.67250
——–——	——	150	———	0.50750
19–27	23	149	0.18625	0.32000
10–18	14	72	0.09000	0.13375
1–9	5	35	0.04375	0.04375

10. Prepare a frequency polygon for the distributions tabled in problems 8 and 9 above. How would you characterize each distribution with respect to skew?

11. Seventy students in Professor Bunsen's chemistry class took a 100-point exam and produced the scores listed below. Summarize this class's performance by constructing a histogram. (Group the data, using seven intervals of size 10. The lowest interval should have limits 31–40.)

66	89	47	63	40	51	62	80	60	97	48	64	67	97
60	67	81	64	74	38	69	50	59	80	57	33	47	72
68	91	74	47	73	50	52	43	54	98	54	63	79	33
51	54	60	44	82	43	32	48	46	57	47	92	72	87
71	79	84	41	40	58	44	61	38	92	71	41	83	84

12. Fifty students in Professor Erlenmeyer's chemistry class also took the exam mentioned in problem 11, producing the list of scores given below. Prepare a separate relative-frequency distribution for each class's scores, and graph both distributions on the same pair of axes. Comparing both distributions with respect to skew, which class seems to have found the exam easier? For both classes, use seven intervals of size 10 and a lowest interval of 31–40. Professor Erlenmeyer's class scored as follows:

81	75	92	64	97	37	86	63	89	71
89	61	72	54	32	64	47	82	76	83
72	78	94	39	73	81	82	88	71	74
77	43	97	54	65	88	67	65	91	50
82	75	64	76	87	96	84	48	51	47

13. The 100 measurements that make up a single distribution are given below:

a. Group the data into eleven intervals of size 40; the lowest interval should have limits of 720–759. Graph the frequency distribution.

b. Now prepare a frequency-distribution table for these data, grouping the data into five intervals of size 100; the lowest interval should have limits of 700–799. Graph the frequency distribution.

c. What differences are there between the graphs for parts a and b above? What characteristic of the distribution appears when eleven intervals are used but disappears when five are used?

1084	1024	812	784	927	761	1110	816	1069	1071
800	781	941	943	819	952	769	979	881	793
1001	971	994	927	1127	774	953	913	805	950
813	787	922	952	913	931	1030	946	949	938
792	943	1002	909	990	960	992	914	1038	784

970	910	921	1009	980	925	926	972	924	979
809	752	956	728	957	917	754	915	821	939
920	971	920	951	885	962	955	932	933	980
778	918	797	942	847	848	729	840	869	981
1145	743	920	922	935	913	1047	915	878	857

14. For the frequency distribution in problem 9 construct a cumulative proportion graph. From the graph estimate the following:

    a. 50th percentile   d. percentile rank of a score of 50

    b. 37th percentile   e. percentile rank of a score of 90

    c. 85th percentile   f. percentile rank of a score of 25

15. Find in a newspaper or news magazine a frequency distribution graph that has either (1) a cut vertical axis, or (2) an axis that is nonlinearly scaled. Prepare a version of the same graph that has two uncut linearly scaled axes. Which version — yours or the published graph — is clearer? Which version is best suited for publication?

3. Central Tendency: In Search of the Typical

After reading this chapter you should be able to do the following:

1. *Explain the common purpose of all measures of central tendency.*
2. *Determine the mean, median, or mode for a group of observations.*
3. *Compute the mean of a grouped-data frequency distribution.*
4. *Specify the level of measurement required to produce each average from data.*
5. *Explain the different interpretations of the three most commonly used measures of central tendency.*
6. *Determine whether a distribution is positively skewed, symmetrical, or negatively skewed when you are given mean, median, and mode.*

The Search for the Typical

According to an Associated Press article published in 1976,[1] the typical wage-earner in the United States made $13,847 that year. Moreover, he[2] was 38 years old, had a 13-year-old son and an 8-year-old daughter, owed $1,500 on a car he bought two years ago, as well as $14,500 more on his mortgage, and $60 more on household furnishings. These estimates were compiled from U.S. Bureau of Labor Statistics data and they are medians rather than arithmetic averages, which are not good measures of "typical" in skewed distributions. The example on pp. 5–6 of the Introduction illustrates the differential effects of skew on mean and median. But what do they say about you or people you know? Or situations that

---

[1]Published in *The Boston Globe*, September 13, 1976, p. 29.

[2]"He" rather than "she," because the modal sex of wage earners is male (see pp. 75–78 for a discussion of the mode). This is just another way of saying that in 1976 more men than women were wage earners.

are likely to affect you in the future? Such statistics say a great deal — that very, very many people are living hand-to-mouth, for instance — but they also do not say a great deal. The purpose of this chapter is to introduce several such measures of "typical" and show their proper uses.

In the preceding chapter, we began to summarize data by putting the observations into table or graph form. It is much easier to evaluate group characteristics from, say, a frequency polygon of 10,000 IQ scores than it is from a list of all 10,000 scores. However, there are times when we need to carry the summarization process a few steps further; we may want to characterize the entire group by one or a very few statistics.

This is the purpose of the various statistics called **measures of central tendency.** This is a family of methods that produce single numbers that can be interpreted as the typical observations. Measures of central tendency are also called **averages,** and there are many of them to choose from. This chapter covers the three generally most useful averages—the **arithmetic mean** (that is, the **arithmetic average**), the **median,** and the **mode.** Appendix 2 also discusses briefly three measures of central tendency that sometimes turn up in specialized applications—the geometric mean, the harmonic mean, and the quadratic mean. Of course, when someone says "average" without indicating which of these statistics is meant, the arithmetic average is almost always intended.

Averages are extremely useful statistics. In *descriptive* statistics they allow us to represent an empirical frequency distribution by a single number. This way we can describe last year's golf scores, 4 years of college grades, or 100 years of weather observations by a single figure. In *inferential* statistics, population averages may be estimated from a sample to answer questions like What is the average age at which American children begin walking? or How many pounds of beef does the average American eat in a year? Some inferential methods allow us to decide from sample data whether certain population averages are the same or different. If, for example, we want to know whether people over 65 years of age have more automobile accidents than people under 65, we can pose a question like this about averages: Does the population of drivers over 65 have a higher average accident rate than the population of drivers under 65 years of age?

Gambling casinos and insurance companies depend heavily on carefully computed averages. Some people collect large amounts of money from both types of institutions, many people pay money to these organizations, and "on the average" the company comes out ahead. Whether a particular person will pay more than he collects cannot be predicted with certainty, but the *average* transaction *can* be predicted with great precision.

The concept of average is a familiar idea, so familiar that we may overlook the fact that there are several types of averages, that each conveys only a limited amount of information about a group, and that each tells you something quite different about the data. To understand differences among averages, it is necessary to consider the computational basis of each; interpretation of each statistic depends upon knowing how the value of the statistic is determined.

The Mode

The **mode** is the simplest measure of central tendency. When the data have been collected and arranged into a frequency distribution, the mode is the most frequently occurring value of the variable.

If a statistics class has ten freshmen, fifteen sophomores, nine juniors, and two seniors, the *modal* class is "sophomore" simply because that observation turned up more often than any other. Notice that the mode is "sophomore" and not "15." If the entire group of cars in a parking lot consists of nineteen Fords, ten Buicks, and two Volkswagens, the modal car type is "Ford."

This should suggest to you one of the primary virtues of the mode as a measure of central tendency: The mode can be determined for data at any level of measurement, including the nominal level. All that is necessary is that the observations can be organized into a frequency table.

As you can see, determining the mode involves little computation. It can be seen immediately as the highest f value in the frequency table. The modal response in the attitude survey summarized in Table 2.2 is "disagree" since that response is accompanied by the highest number in the f column.

When the data have been grouped, the mode is taken to be the *middle* of the interval with the highest frequency. Thus, modal IQ shown in Table 2.3 is 104.5, the score value halfway between the upper and lower limits of the most frequently occurring interval.

Determining the mode from a frequency graph (bar graph, histogram, or frequency polygon) is also quite easy. When graphs are drawn as described in Chapter 2, look for the highest bar or the highest point on the polygon. The mode will be the value on the horizontal axis below that point. Figure 3.1 is a frequency polygon describing the ages of students at State University. The highest frequency is easily recognized as 1,500 students; the modal age is thus 20 years.

The primary disadvantage of using the mode to describe the typical observation is that its value is unaffected by frequency observations outside the modal interval. Thus, many situations can occur where the mode is hardly the typical observation. In distributions (a), (b), (c), and (d) of Figure 3.2 the mode is the same value, 75, even though the distributions differ radically in shape. On the other hand, distributions (d) and (e) are identical except for very slight frequency differences at the values 50 and 75. Although small, these differences are enough to shift the mode from one end of the distribution to the other. The illustrations in Figure 3.2 are extreme examples of what happens in actual frequency distributions when the frequencies in different categories are nearly equal.

Another way to describe this disadvantage is to say that the mode uses

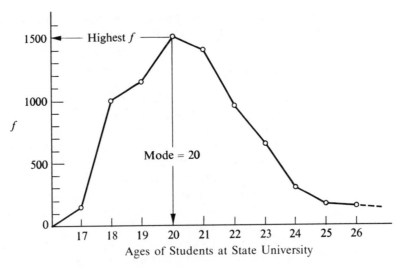

Figure 3.1 Frequency polygon summarizing ages of students at State University. The mode is the value of the variable with the highest frequency, 20 years.

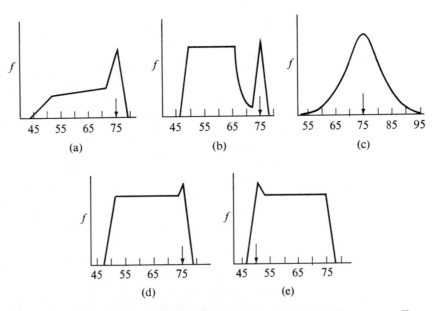

Figure 3.2 The modes of five different frequency polygons are indicated by arrows. Even though the shapes of (a), (b), (c), and (d) are radically different, all four distributions have the same mode. Distributions (d) and (e) are nearly identical frequency distributions with radically different modes.

very little of the information in a distribution; for this reason, the mode is not usually chosen as the single statistic to stand for the entire group when it is appropriate to use the mean or the median. When the observations represent nominal-level data, however, the mode is the only available measure of the typical observation.

You may also notice that the mode is not very informative in small groups of data. In a group of, say, ten test scores, it may occur that no two scores are alike. A related problem occurs for continuous data that have not been grouped; again, it is possible that no two observations are equal.

The mode is perhaps the most useful average for describing the typical member of a distribution like the one in Figure 3.3. This polygon summarizes the ages of people in a certain neighborhood and suggests that many young children and many relatively young parents live there, but not many other people. Computing other averages, such as the mean or median (see pp. 77–86), would produce descriptions of typical as being around 16 or 17 years of age, a group that is hardly represented at all in this distribution. In a frequency polygon like this, we can define *two* modes and say that the age distribution is **bimodal;** modal ages are 4 and 24 years. Distributions may also be **trimodal** (having three modes) or, in general, **multimodal** if more than one mode is defined.

Since there can be more than one mode per distribution, we must redefine the mode as the most frequently occurring value *in its vicinity* on the variable of interest. In Figure 3.3, age 4 is the most frequent value in its vicinity on the age scale and 24 is the most frequent in its vicinity.

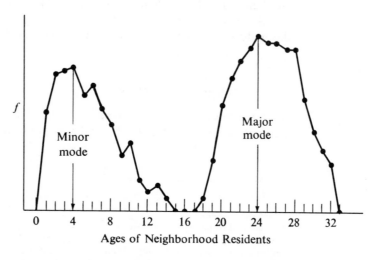

Figure 3.3   Bimodal distribution of ages of residents of a neighborhood, showing major and minor modes. Age categories 4 and 24 are modes, because they have the highest frequencies in their vicinities of the age scale; 23 is not a mode, even though it has a higher *f* than 4, because there is a higher *f* located in the immediate vicinity of 23.

Notice that even though there are more 23-year-olds than 4-year-olds, 23 is not a mode because there is a higher frequency in its immediate vicinity.

Sometimes the most frequently occurring value is called the **major mode** and other modes are referred to as **minor modes.**

The Median

A small class of eleven students took a ten-point quiz and produced the following eleven quiz scores:

9, 9, 6, 4, 7, 6, 8, 2, 9, 10, 3

What is the typical score? the average performance? The modal value is easily determined as 9, but we have seen that the mode can convey very little information about the entire group. We may instead choose the **median** as our average because the value of this statistic depends on the *order* among score values and thus tells us more about the group than the mode does. Strictly speaking, the median is the value on the measurement scale with half of the observations falling above it and half of the observations falling below it. That is, the median is simply the 50th percentile.

To find the median of the quiz scores, first line them up in order from high to low or low to high:

2, 3, 4, 6, 6,	7,	8, 9, 9, 9, 10
Five scores below	Middle value	Five scores above

Then find the middle value, the one with an equal number of scores above and below it. The median is taken to be the middle value.[3]

If there is an even number of scores in the group, you will have to find the middle *pair* of values. Suppose we had the following ten scores:

2, 3, 4, 6, 6, 7, 7, 9, 9, 10

Middle pair of values

Here the median may be taken as the value halfway between each member of the middle pair. To find this point, add the two values and divide the total by 2:

$$\text{Median} = \frac{6 + 7}{2} = \frac{13}{2} = 6.5$$

---

[3]This method of finding the median will generally be acceptable in most situations. Occasionally, however, when more than two scores share the middle value, and there are relatively few observations involved, computing the median becomes a slightly more complicated matter. Consider the distribution: 2, 3, 3, 4, 7, 7, 7, 7, 7, 8, 9. The middle score is 7, but four other scores also share that value. In such cases, one has to determine where the median lies between the upper and lower true limits of the value 7, that is, where exactly the median falls between 6.5 and 7.5. Here, the middle score stands second among five scores that are equal. The median is thus two-fifths the difference between 6.5 and 7.5 above the lower true limit of 6.5. The true median is thus 6.5 + 0.4 = 6.9.

It may come as a surprise that the median can assume a value that does not actually appear in the group of observations. Remember that the statistic is defined as the point on the measurement scale with half of the values above it and half of the values below it, *not* the observed score that meets this criterion.

Since the median is determined by the rank order of observations, it cannot be computed for data unless the observations can be arranged in an order. Thus, a median cannot be computed for nominal-scale data. Ordinal-, interval-, ratio-, and absolute-scale data *can* be characterized by a median. Ten soldiers lined up in order of rank might look like this:

sgt, sgt, sgt, cpl, cpl, cpl, pfc, pfc, pvt, pvt
\/
Middle pair of values

sgt = sergeant
cpl = corporal
pfc = private,
    first class
pvt = private

The ranks are ordinal-level data with a median rank of corporal. If the two middle values had been different ranks, the median might have been specified as sgt–cpl. When the data are at the interval, ratio, or absolute levels, the point between two middle values can be computed, as shown with the quiz scores.

The median has one characteristic that makes it the best average for describing some distributions: Its value is unaffected by the presence of a few extremely high or extremely low values in the distribution. The median is used quite often when the distribution is heavily skewed, that is, when many observations are at one part of the scale and only a few values appear far above or far below that group.

In a sociological study the annual incomes of four groups of people were determined. The incomes of the people in each group were lined up in order, and the median was taken as the middle value in each case (see Table 3.1). Groups 1 and 2 form symmetrical distributions; each has observations at the same distances above and below the middle point. Group 3 is positively skewed (one extremely high value), and group 4 is negatively skewed (one extremely low value). All groups have exactly the same median because its value is determined entirely by the value of the middle score; the rank orders of the observations determine which score

Table 3.1   *Annual Incomes of Four Groups of People*

GROUP 1	GROUP 2	GROUP 3	GROUP 4
$ 3,000	$6,000	$ 6,203	$ 100
5,000	6,500	6,872	6,789
7,000	7,000	7,000	7,000
9,000	7,500	11,500	7,210
11,000	8,000	100,000	7,595
	Median in all groups = $7,000		

is the middle one. Values of observations above and below the middle are not reflected in the median's value.

Teachers quite often use the median test score to specify the average, because it is not uncommon for one or a few students to score far above or far below the rest of the group. Unlike the arithmetic mean (pp. 80–86), the median is relatively insensitive to the presence of these extreme values. The median is not the average to use, however, if you *do* want a measure of central tendency that reflects the value of every group member.

Another advantage to the median is that it can be computed when some values are missing from the high or the low end of the distribution. In such cases, the distribution is said to be open-ended. Suppose, for example, that a group of ten people were asked to list their ages so that the group's median age could be determined. Two of the older adults in the group, however, wish to keep their ages secret. Since they are obviously older than the median age of the group, the median can still be computed. The unknown ages are simply registered as ? when the observations are lined up in order:

7, 8, 9, 14, 17, 19, 29, 37, ?, ?

Median age $= \dfrac{17 + 19}{2} = 18$ years

In such cases, it is only necessary to know whether the missing values lie above or below the median; the value of the median is unchanged regardless of the missing values.

When the group contains many observations, lining them up in order and finding the middle value can be impractical. Since the median is equivalent to the 50th percentile, you may prepare a cumulative-proportion table for the distribution and estimate the median from a cumulative-proportion graph (pp. 55–61) or compute the estimate directly from the table in Appendix 3.

The Arithmetic Mean

The **arithmetic mean** is also called the **arithmetic average** by statisticians and simply the average by most of us. Here it will be called the **mean,** even though that designation also belongs to several other statistics.[4] It is by far the most popular and generally useful measure of central tendency. Although it is a familiar statistic, it has some strong points and some limitations of which you may not be aware.

Several different symbols are commonly used to represent the mean, including the following:

---

[4]See Appendix 2 for a brief introduction to the harmonic mean, the geometric mean, and the quadratic mean.

μ — (Pronounced "mue") Represents the mean of a population in inferential applications. Calculation of μ involves every member of the population.

$\bar{X}$ — (Bar X) In inferential statistics $\bar{X}$ is the mean of a sample. In descriptive statistics it is the mean of a small group.

M — Another symbol used by some authors to indicate $\bar{X}$.

The mean is usually defined as the sum of all values divided by the number of values. Thus, if a halfback carries the football three times and runs 7.1 yd, 5.2 yd, and 3.4 yd on the three carries, the mean yards per carry for this group of observations is:

$$\bar{X} = \frac{7.1 + 5.2 + 3.4}{3} = \frac{15.7}{3} = 5.23 \text{ yd}$$

He runs a mean 5.23 yd/carry.

Notice that every value in the group contributes to the value of the mean. If nine families each have one child, the mean number of children per family is computed as follows:

$$\bar{X} = \frac{1 + 1 + 1 + 1 + 1 + 1 + 1 + 1 + 1}{9} = \frac{9}{9} = 1$$

If, however, another family with eleven children is added to the group, the mean number of children becomes:

$$\bar{X} = \frac{1 + 1 + 1 + 1 + 1 + 1 + 1 + 1 + 1 + 11}{10} = \frac{20}{10} = 2$$

Adding one value of 11 in this case drastically changes the picture of the typical observation specified by the mean. Unlike the median and the mode, the mean is sensitive to the presence of a few high or low values in the distribution.

Computing the Mean: Subscripts and Summation

One reason for the popularity of the mean as a measure of central tendency is that it can be symbolized conveniently. This makes it easy to use as part of formulas for more complex statistical operations. In order to appreciate this virtue and — more importantly — in order to produce a formula for computing the mean, we must briefly cover here the notions of **subscripted variables** and **summation notation.** If you are not familiar with these concepts, it is important that you learn them now before moving on; they will turn up many times in the chapters that follow.

Why is the mean so often symbolized as $\bar{X}$? The X in the symbol indicates that we are dealing with the mean of a number of observations, all of which represent values of some variable "X." The bar above designates the mean of the "X" observations.

Table 3.2   *Example Data for Subscripted Variables X_i and Y_i*

PERSON i	APTITUDE SCORE X_i	ACHIEVEMENT SCORE Y_i
1	75	81
2	84	81
3	93	96
4	76	75
5	85	89
6	71	69
7	87	86
8	64	67
9	78	82
10	82	89

To make an example, let us say that X represents the scores of ten people on an aptitude test, and Y represents their scores on an achievement test. Table 3.2 presents these scores.

Now we need a way to refer to individual scores. The symbol i will become a subscript and X and Y will become **subscripted variables.** Any single aptitude score can be referred to as X_i, for instance. If we want person 1's aptitude score, we want X_1; if we want the aptitude score for person 9, then we want X_9. Similarly, if we want the achievement score for person 1, we want Y_1 and if we want person number 8's achievement score, we want Y_8. From the data in Table 3.2, then, $X_7 = 87$, $X_4 = 76$, $Y_6 = 69$, and $Y_{10} = 89$.

Next, we need a way to indicate that a certain group of observations are to be added up; this is the purpose of **summation notation.** What follows may look complex if you have never seen it before, but it is simply a set of instructions for adding up data points. If, for instance, we want to add up all of the ten X_i values, we write:

$$\sum_{i=1}^{10} X_i$$

This group of symbols has a number of parts, and each part specifies something about the addition process we are to do. The $\sum$ is a capital Greek sigma, and it means, "Take the sum of some values." The X_i to the right means we will be summing individual values of the variable X. Notice that $i = 1$ appears below the $\sum$ and 10 appears above the sigma. The notation $i = 1$ specifies that we start with X_1 and the 10 indicates that we use all values up to and including X_{10}. Thus, X_{10} is the last value to be added. For the data in Table 3.2:

$$\sum_{i=1}^{10} X_i = X_1 + X_2 + X_3 + X_4 + X_5 + X_6 + X_7 + X_8 + X_9 + X_{10}$$
$$= 75 + 84 + 93 + 76 + 85 + 71 + 87 + 64 + 78 + 82$$
$$= 795$$

Similarly:

$$\sum_{i=1}^{10} Y_i = Y_1 + Y_2 + \cdots + Y_{10} \quad \text{(The symbols} \cdots \text{are also a way}$$
of saying that we "add Y_1 and Y_2 *and so on* until we reach Y_{10}.)

$$= 815$$

The summation notation can also specify that we add up only some of the values. For instance:

$$\sum_{i=1}^{4} X_i = X_1 + X_2 + X_3 + X_4 = 75 + 84 + 93 + 76 = 328$$

$$\sum_{i=4}^{8} Y_i = Y_4 + Y_5 + Y_6 + Y_7 + Y_8 = 75 + 89 + 69 + 86 + 67 = 386$$

In each of these cases, the first value to be added is specified under the summation sign, and the last value to be added is indicated by the subscript number above the summation sign.

Now, if we use the symbol N to stand for the number of observations in a group of scores, we can describe computation of the mean with the symbols:

$$\bar{X} = \frac{\sum_{i=1}^{N} X_i}{N}$$

This says, "Add up all of the X values, beginning with X_1 and ending with X_N (remember that here N = 10), and then divide this sum by N." For these data:

$$\bar{X} = \frac{\sum_{i=1}^{N} X_i}{N} = \frac{795}{10} = 79.5$$

$$\bar{Y} = \frac{\sum_{i=1}^{N} Y_i}{N} = \frac{815}{10} = 81.5$$

⊡ The Mean From Grouped Data

When the data have been grouped, the exact mean cannot be computed because not every value in the distribution is specified. All values within each interval are treated as equal, even though some may have fallen near the upper limit and others near the lower limit. This is one characteristic of the grouping process: We lose some of the original information about the distribution because the frequency table does not allow us to recover every original observation. Nonetheless, the mean can still be *estimated* from the grouped-data table.

Table 3.3 *Summary of Dollar Values of Grocery Sales for Grouped-Data Computational Example*

INTERVAL	VALUE OF SALE ($)	f_i	X_{mi}
5	37–45	100	41
4	28–36	150	32
3	19–27	300	23
2	10–18	500	14
1	1–5	400	5

Suppose you wish to find the mean dollar value of a day's grocery sales at a nearby supermarket by examining the accumulated cash-register slips for that day. The dollar values listed on the slips are summarized in Table 3.3. An extra column has been added, called X_{mi}. Values in this column show the **midpoints** of the dollar intervals, that is, the values halfway between the upper and lower limits of each interval.[5] When you compute the mean for grouped data, X_{mi} will stand for every observation in the interval.

The algebraic formula for computing the mean from grouped data is simpler to use than it looks:

$$\bar{X} = \frac{\sum_{i=1}^{k} f_i X_{mi}}{\sum f_i}$$ (Notice that with grouped data, i stands for an interval and not for an individual data point.)

Let us look at the parts of this entity.

f_i represents the frequency in interval i

X_{mi} represents the midpoint of interval i

k represents the number of intervals; here $k = 5$

Thus, to follow the instructions for the grouped-data mean, we first deal with a subscript value of $i = 1$. In the numerator of the fraction, we are thus told to multiply the frequency of interval 1 (f_i) by the midpoint of interval 1 (X_{mi}); then we change i to 2 and do the same thing for interval 2, and so on, until we have added up all the products. In the denominator, we simply add up the total frequencies. In numbers:

$$\bar{X} = \frac{\sum_{i=1}^{k} f X_{mi}}{\sum_{i=1}^{k} f_i}$$

---

[5]To find each X_{mi} add the interval's lower limit to the upper limit and divide by 2.

$$= \frac{(400)(5) + (500)(14) + (300)(23) + (150)(32) + (100)(41)}{400 + 500 + 300 + 150 + 100}$$

$$= \frac{2000 + 7000 + 6900 + 4800 + 4100}{1450}$$

$$= \frac{24,800}{1450}$$

$$= \$17.103 = \$17.10$$

This would probably differ slightly from the value of $\bar{X}$ that would appear if we used the actual value of each grocery sale in the computations. However, when the data have been grouped, it is the best estimate of the mean that we can produce.

As another example, Table 3.4 summarizes computation of the grouped-data mean for the 10,000 IQ scores listed in Table 2.3. Before

Table 3.4 *Computation of Grouped-Data Mean for 10,000 IQ Scores*

i	IQ	X_{mi}	f_i	$f_i X_{mi}$
14	160–169	164.5	3	3(164.5) = 493.50
13	150–159	154.5	20	3,090.00
12	140–149	144.5	110	15,895.00
11	130–139	134.5	310	41,695.00
10	120–129	124.5	820	102,090.00
9	110–119	114.5	1812	207,474.00
8	100–109	104.5	2352	245,784.00
7	90–99	94.5	2300	217,350.00
6	80–89	84.5	1450	122,525.00
5	70–79	74.5	560	41,720.00
4	60–69	64.5	200	12,900.00
3	50–59	54.5	40	2,180.00
2	40–49	44.5	20	890.00
1	30–39	34.5	3	103.50

$$\sum_{i=1}^{14} f_i = 10,000 \qquad \sum_{i=1}^{14} f_i X_{mi} = 1,014,190.00$$

(Each $X_{mi} = \dfrac{\text{lower limit} + \text{upper limit}}{2}$. For example, for

interval 70–79, $X_{mi} = \dfrac{70 + 79}{2} = 74.5$.)

$$\frac{\displaystyle\sum_{i=1}^{14} f_i X_{mi}}{\sum f_i} = \frac{1,014,190.00}{10,000} = 101.42$$

computing the mean, it is important to be sure that the data represent at least interval-level measurements. The value of the mean is sensitive to *distances* on the measurement scale, and distances on nominal- and ordinal-level scales do not have meaning.

Choosing a Measure of Central Tendency

The mean, the median, and the mode all have different virtues and different limitations. The following summary of the preceding sections may help you to choose among these averages for specifying the typical group member.

>   Mode — (Most frequently occurring value.) The only measure of central tendency available when the observations represent nominal-scale data.
>
>   — Appropriate when the distribution has two or more modes and you wish to describe members of each modal group as typical.
>
>   — Not very informative with small numbers of observations or ungrouped continuous data, situations where every observation is likely to be different from every other observation.
>
>   Median — (Value with half the distribution falling above it and half the distribution falling below it.) Appropriate only for data at the ordinal level or higher.
>
>   — Its value is not influenced by the presence of a few extremely high or low values in the distribution; it is thus used in heavily skewed distributions.
>
>   Mean — (Sum of all values divided by the number of values.) Requires at least interval-level data.
>
>   — Reflects the value of every member of the group.
>
>   — Its value is sensitive to the presence of a few high or low values in the group.

Since each measure conveys different information about a group, the most comprehensive description of the typical observation can be provided by specifying all three averages for a single group.

Mean, Median, and Mode in Skewed Distributions

Sometimes it is very difficult to detect skew from a frequency-polygon graph; empirical frequency distributions are usually jagged, with an overall shape that may not resemble the textbook shapes shown in Chapter 2. At other times you may want to examine a distribution for skew without actually graphing a polygon. In these cases, skew can be determined by comparing the mean, the median, and the mode.

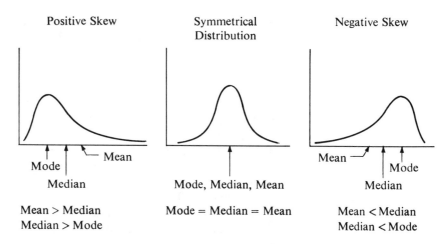

Figure 3.4   Relations among mean, median, and mode in distributions of three different shapes.

Remember that the mean is highly sensitive to the presence of a few extreme scores in a distribution. Of all the measures of central tendency, the mean will be pulled most toward the tailed end of a skewed distribution. The median is much less sensitive to a few extreme scores, and the mode is not influenced at all by observations outside the modal category. This means that in a *positively* skewed distribution the mean will be greater than the median and the median will, in turn, be greater than the mode. Conversely, in a *negatively* skewed distribution, the mean will have a lower value than the median, which will have a lower value than the mode. In a perfectly symmetrical distribution all three measures of central tendency will be the same. These relations are illustrated in the frequency polygons shown in Figure 3.4.

Can you tell from looking at Table 2.4 whether the distribution of chemistry exam scores from Adams H.S. students is positively skewed, symmetrical, or negatively skewed? The modal value for this group is the midpoint of the interval with the highest frequency, 72.5. The median for this group was estimated earlier to be 67. From the grouped-data table the value of the mean is computed to be 64.8. Thus, the mode is greater than the median, and the median is greater than the mean, suggesting a *negatively* skewed distribution.

SUMMARY

Measures of central tendency are methods that produce single numbers specifying the typical observation. These statistics are also called averages. The most common averages are the mode, the median, and the arithmetic mean. Each has its strong points and its limitations.

The mode is simply the most frequently occurring observation in the group. It is easy to determine from a frequency table or graph, and it can be applied to data from all levels of measurement. When a frequency distribution has two peaks, it may be useful to describe the distribution as bimodal; in this case, each mode is the most frequently occurring value in its vicinity on the measurement scale. The major weakness of the mode is that its value is insensitive to observations falling outside of the modal interval.

The median is defined as the point on the measurement scale with half of the observations falling above it and half falling below it. It is thus appropriate only for data that can be rank-ordered, that is, for ordinal-level data or higher. For an odd number of ungrouped observations lined up in order, the median is taken as the middle value. When there are an even number of values, the median is halfway between the two middle values. If the data have been grouped, the median can only be estimated. A simple way to estimate it is to find the point on a cumulative proportion graph corresponding to a cumulative proportion of 0.50.

By far the most familiar and useful measure of central tendency is the arithmetic mean (or arithmetic average). The mean can be defined as the sum of the values divided by the number of values in the group. It differs from the median and mode in that every member of the group contributes to the value of the mean. This characteristic makes its value sensitive to the presence of a few extremely high or extremely low values in the group — a property that may or may not be desirable, depending on the purpose of the statistician. Data must be at the interval level or higher to be appropriate raw material for the mean.

Since each measure of central tendency conveys a different kind of information about the group, the most comprehensive picture of the typical observation is provided by specifying all three averages for the group.

Comparing the relative values of mean, median, and mode is a useful way of detecting skew in a distribution. In positively skewed distributions the mean is greater than the median, and the median is greater than the mode. This order is reversed in negatively skewed distributions. Because of its sensitivity to extreme values, the mean is pulled furthest in the direction of the skew.

KEY CONCEPTS

measures of central tendency	arithmetic mean (arithmetic average)	subscripted variable
averages	mode	summation notation
	median	midpoint (of an interval)

SPOTLIGHT 3   **On the Average**[6]

*The figure 2.2 children per adult female was felt to be in some respects absurd,
and a Royal Commission suggested that the middle classes be paid money to
increase the average to a rounder and more convenient number.*

— PUNCH

In former times, when the hazards of sea voyages were much more
serious than they are today, when ships buffeted by storms threw a
portion of their cargo overboard, it was recognized that those whose
goods were sacrificed had a claim in equity to indemnification at the
expense of those whose goods were safely delivered. The value of the lost
goods was paid for by agreement between all those whose merchandise
had been in the same ship. This sea damage to cargo in transit was known
as *havaria* and the word came naturally to be applied to the compensation
money which each individual was called on to pay. From this Latin word
derives our modern word "average." Thus the idea of an average has its
roots in primitive insurance. Quite naturally with the growth of shipping,
insurance was put on a firmer footing whereby the risk was shared, not
simply by those whose goods were at risk on a particular voyage, but by
large groups of traders. Eventually the carrying of such risks developed
into a separate skilled and profit-making profession. This entailed the
payment to the underwriter of a sum of money which bore a recognizable
relation to the risk involved.

The idea of an average is common property. However scanty our
knowledge of arithmetic, we are all at home with the idea of goal averages, batting and bowling averages and the like. We realize that the
purpose of an average is *to represent a group of individual values* in a
simple and concise manner so that the mind can get a quick understanding
of the general size of the individuals in the group, undistracted by fortui-
tous and irrelevant variations. It is of the utmost importance to appreciate
this fact that the average is to act as a *representative*. It follows that it is
the acme of nonsense to go through all the rigmarole of the arithmetic to
calculate the average of a set of figures which do not in some real sense
constitute a single family. Suppose a prosperous medical man earning
£3,000 a year had a wife and two children none of whom were gainfully
employed and that the doctor had in his household a maid to whom he
paid £150 a year and that there was a jobbing gardner who received £40 a
year. We can go through all the processes of calculating the average
income for this little group. Six people between them earn £3,190 in the
year. Dividing the total earnings by the number of people, we may
determine the average earnings of the group to be £531 13s. 4d. But this
figure is no more than an imposter in the robes of an average. It represents
not a single person in the group. It gives the reader a totally meaningless

---

[6]From M. J. Moroney, *Facts from Figures* 2d ed. (London: Penguin Books, 1953), pp.
34–35. © M. J. Moroney, 1951. Reprinted by permission of Penguin Books Ltd.

figure, because he cannot make one single reliable deduction from it. This is an extreme example, but mock averages are calculated with great abandon. Few people ask themselves: What conclusions will be drawn from this average that I am about to calculate? Will it create a false impression?

The idea of an average is so handy that it is not surprising that several kinds of average have been invented so that as wide a field as possible may be covered without misrepresentation. We have a choice of averages; and we pick out the one which is appropriate both to our data and our purpose. We should not let ourselves fall into the error that because the idea of an average is easy to grasp there is no more to be said on the subject.

PROBLEMS

1. Ten six-year-olds were given an intelligence test, producing the following scores:

Child	Score	Child	Score
1	120	6	110
2	97	7	107
3	134	8	115
4	106	9	94
5	85	10	106

If the score for child i is called X_i, what are the values of

a. $\sum_{i=1}^{N} X_i$ b. $\sum_{i=1}^{5} X_i$ c. $\sum_{i=7}^{N} X_i$ d. $\sum_{i=1}^{N} X_i^2$

2. Using the data in Table 3.2, give the numerical values of

a. $\sum_{i=1}^{5} X_i$ d. $\sum_{i=1}^{4} X_i Y_i$

b. $\sum_{i=3}^{9} X_i^2$ e. $\sum_{i=1}^{5} X_i + \sum_{i=1}^{4} Y_i^2$

c. $\sum_{i=1}^{5} (X_i + Y_i)$ f. $\left(\sum_{i=2}^{5} X_i \right)^2$

3. For the scores given in problem 1 above, what is the mean? The median?

4. For the following set of numbers, determine the mean, median, and mode:

$$19 \quad 26 \quad 37 \quad 14 \quad 21 \quad 18 \quad 26 \quad 7 \quad 24 \quad 22 \quad 21$$
$$26 \quad 18 \quad 13 \quad 21 \quad 19 \quad 27 \quad 23 \quad 31 \quad 37 \quad 31 \quad 29$$

5. Which measure of central tendency
   a. May be applied to nominal-scale data?
   b. Is most easily influenced by a few extreme score values?
   c. Has half of the score values in a distribution below it and half above it?
   d. Is easiest to manipulate algebraically?
   e. Should be chosen if you want to show the typical score value but do not want a few extremely high scores at one end of a distribution to influence the value of the statistic? (There is only one of these per distribution.)
   f. Should be chosen if you want a statistic to reflect the value of every group member?

6. For each group of measurements below, determine the mean, median, and mode(s). Also indicate any measures of central tendency that should not be used for a particular distribution because they might be misleading.
   a. Ages of people in class:
      17 18 21 25 23 24 19 24 36
   b. Noontime temperatures for one week:
      62° 78° 61° 71° 52° 80° 61°
   c. Scores on one class's statistics exam:
      54 69 76 52 49 95 41 52 68 87
      58 78 62 74 86 49 61 98 75 84
      89 75 88 68 89 75 14 86 71 64

7. For each group of observations in problem 6, "throw out" the highest and lowest values. Now recompute each mean and median. Show the change (if any) in mean and median that occurred as the result of excluding the highest and lowest values.

8. What are the grouped data means for the following distributions?

a. Interval	f	b. Interval	f
46–54	22	601–700	620
37–45	31	501–600	975
28–36	54	401–500	1,140
19–27	27	301–400	1,270
10–18	19	201–300	372
1–9	3		

9. A football team's members have the weights given below (in pounds). Compute the mean and median:
   191 250 224 296 241 187 301
   182 221 280 178 237 213 204
   191 212 217 240 206 204 198

10. Using the summary tables in problems 8 and 9 in Chapter 2, compute the grouped-data means.

11. Prepare a cumulative-proportion graph for the IQ scores tabulated in Table 2.3. Use the graph to estimate the median IQ. Compute the grouped-data mean for these scores. For this distribution, would you consider either mean or median as acceptable measures of the typical observation?

12. Prepare a cumulative-proportion graph for the data in problem 9, Chapter 2. Use the graph to estimate this group's median. Comparing the median and grouped-data mean, decide whether this group is positively or negatively skewed.

13. For some data, a process called *coding* makes computation of the mean easier. One kind of coding involves subtracting the same value from every score and then computing the mean of the remainders; then, the same value is added to the mean of the remainders to find the original score mean. To demonstrate coding:

 a. First compute the mean of these data in the ordinary manner:

    1,010   1,024   1,036   1,029   1,045

    1,020   1,027   1,027   1,031   1,037

    1,047   1,018   1,032   1,024   1,029

 b. Subtract 1,000 from each value and compute the mean of the remainders. Then add 1,000 to your computed value. Does this produce the same mean that you found in part *a*?

14. Fifty students took a chemistry exam and produced the list of scores given in problem 12, Chapter 2 (p. 71). These scores are summarized in the grouped-data frequency distribution:

SCORES	f
91–100	6
81–90	13
71–80	12
61–70	8
51–60	3
41–50	5
31–40	3

 a. Compute the grouped data mean for this distribution.

 b. Using the scores listed on p. 71, compute the raw-data mean.

 c. Explain why the grouped-data mean and raw-data mean are not exactly equal to each other.

15. Prepare a cumulative-proportion graph for the grouped-data test score distribution given in problem 14 above. From the graph, estimate this group's median score and compare it to the grouped-data mean. On the basis of this comparison, how would you evaluate this group with respect to skew?

4. Variability: How Much Do Group Members Differ from Each Other?

After reading this chapter you should be able to do the following:

1. *Explain the concept of variability and the purpose of measures of variability.*

2. *Select, compute, and interpret the appropriate measure of variability for a group of data.*

3. *Discuss the virtues of standard scores.*

4. *Compute standard scores for raw scores and convert them to transformed standard scores.*

Variability

The big football game between the Zephyrs and the Zebras is scheduled for next Saturday. Neither coach will release statistics on individual players, but we do know that both teams have approximately equal offensive units and we have sent spies into the training camps to try to find out something about the defensive units. The spies report that the mean weight of the Zephyrs' defensive line is 200 lb and the mean weight of the Zebras' defensive line is 201 lb. It looks like we're in for a close game, and with nothing else to go on we put our money on the Zebras.

When Saturday arrives, however, we discover another shortcoming of measures of central tendency like the mean. Searching through the program, we find the weights of the defensive linemen:

ZEPHYRS		ZEBRAS	
Arlo	215 lb	Ed	146 lb
Bob	180 lb	Frank	166 lb
Chuck	200 lb	George	191 lb
Dick	205 lb	Hank	301 lb

A quick check shows the spies' information to be accurate:

$$\bar{X}_{\text{Zephyrs}} = \frac{800}{4} = 200 \text{ lb}; \qquad \bar{X}_{\text{Zebras}} = \frac{804}{4} = 201 \text{ lb}$$

As the Zephyr offense proceeds to avoid Hank's end of the line and run all its plays through Ed's position, we realize that some important characteristics of a distribution are not specified by the mean. Both groups have nearly the same mean weights, but Zephyr weights are all close to the typical value while most Zebra weights are not. The two groups differ with respect to **variability** — the extent to which each group's members differ among themselves.

Often it is useful to specify the degree of variability within a group, and statistics designed for this purpose are called, naturally, **measures of variability.** Specifying one of these *and* a measure of central tendency conveys a great deal of information about a group. Together, the two statistics describe both the typical member and the spread of the group.

Figure 4.1, showing three different frequency distributions, illustrates the need for variability indices. It shows mathematics exam scores for three groups of students. All have the same mean score (60), but all differ in variability. What is needed is a single statistic to indicate that class A's scores were spread all over the test-score scale, while class C's were bunched together.

The figure also suggests another use for measures of variability. Suppose you took the math exam and scored a 75. How would you stand with respect to your classmates? That would depend on which class you were in. A score of 75 is above the mean by fifteen points in all classes, but in Class C it is above almost all the other students. In class A many people scored above 75. Thus, 75 has a different *relative* standing in each group. In the discussion on z-scores we will use a measure of variability together with a measure of central tendency to specify an individual's standing within the group.

A measure of variability is a single statistic whose value is relatively large when the group members are relatively scattered and relatively small when the values are bunched together. We will consider four such measures: (1) the **range;** (2) the **interquartile range;** (3) the **mean deviation;** and (4) the **standard deviation** (and its relatives, **variation** and **variance**).

Just as measures of central tendency, each of these measures of variability has advantages and limits. Choosing the best measure depends on the characteristics of the data and the message you wish to convey. Understanding and interpreting each requires a knowledge of the computational basis of the statistic.

The Range

The **range** is already familiar to us from the discussion of grouped data (pp. 45–48), and it is a concept that you probably understood long before reading Chapter 2. We should, however, take a look at its precise computation and some of its weaknesses.

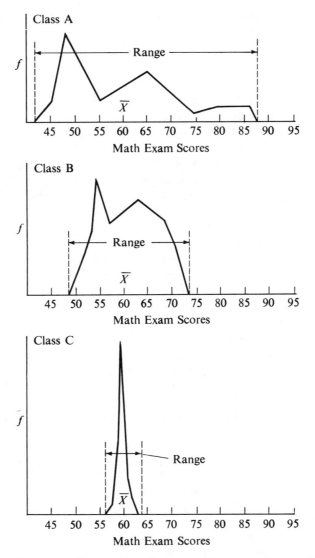

Figure 4.1   Frequency distributions of mathematics exam scores produced by three classes. Even though all groups have the same mean score, the three distributions are quite different. One important difference among classes is the extent to which class members differ among themselves, that is, variability. One measure of variability for a group is the range of scores obtained; ranges obtained by each class are shown as distances on the figure.

Essentially, the range is the *distance on the measurement scale* between the highest and lowest values in the group. Figure 4.1 shows these distances for the three distributions of exam scores. When the group of observations has been collected, the range is computed as follows:

Range = Highest value − lowest value

+ 1(smallest measurement increment specified)

For the two groups of football players, the ranges of defensive linemen weights are:

$$\text{Range}_{\text{Zephyrs}} = 215 - 180 + 1(1) = 35 + 1 = 36 \text{ lb}$$
$$\text{Range}_{\text{Zebras}} = 301 - 146 + 1(1) = 155 + 1 = 156 \text{ lb}$$

If only our spies had informed us that the Zephyr line had a mean weight of 200 lb *and* a range of 36 lb, while the Zebras had a mean of 201 lb *and* a range of 156 lb. That news would have been a more informative description of the two groups.

You may be wondering why the last term in the expression is included: 1(smallest measurement increment specified). This is a very important point, since we are concerned with the distance (on the weight scale) covered by the observations. For the football players, the smallest weight unit specified was 1 lb, that is, there were no readings that included fractions of pounds. The range of weights is the distance between the lower true limit of the lowest weight reported and the upper true limit of the highest weight reported; since the weights are reported in 1-lb units, these true limits are located 0.5 lb below the lower limit and 0.5 lb above the upper limit. Thus, it is necessary to add 1(1 lb) to the difference between high and low values to cover the appropriate range. Figure 4.2 also illustrates why the addition is necessary.

Table 4.1 shows computation of the range for some groups where the data are reported with varying degrees of precision.

All the preceding applies, of course, only to data that come from a measurement scale where distances have meaning, namely, interval, ratio, or absolute scales. When the data are measurements on an ordinal

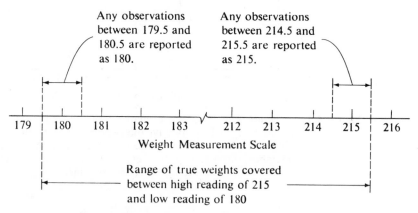

Figure 4.2   Demonstration of the range as a distance on the measurement scale. The distance cannot be computed simply by subtracting the lowest observation from the highest, because the lowest reported value might include some observations with true values below it, and the highest reported value might include some observations with true values above it. These extra distances are included in the value of the range by adding 1 (smallest measurement increment reported) to the difference between high and low observations.

Table 4.1 *Smallest Measurement Increment and Range Obtained for Data Reported with Varying Degrees of Precision*

TYPICAL OBSERVATIONS	HIGH VALUE	LOW VALUE	SMALLEST MEASUREMENT INCREMENT	RANGE
6.8, 4.9, 8.1, 3.7	10.6	3.3	0.1	7.4
6.85, 4.93, 8.18	10.64	3.34	0.01	7.31
4,000, 6,000, 3,000	10,000	3,000	1,000	8,000
$98\frac{4}{5}$, $102\frac{1}{5}$, $57\frac{2}{5}$	$115\frac{3}{5}$	$55\frac{1}{5}$	$\frac{1}{5}$	$60\frac{3}{5}$

scale, the procedure is much simpler — list the high and low values. We could say that a group of soldiers ranged in rank from captain to private. The range, of course, is not used with nominal-scale data, since nominal-scale observations cannot be rank-ordered.

Although easily computed and easily understood, the range is limited as a measure of variability because it has some serious weaknesses. Its value is based only on the highest and lowest values; it tells us nothing about variability among group members between the extremes. Consider, for example, a class of sixty-three students who reported the number of miles they traveled from home to attend college. Most had come between 5 and 100 miles, but one person had come 1,800 miles and one person only 1 mile. The latter two values represented the highest and lowest values reported, so the range of distances represented was:

Range = 1,800 − 1 + 1(1) = 1,800 miles

Then one more student transferred into the class who had traveled 3,000 miles from home; the range then became:

Range = 3,000 − 1 + 1(1) = 3,000 miles

Adding one extreme value drastically alters the value of the range in this case. While the range accurately portrays the difference between high and low values, it can be a misleading statistic if the intent is to characterize dispersion of the entire group. Other statistics, such as the interquartile range and the standard deviation, do not change values quite so drastically when the highest or lowest values are altered.

The Interquartile Range and the Semi-Interquartile Range

Instead of measuring the distance between highest and lowest values to obtain an idea of a distribution's spread, we can measure the distance between the 75th and 25th percentiles, the third and first quartiles, respectively. This distance is known as the **interquartile range,**[1] and as a

---

[1] The interquartile range is often called simply the *quartile range*.

measure of variability it is more stable than the simple range, because one can change the highest and lowest values without altering its value.

When Q_3 and Q_1 have been determined (either through computation or estimation from a cumulative graph), the interquartile range is found by direct subtraction:

Interquartile range $= Q_3 - Q_1$

During the year 1967 the 25th percentile (Q_1) for annual family incomes in the United States was about \$4,900; the 75th percentile (Q_3) was about \$12,150.[2] Thus, the interquartile range of family incomes in 1967 was:

\$12,150 $-$ 4,900 $=$ \$7,250

The interquartile range is similar in some respects to the median; it is applicable with data at the ordinal level or higher, and like the median, it is particularly useful in very skewed distributions where a few extreme values might distort the value of other variability indices such as the standard deviation (pp. 101–103). The median family income in the United States in 1967 was about \$8,000,[3] and this information combined with the interquartile range of \$7,250 gives a more accurate picture of the way family incomes were distributed than the mean and standard deviation might.

Often, statisticians divide the interquartile range by 2 to produce the **semi-interquartile range.** For the distribution of annual incomes, this would be:

$$\text{Semi-interquartile range} = \frac{Q_3 - Q_1}{2} = \frac{\$12,150 - 4,900}{2} = \$3,625$$

Of course, the semi-interquartile range has the same strong and weak points as the interquartile range. It is the statistician's personal preference that dictates which of the two is used.

The Mean Deviation

The range and interquartile range each have advantages and are useful in certain situations, but neither statistic reflects the value of every group member. When we want a measure of variability that is determined by all the observations, we may choose either the **mean deviation** or the **standard deviation.** The standard deviation is by far the most common and useful variability index, while the mean deviation is rarely used. However, understanding the mean deviation makes the logic underlying the standard deviation a little more understandable, so we will discuss it here, even though you may never meet it in print or be called upon to compute it.

---

[2] U.S. Bureau of the Census figures.
[3] U.S. Bureau of the Census figures.

Instead of looking only at two extreme points in a distribution, we can approach the problem of describing variability by indicating the degree to which group members differ, or *deviate*, from some fixed point on the measurement scale. The more that values tend to be scattered from each other, the more they will tend to differ, overall, from any single reference value we choose. But what value shall we choose as the reference point? And how shall we specify how much group members differ from it?

We could measure each observation's deviation from the value zero, or each observation's deviation from the highest value in the group, or any other point. The group's arithmetic mean, however, is the ideal reference point for this purpose for two mathematical reasons: (1) As we shall see, the sum of the deviations about the mean is zero (which says that the mean is, in effect, the group's center of gravity); and (2) the sum of the squared deviations about the mean is lower than the sum of squared deviations about any other point on the measurement scale. These properties are important because, in effect, they ensure that the value of our variability index will reflect only *differences* among group members and not their absolute values.

Using the mean as the group's reference point, we might describe variability by taking the mean distance between the scores and the mean. Let's return to the Zephyrs' weights and see how this would be done. Table 4.2 lists each observation (each X_i); the group mean ($\bar{X}$) is listed four times, and **deviation scores** are listed in column 3 (one for every X_i). A deviation score for any X_i is simply the difference between the observation and the group mean. To compute one, subtract the mean from the X_i, even if $\bar{X}$ is greater than X_i:

$$\text{Deviation score of } X_i = \text{Raw Score} - \text{Mean}$$
$$= X_i - \bar{X}$$

By convention it is *never* $\bar{X} - X_i$.

On Figure 4.3 some deviation scores are represented as distances on the measurement scale.

Now, if we try to describe variability by taking the mean of these

Table 4.2 *Computation of Deviation Scores for Zephyr's Weights*

X_i	$\bar{X}$	DEVIATION SCORE $(X_i - \bar{X})$
215	200	15
180	200	-20
200	200	0
205	200	5
		$\sum(X_i - \bar{X}) = 0$

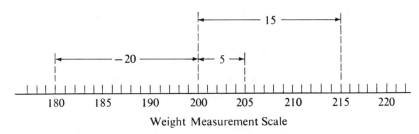

Weight Measurement Scale

Figure 4.3 Illustration of deviation-score concept for the group of four football players' weights. For each X_i there is also an $X_i - \bar{X}$ represented as a distance on the measurement scale. For $X_i = \bar{X}$, of course, $X_i - \bar{X} = 0$.

deviation scores, we run into a problem: The sum of the deviation scores about the mean is always equal to zero.[4] And dividing this sum by N always yields a quotient of zero.[5]

$$\frac{\sum(X_i - \bar{X})}{N} = \frac{15 + (-20) + 0 + 5}{4} = \frac{0}{4}$$

We can still compute a mean deviation, however, by taking the absolute values of the deviation scores.[6] That way we are taking the mean of deviation *sizes*, forgetting about the direction of the deviation. In symbols and numbers:

$$\text{Mean deviation} = \frac{\sum |X_i - \bar{X}|}{N} = \frac{|15| + |-20| + |0| + |5|}{4}$$

$$= \frac{15 + 20 + 5}{4} = \frac{40}{4} = 10$$

This value, 10 lb, is the typical deviation of Zephyr weights from the Zephyr mean; it can be compared to the mean deviation of Zebra weights (50.0 lb) to determine which group has more variability. The concept of mean distance from the mean is easily understood by most people.

Why then is the mean deviation rarely used? One important reason is that statisticians need a measure of variability that will fit conveniently into the formulas for more complex statistics (such as the z-score or the standard error of the difference between two means). The absolute value sign in the formula for mean deviation would cause problems in the algebraic operations necessary to solve these. Thus, statisticians usually specify a group's variability by a statistic that fits easily into more com-

---

[4]This is, in fact, one definition of the mean: The mean is the point about which the sum of the deviation scores equals zero.

[5]In this example, and many that follow in this book, the starting point and the end point of the summation process will not adorn the $\sum$ when it is clear that all members of the group are to be added.

[6]The **absolute value** of a number is its magnitude considered independently of its sign. Converting a group of numbers to their respective absolute values is essentially a process of throwing away the minus signs. The absolute value of $+3$ is 3, the absolute value of -3 is also 3. In symbols, $|3| = |-3| = 3$.

plex formulas and reflects the value of every group member: the **standard deviation.** Since we are required to share a planet with the statisticians, and sometimes understand them, we should take a close look at this favorite statistic of theirs.

The Standard Deviation

The standard deviation serves the same purpose as the mean deviation: It describes the distance on the measurement scale by which the typical group member differs from the mean. The standard deviation is represented by several different symbols, which become important when we move beyond descriptive statistics and talk about samples and populations involved in inferential statistics. Some of these are as follows:

s — Standard deviation in descriptive situations, where the purpose is simply to describe variability of the group at hand.

$\hat{s}$ — Standard deviation of a sample in inferential statistics, where the purpose is to make inferences about population characteristics from sample data. Computed slightly differently from the descriptive s, $\hat{s}$ will be discussed in Chapter 10. ($\hat{s}$ may be called s–hat.)

σ — (Lowercase sigma) Standard deviation of the population in inferential statistics. Every member of the population contributes to the value of σ.

Instead of taking the absolute values of deviation scores to eliminate negative values, we base the standard deviation on *squared* deviation scores. Any number (positive or negative) squared, of course, yields a positive product. Specifically, the standard deviation is the square root of the mean of the squared deviation scores about the mean.[7] This definition is quite easy to digest if taken in pieces:

$$s = \sqrt{\frac{\sum (X_i - \bar{X})^2}{N}}$$

where $\sqrt{}$ signifies "the square root of,"

$\dfrac{\sum}{N}$ signifies "the mean of,"

$(X_i - \bar{X})^2$ signifies "the squared deviation scores about the mean."

This statistic is based on deviation scores — distances on the measurement scale. This should indicate to you that the only data appropriate for this statistic are those representing interval-, ratio-, or absolute-scale levels of measurement.

---

[7]Those who have read Appendix 2 may recognize the standard deviation as the quadratic mean of the deviation scores.

Table 4.3 *Computation of the Standard Deviation (s) for the Football Data*

		ZEPHYRS				ZEBRAS	
X_i	$\bar{X}$	$X_i - \bar{X}$	$(X_i - \bar{X})^2$	X_i	$\bar{X}$	$(X_i - \bar{X})$	$(X_i - \bar{X})^2$
215	200	15	225	146	201	-55	3025
180	200	-20	400	191	201	-10	100
200	200	0	0	166	201	-35	1225
205	200	5	25	301	201	$+100$	10,000

1. Find the sum of squared deviations, $\sum (X_i - \bar{X})^2$

   650   14,350

2. Find mean squared deviation, $\dfrac{\sum (X - \bar{X})^2}{N}$

   $\dfrac{650}{4} = 162.5$ $\dfrac{14,350}{4} = 3587.50$

3. Find standard deviation,

   $s = \sqrt{\dfrac{\sum (X - \bar{X})^2}{N}}$

   $\sqrt{162.5} = 12.75$ lb $\sqrt{3587.0} = 59.90$ lb

An old name for the standard deviation is the root mean square, a term that may help you remember the computational basis of the statistic.

Computation of s can be demonstrated by finding the standard deviations for both football lines. The line having the greatest variability should produce the largest s. Table 4.3 shows the computations. The table is similar to Table 4.2. Separate columns list each X_i, $\bar{X}$, and each $(X_i - \bar{X})$. In addition, a fourth column shows the squared deviation scores $(X_i - \bar{X})^2$. In each case, the squared deviation scores are added up and divided by N to find the mean squared deviation. The square root of this value is s. Thus, the Zephyr line has a mean weight of 200 lb and a standard deviation of 12.75 lb, much less than the Zebras' standard deviation of 59.90.

---

[c] The Standard Deviation from Grouped Data

When the data have been grouped, we can compute an *estimated* standard deviation from the frequency table. Here, just as with the grouped-data mean, we must let the interval midpoint (X_{mi}) stand for every individual in the interval, and that is why our computed s is an estimate and not the exact standard deviation for the whole group. The only way to get the true standard deviation (or true mean) is to enter every value in the group into the computation. The grouped-data estimate of s is computed as follows:

$$s = \sqrt{\frac{\sum f_i (X_{mi} - \bar{X})^2}{\sum f_i}}$$

where f_i = frequency in interval i,

$(X_{mi} - \bar{X})$ = deviation score associated with the midpoint of interval i.

Table 4.4 (p. 104) illustrates the computations involved in obtaining the grouped-data s. A deviation score is computed for each interval's midpoint. These are first squared and then multiplied by the frequencies in their respective intervals. The products are added, divided by the total frequency, and the square root of this quotient is taken as s.

Choosing a Measure of Variability

The following list is a summary of points you may want to consider in choosing among range, interquartile range, and standard deviation to describe the spread of your data.

1. *Level of measurement.* Nominal-level data can be characterized with a mode, but there is no corresponding measure of variability applicable to it. If the data represent ordinal-level observations, only the range or interquartile range are usable. These and the standard deviation are available for interval-, ratio-, or absolute-scale data.

2. *The companion measure of central tendency.* Often an index of variability is used in conjunction with a measure of central tendency. If you were describing the typical observation with a median, the variability measure of choice would probably be the interquartile range, since both statistics are based on percentiles. The mean would probably be accompanied by the standard deviation, since both statistics reflect the value of every group member and both are useful in more complex analyses that may follow.

3. *Skew.* The interquartile range is less sensitive than the standard deviation to extreme skew in a distribution; hence, it will usually be chosen over the standard deviation to describe variability in heavily skewed distributions (just as the median is sometimes chosen over the mean to describe such distributions).

Relatives of the Standard Deviation

Two parts of the standard deviation formula are useful as statistics in their own right. They are not used so much to describe variability per se, but rather as parts of more complex operations involving variability. You already know how to compute them (from the previous section), and they will simply be identified here. It is well to learn the parts of the standard deviation formula now, and their relations to each other, because we will

Table 4.4 *Computation of the Standard Deviation (s) for Grouped Data*

i	INTERVAL	INTERVAL MIDPOINT (X_{mi})	f_i	$(X_{mi} - \bar{X})$	$(X_{mi} - \bar{X})^2$	$f_i(X_{mi} - \bar{X})^2$
11	96–100	98	6	19.1129	365.3031	2191.8184
10	91–95	93	5	14.1129	199.1740	995.8702
9	86–90	88	11	9.1129	83.0450	913.4951
8	81–85	83	11	4.1129	16.9160	186.0757
7	76–80	78	7	−0.8871	.7869	5.5086
6	71–75	73	6	−5.8871	34.6579	207.9475
5	66–70	68	7	−10.8871	118.5289	829.7021
4	61–65	63	4	−15.8871	252.3998	1009.5994
3	56–60	58	0	−20.8871	436.2708	0.0000
2	51–55	53	1	−25.8871	670.1418	670.1418
1	46–50	48	4	−30.8871	954.0127	3816.0510

$$\sum f_i = 62 \qquad\qquad \sum f_i(X_{mi} - \bar{X})^2 = 10,826.2098$$

1. Find $\bar{X}$ as described in Chapter 3. (This table does not show computation of the necessary $\sum f_i X_{mi}$.) If in doubt, see Chapter 3.

$$\bar{X} = \frac{\sum f_i X_{mi}}{\sum f_i} = \frac{4891}{62} = 78.8871$$

2. Compute deviation score for each X_{mi}. For the highest interval:

$$X_{mi} - \bar{X} = 98 - 78.8871 = 19.1129$$

3. Square each $(X_{mi} - \bar{X})$. For the same interval:

$$(X_{mi} - \bar{X})^2 = (19.1129)^2 = 365.3031$$

4. Multiply the frequency for each interval by the squared deviation score. For the first interval:

$$f_i(X_{mi} - \bar{X}_{mi})^2 = 6(365.3031) = 2191.8184$$

5. Add the quantities obtained in step 4, divide by the total frequency ($\sum f_i$), and extract the square root of the quotient. That is,

$$s = \sqrt{\frac{\sum f_i(X_{mi} - \bar{X})^2}{\sum f_i}} = \sqrt{\frac{10,826.2098}{62}} = \sqrt{174.6163} = 13.21$$

(If you reproduce these calculations on your own, you may obtain *slightly* different values for entries in the two columns at right, depending on how many digits you round off after different computational steps.)

be using them regularly in the chapters on regression and hypothesis testing.

The **variation** is the sum of the squared deviation scores, $\sum (X_i - \bar{X})^2$. The variation can, of course, be identified within the formula for s:

$$s = \sqrt{\frac{\overbrace{\sum (X_i - \bar{X})^2}^{\leftarrow \text{Variation}}}{N}}$$

This statistic is also known as the **sum of squares,** or SS, which is more descriptive of its origin. In summary,

$$\text{Variation} = \text{Sum of squares} = SS = \sum (X_i - \bar{X})^2$$

Variance is the name of another statistic, the squared standard deviation. Variance is also known as the **mean square,** or MS. Thus:

$$\text{Variance} = s^2 \text{ (or } \sigma^2 \text{ for populations)}$$

$$= \frac{\text{Variation}}{N} = \text{Mean square} = MS$$

It is a good idea to learn the relations among standard deviation, variation, and variance well enough so that given the value of one of them, and the group size, you could find the values of the other two. For example, a group of ten observations has a *variation* of 360. What are the variance and the standard deviation?

$$\text{Variance} = \frac{\text{Variation}}{N} = \frac{360}{10} = 36$$

$$\text{Standard deviation} = \sqrt{\text{Variance}} = \sqrt{36} = 6$$

---

[c] Computational Procedures for Standard Deviation

At this point you should be able to compute the standard deviation, variation, and variance from a group of data. The last parts of this chapter will cover some of the things you can do with these statistics once you get them — ways to specify characteristics of the group and individuals within the group. Before proceeding, however, it will again be worthwhile to pause and learn another method of computing these statistics.

We have just covered the definitional formulas for standard deviation, variance, and variation; we will now present computational formulas for these same statistics.

Why bother learning two ways to compute the same statistics? One reason is that the second set of formulas gives you a means of checking your computations on the first set. You should be able to compute variance for a group one way and then check your work by getting exactly the same value from the other method. Another reason is that each kind of formula has a different virtue. The definitional formulas show you what deviation scores are and how they produce your statistics. However, they are more laborious to use than the computational formulas, particularly when there are many observations. The computational formulas do not require computation of individual deviation scores, nor do they require you to compute the group mean; they are easier to use in hand calculations and small calculator operations. Finally, students who continue into more advanced statistics will find a special benefit from learning the computational formula for variation: practically all computations in complex applications of the *analysis of variance* can be handled with simple variations of the formula for variation.

Computational formulas are presented side by side with the corresponding definitional formulas. We shall use the group we have already seen in Table 4.3. Here, we begin the computational approach with a two-column table (Table 4.5) having only X_i and X_i^2 and the sums of these values.

Table 4.5   *Table Squared Data Values and Necessary Sums for Computing Variability Statistics from Computational Formulas*

X_i	X_i^2
215	46,225
180	32,400
200	40,000
205	42,025
$\sum X_i = 800$	$\sum X_i^2 = 160,650$

Begin by computing the *variation* (sums of squares):

Computational Formula

$$SS = \sum X_i^2 - \frac{(\sum X_i)^2}{N}$$

$$= 160,650 - \frac{(800)^2}{4}$$

$$= 160,650 - 160,000$$

$$= 650$$

Definitional Formula

$$SS = \sum (X_i - \bar{X})^2$$

$$= (215 - 200)^2 + (180 - 200)^2$$
$$+ (200 - 200)^2 + (205 - 200)^2$$

$$= 225 + 400 + 0 + 25$$

$$= 650$$

Now obtain the variance (mean square):

$$MS = \frac{SS}{N} = \frac{650}{4} = 162.50$$

The standard deviation is taken as the square root of the variance:

$$s = \sqrt{MS} = \sqrt{162.50} = 12.75$$

To summarize, then, we can write the complete set of computational operations for the standard deviation as:

$$s = \sqrt{\frac{\sum X_i^2 - \frac{(\sum X_i)^2}{N}}{N}}$$

The same computational formula can be modified slightly to compute standard deviations from grouped data. Just as when computing the grouped-data mean, we must have the midpoint for each interval (X_{mi}) and the frequency (number of observations, f_i) in each interval. The data from Table 4.4 are reprinted in Table 4.6, along with some intermediate computations involved in computing the grouped-data standard deviation.

There are two operations to note carefully in Table 4.6. First, each midpoint is multiplied by the frequency for its interval (as shown in the third column), each midpoint is squared, and these squared values are also multiplied by the corresponding frequency (fifth column).

Now we can compute variation (SS):

Computational formula

$$SS = \sum f_i X_{mi}^2 - \frac{(\sum f_i X_{mi})^2}{\sum f_i}$$

$$= 396{,}663 - \frac{(4891)^2}{62}$$

$$= 396{,}663 - 385{,}836.7903$$

$$= 10{,}826.2097$$

Definitional formula

$$SS = \sum f_i (X_{mi} - \bar{X})^2$$

$$= 10{,}826.2098$$

(See Table 4.4 for this definitional formula calculation)

(The slight difference between computational and definitional results is due to rounding error in the definitional computations.) As with the ungrouped approach:

$$MS = \frac{SS}{N} \qquad \text{here } N = \sum f_i$$

For our data:

$$MS = \frac{10{,}826.2098}{62} = 174.6163$$

Finally, the standard deviation is taken as the square root of the variance:

$$s = \sqrt{MS}$$

$$s = \sqrt{174.6163} = 13.21$$

Table 4.6 *Data and Preliminary Computations for Calculating the Grouped-Data Standard Deviation, Using the Computational Formula. The Data are the Same as in Table 4.4.*

FROM TABLE 4.4				
X_{mi}	f_i	$f_i X_{mi}$	X^2_{mi}	$f_i X^2_{mi}$
98	6	(98)(6) = 588	$(98)^2$ = 9604	(6)(9604) = 57,624
93	5	465	8649	43,245
88	11	968	7744	85,184
83	11	913	6889	75,779
78	7	546	6084	42,588
73	6	438	5329	31,974
68	7	476	4624	32,368
63	4	252	3969	15,876
58	0	0	3364	0
53	1	53	2809	2,809
48	4	192	2304	9,216
Need sums:	$\sum f_i$ = 62	$\sum f_i X_{mi}$ = 4891		$\sum f_i X^2_{mi}$ = 396,663

We summarize these operations in a single computational formula:

$$s = \sqrt{\frac{\sum f_i X^2_{mi} - \dfrac{(\sum f_i X_{mi})^2}{\sum f_i}}{\sum f_i}}$$

The computational formulas for grouped and ungrouped data look more complex than the definitional formulas, but they will save you a great deal of time if you must calculate a number of standard deviations.

Standard Scores

One important use of the standard deviation is in evaluating the relative positions of individual observations within the group. Remember that this was also the purpose of percentiles — showing the percentage of a distribution below an observation — but percentiles are very limited statistics. They cannot be added, multiplied, averaged, or combined in complex operations (see Chapter 2). The standard score provides an alternative method of specifying a single observation's standing in the group; it *can* be averaged or used in more complex operations.

During a one-semester course in biology, two examinations were given to a class, a midterm and a final. Both exams were to be counted equally in determining each student's grade, and the class produced a mean score of 70 on both exams. On the midterm John received a 90, but on the final

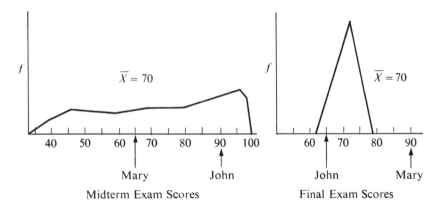

Figure 4.4   Frequency distributions of scores obtained by the same class on a midterm exam and a final exam. Relative to the rest of the class, Mary's score of 90 on the final is a better performance than John's score of 90 on the midterm, even though the class produced a mean score of 70 on both exams. The reason is that there was substantially more variability among midterm scores than among final exam scores.

he managed only to score a 65. Mary, on the other hand, scored only 65 on the midterm but managed to obtain a 90 on the final. Did John and Mary do equally well overall? Should they get the same grade?

If the grades were based on absolute test scores, the answer is yes; both received a total of 155 points and a mean score of 72.5 for the two exams. But if *relative performance within the class* were to be considered, the answer is no. Consider the frequency distributions of scores for each exam as shown in Figure 4.4. Both people scored twenty points over the mean on one test and five points under the mean on the other, but these deviations from the mean were produced in frequency distributions with different spreads, different standard deviations. We need a way to specify that twenty points over the mean is a much better relative performance on the final than on the midterm. The best way to indicate the relative standing of each score, taking into account the different standard deviations of the distributions, is to compute a standard score for each individual raw score.

A **standard score** — or equivalently, a **z-score** — indicates directly how many standard deviations above or below the mean a given score lies. If the mean is 70 and the standard deviation is 20, a score of 90 is 1 standard deviation above the mean. If the mean is 70 in a distribution where $s = 5$, a score of 90 is 4 standard deviations above the mean.

In formula form, z-scores are computed as follows[8]: The z-score (stan-

---

[8]z-scores are "pure numbers," in the sense that the units of measurement cancel out in the conversion process. That is:

$$z = \frac{(X \text{ points} - \bar{X} \text{ points})}{s \text{ points}}$$

Table 4.7 *Computation of z-Scores for John's and Mary's Raw Scores*

	MIDTERM EXAM $\bar{X} = 70; s = 20$	FINAL EXAM $\bar{X} = 70; s = 5$
John	Raw score = 90	Raw score = 65
	$z = \dfrac{90 - 70}{20} = +1.0$	$z = \dfrac{65 - 70}{5} = -1.0$
Mary	Raw score = 65	Raw score = 90
	$z = \dfrac{65 - 70}{20} = -0.25$	$z = \dfrac{90 - 70}{5} = +4.0$

dard score) (z_i) corresponding to raw score X_i is equal to the difference between that score and the distribution mean ($X_i - \bar{X}$) divided by the standard deviation of the distribution (s).

$$z_i = \frac{X_i - \bar{X}}{s}$$

Every score in the distribution has a z-score, and the value of that standard score depends on the group's mean and standard deviation. Table 4.7 shows how z-scores were obtained for each of John's and Mary's raw scores. Thus, Mary's z-score of 4.0 on the final indicates she scored 4 standard deviations above the mean with her 90, while John's z-score of 1.0 shows that his 90 on the midterm was only 1 standard deviation above the mean. Relative to the rest of the group in each case, Mary's 90 was a far better performance than John's. Notice that z-scores can be negative as well as positive — both students received one positive and one negative z-score. A negative z-score simply represents an observation below the mean.

Advantages of z-Scores over Raw Scores

In order to get a job as a technician in a certain plant, an applicant must do well on several job-related aptitude tests. Suppose these include a manual dexterity test (scored on a 100-point scale), a memory test (scored on a 500-point scale), and a color vision discrimination test (scored on a 3-point scale). One applicant took the test and scored 60 on the manual dexterity test; 370 on the memory test; 1.6 on the color discrimination test. How did he do overall? How do you compare 60, 370, and 1.6 when they come from different scales? To what degree are these scores better or worse than those of other people?

z-scores provide answers to these questions. If we know the means and the standard deviations of scores produced by all applicants on each exam, this applicant's scores can be converted to z-scores. Suppose these turned out as follows:

$z = -0.5$ on the manual dexterity test

$z = 0.1$ on the memory test

$z = -2.0$ on the color discrimination test

Relative to other applicants, he scored below the mean on two of the three exams and only 0.1 standard deviation above the mean on another. Here we can compare observations on one variable directly with observations on another.

Characteristics of z-Scores

If you compute a z-score for every X in a distribution and plot a frequency distribution of z-scores, it will have exactly the same shape as the original graph of raw scores. Only the numbers on the horizontal axis will have changed in the transformation to standard scores. A distribution of raw scores is shown in Figure 4.5, together with the corresponding distribution of z-scores. The figure also suggests some other important characteristics of z-scores: (1) The group of z-scores for any distribution will always have a mean ($\bar{z}$) of zero and a standard deviation of 1, and (2) when the distribution is relatively "bell shaped," almost all z-scores will have values between -3 and $+3$. This second characteristic, however, is only a rule of thumb and not an ironclad law. Scores *can* have z values much greater than $+3$ or much less than -3. They just do not appear very often in bell-shaped frequency distributions.

Transformation of z-Scores

Teachers are probably among the world's greatest users of standard scores. Recording each student's exam, quiz, homework, and classroom performances as z-scores offers a concise and useful record. Since most z-scores tend to fall between -3 and $+3$ (with the majority between -1 and $+1$ in many distributions), the teacher's grade-book can become crowded with numbers such as $+0.2$, -0.13, $+2.71$, $+0.08$, -1.06, and so on. Many teachers need to use z-scores but find arrays of such numbers hard to work with. One must be very careful not to lose track of the plus and minus signs; a z of $+2$ is very different from a z of -2. Thus, many teachers turn their z-scores into an alternative form of standard score, the **transformed standard score.**

Similarly, large testing organizations use standard scores to specify how students do on such exams as the Scholastic Aptitude Test or the Graduate Record Examination relative to all other students in the country who take the test. If you have taken such an exam, you may remember that your score was a three-digit number (or a pair of them) such as 490 or 800. These, too, are standard scores, but they have been converted to the alternative form.

A distribution of z-scores may be transformed into another distribution

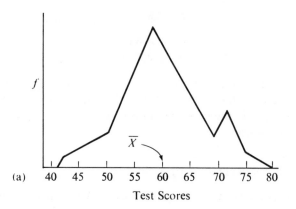

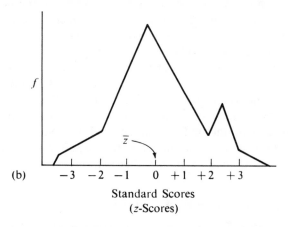

Figure 4.5 Frequency distribution for (a) a group of raw scores and (b) the same scores converted to z-scores. Notice that the transformation to standard scores does not alter the *shape* of the original frequency distribution, only the scaling of the horizontal axis. Regardless of the original values, the z-distribution will have a mean of zero and a standard deviation of 1.

with a mean and standard deviation arbitrarily chosen by the statistician. You may change a group of raw scores into z-scores and then, if you find z-values difficult to work with, perform a second transformation and convert the z-scores into transformed standard scores. Transformed standard scores are sometimes symbolized z'. The new z' distribution may have any mean and standard deviation you choose. The transformation to z'-scores will not alter the shape of the distribution nor change the rank order among individual scores nor alter the relative distances between scores.

Production of z'-scores can be divided into three steps:

1. Determine raw-score mean and standard deviation for the original distribution.

2. Determine the z-score for each raw score.

3. Convert each z-score to a z'-score in the following way:

$$z_i' = \bar{z}' + (s_{z'})(z_i)$$

where $z_i' = z'$-score for an individual observation X_i,

$\bar{z}'$ = arbitrarily chosen mean of the new z'-distribution,

$s_{z'}$ = arbitrarily chosen standard deviation of the z'-distribution,

$z_i = z$-score for the observation X_i.

Two examples should suffice to illustrate this process. In a previous section, we had determined that Mary's test scores corresponded to z-scores of -0.25 and $+4.0$. Suppose that the teacher found z-values difficult to work with and decided instead to specify these performances as standard scores within a distribution with a mean of 500 and a standard deviation of 100. Then, according to the preceding formula, the z'-score for Mary's midterm is:

$$
\begin{aligned}
z' &= 500 + 100(-0.25) \\
&= 500 - 25 \\
&= 475
\end{aligned}
$$

On her final, her z'-score is:

$$
\begin{aligned}
z' &= 500 + 100(+4.0) \\
&= 500 + 400 \\
&= 900
\end{aligned}
$$

Of course, the teacher could have chosen any other value for the mean or standard deviation of the z'-distribution; he might have preferred to change the class's z-scores to z'-scores distributed with a mean of 100 and a standard deviation of 20. In this case, Mary's midterm z'-score would have been:

$$z' = 100 + 20(-0.25) = 100 - 5 = 95$$

and her final exam score would have been:

$$z' = 100 + 20(4.0) = 100 + 80 = 180$$

You may wonder how it is possible to use *any* $\bar{z}'$ and $s_{z'}$ we wish and still produce z'-scores with meaning. Remember that a z'-score still indicates how many standard deviations above or below the mean a given observation falls, if we know the mean and standard deviation of the z'-distribution; otherwise, the z'-scores are meaningless. Those with mathematical background will recognize the z' formula (and the original formula for z as well) as a linear transformation; if the same linear transformation is applied to all scores in a group (that is, if the same $\bar{z}'$ and

$s_{z'}$ are used for everybody), the transformed scores will fall in the same rank order, at the same relative distances as the original scores.[9]

In some kinds of educational tests the symbol T is used to refer to a transformed standard score that belongs to a distribution with a mean of 50 and standard deviation of 10.[10] The formula for converting each z-score into a **T-score** is:

$$T_i = 50 + (10)(z_i)$$

where T_i = T-score corresponding to individual observation X_i,

z_i = z-score corresponding to individual observation X_i.

According to this formula, a raw score that produces a z of +1 has a T-score of 50 + 10(1), or 60.

There is no mathematical advantage to using z'-scores instead of z-scores. Those teachers and testing companies who do so simply prefer to record large two- or three-digit numbers instead of z-scores, which are small decimal values with signs.

SUMMARY

Since different-shaped distributions can have the same mean, median, or mode, it may be necessary to specify a measure of central tendency *and* a measure of variability to describe them completely. A measure of variability is a statistic whose value is relatively large if the observations are relatively scattered on the measurement scale and relatively small if the observations are relatively bunched together. Four such measures were considered: the range, the interquartile range, the mean deviation, and the standard deviation. Which one is best depends on the characteristics of the data and the message you wish to convey. Understanding and interpreting each requires a knowledge of the computational basis of the statistics.

The range is essentially the distance between the distribution's highest and lowest observations. It can be used with ordinal- or higher-level data to describe the width of a distribution. Although easy to compute and

---

[9]The usual form of the linear relationship is given as

$$Y = bX + a$$

where b = a constant (the slope),
a = a constant (the intercept).

The z-score formula may be written in this form when s and $\bar{X}$ are constants:

$$z = \frac{X - \bar{X}}{s} = \frac{X}{s} - \frac{\bar{X}}{s} = \left(\frac{1}{s}\right)X + \frac{\bar{X}}{s}$$

where $\frac{1}{s}$ = slope,

$\frac{\bar{X}}{s}$ = intercept.

Linear relationships will be covered in Chapter 7 when we deal with linear regression.

[10]T-scores should not be confused with the t-distribution (discussed in Chapter 10).

easily interpreted, it is an unstable statistic because its value depends entirely on the values of two observations.

In skewed distributions a useful measure of variability is the interquartile range — the difference between the 75th and 25th percentiles (between Q_3 and Q_1). Although more stable than the range, and applicable to ordinal- or higher-level data, it does not reflect the values of all observations in the distribution.

Another way to describe variability is to indicate the degree to which observations differ from some fixed point on the measurement scale. For mathematical reasons, the reference point chosen for this purpose is the distribution's mean. Two measures of variability based on deviations from the mean are the standard deviation and the mean deviation. In practice, the mean deviation is rarely used. However, the logic underlying its derivation helps to explain the more useful standard deviation. The mean deviation is the mean distance between each observation and the group mean.

The most commonly used index of variability is the standard deviation. Defined verbally, it is the square root of the mean of the squared deviation scores. It is highly useful for describing a distribution's spread because (1) it is expressed in the same units as the original data, (2) it reflects the value of every observation in the distribution, (3) it is more stable than other measures of variability, and (4) its mathematical properties permit its use in complex statistical operations.

Sometimes we may wish to do more than describe characteristics of a distribution as a whole; we may want to specify the relative standing of one observation within the distribution. Standard scores are used for this purpose.

Standard scores (or z-scores) specify how many standard deviations above or below the mean an individual score stands. When a group of observations are transformed to standard scores, the shape of the new distribution is the same as the shape of the raw-score distribution. The z-score distribution, however, has a mean of zero and a standard deviation of 1. z-scores are useful because they allow comparison of an individual's performance on one variable with his performance on another, even though the variables may be measured on quite different scales.

Standard scores need not be expressed only in distributions with a mean of zero and a standard deviation of 1. They may be transformed into a distribution of scores with any mean and standard deviation we choose, known as transformed standard scores or z'-scores. The transformation process is linear, which means that the z'-distribution's shape and the relative standing of individual observations within the distribution are unchanged from the original distribution.

z-scores and z'-scores are evaluated in the same way; both specify an observation's standing within a group in terms of standard deviations above or below the mean. z-scores, however, are difficult for some people to work with, simply because they are small decimal numbers with signs.

KEY CONCEPTS

variability	standard deviation	absolute value
measures of variability	variation (sum of squares)	standard score (z-score)
range	variance (mean square)	transformed standard score
interquartile range	semi-interquartile range	(z'-score)
mean deviation	deviation scores	T-score

SPOTLIGHT 4   **Sir Francis Galton Discovers Variability**

This chapter discusses two important uses for measures of variability: (1) describing the spread of distributions, and (2) describing an individual's standing within a group *independent of the units of measurement*. z-scores could tell us that a person's height is 1 standard deviation above the mean and his weight 0.5 standard deviations above the mean without mentioning inches or pounds. These ideas may seem obvious and indispensable now, but this was not always so.

While the concept of average originated at some uncertain period in antiquity, the idea of a variability measure has a much shorter history. During the first part of the nineteenth century, mathematicians described dispersion by *probable error* and the *interquartile deviation*.[11] The science of statistics was well underway at the time, but statisticians were slow to appreciate the utility of these mathematical curiosities. The idea of using variability indices, as we use them here, originated primarily with the English biologist, anthropologist, and psychologist Sir Francis Galton (1822–1911).

Galton's overriding interest was in uncovering the laws of inheritance; to this end he spent an active lifetime collecting and analyzing massive numbers of measurements on all sorts of phenomena. Sweet peas, race horses, dogs, and all classes of Englishmen were brought to his laboratory and measured in dozens of ways. When available statistical methods of his day would not suffice, he invented his own. The concepts of correlation and linear regression were developed by Galton as means of describing the relation between characteristics of parents and their offspring. He also pioneered the use of the normal curve in analyzing frequency distributions of biological and psychological characteristics. All these contributions, in turn, would not have been possible had he not brought variability into the laboratory.

It is hard to believe now, but the idea of describing variability in groups was quite foreign to Galton's contemporaries. In 1890 he wrote:

---

[11]The probable error is very similar to our modern standard deviation; the standard deviation was brought into general use by Karl Pearson in the 1890s.

It seems to be a great loss of opportunity when, after observations have been laboriously collected, and been subsequently discussed in order to obtain mean values from them, that the small amount of extra trouble is not taken, which would determine other values whereby to express the variety of all the individuals in those groups. . . . There are numerous problems of special interest to anthropologists that deal solely with variety.

There can be little doubt that most persons fail to have adequate conceptions of the orderliness of variability, and think it is useless to pay scientific attention to variety, as being in their view, a subject wholly beyond the powers of definition. They forget that what is confessedly undefined in the individual may be definite in the group, and that uncertainty as regards the one is in no way incompatible with statistical assurance as regards the other. . . .

Greater interest is attached to individuals who occupy positions towards the middle of a marshalled series than to those who stand about its middle. An average man is morally and intellectually an uninteresting being.[12]

Just as important, moreover, was his realization that distributions of different variables can be compared to one another if the observations are specified in units of variability. Galton's term for measures of variability was ''statistical units,'' whose ''office,'' he said,

is to make the variabilities of totally different classes, such as horses, men, mice, plants, proficiency in classics, etc. etc., comparable on equal terms. The statistical unit of each series is derived from the series itself.[13]

He no doubt recognized that his discovery of a use for measures of variability was of major consequence, and he described the moment of insight with a scene that brings to mind Newton's apple tree:

As these lines are being written, the circumstances under which I first clearly grasped the important generalisation that the laws of Heredity were solely concerned with deviations expressed in statistical units, are vividly recalled to my memory. It was in the grounds of Naworth Castle, where an invitation had been given to ramble freely. A temporary shower drove me to seek refuge in a reddish recess in the rock by the side of the pathway. There the idea flashed across me, and I forgot everything else for a moment in my great delight.[14]

Applications of Galton's ''important generalisation'' go far beyond z- and T-scores. Without the ability to measure variability, and to use that index to specify an individual's standing in a group, we would have to end the book at this point. Correlation, regression, and most inferential statistics require a means of specifying an individual's place in a group independently of the original units of measurement.

---

[12] From Galton's *Anthropometric Laboratory, Notes and Memoirs*, No. 1, quoted in Karl Pearson, *The Life, Letters and Labours of Francis Galton* (Cambridge: Cambridge University Press, 1924), II, 384–385. Reprinted by permission of Cambridge University Press.
[13] Sir Francis Galton, *Memories of My Life*, 3d ed. (London: Methuen, 1909), p. 298.
[14] Ibid., p. 300.

PROBLEMS

1. An instructor has two different statistics classes. Both classes take the same exam and produce the following scores:

78	83	91	60	77		71	68	71	79	64
87	56	92	78	64		64	74	95	72	73
56	29	85	76	31		69	32	75	72	69

a. Compare these two classes' performances by comparing their means and their ranges.

b. Now compare these groups by comparing their means and standard deviations.

c. Which comparison, a or b above, is more informative? Why do the standard deviations tell you something different about variability than the two ranges tell you?

2. In measuring the amount of time it takes for each of 12 drivers in a driving-simulator device to get his or her foot on the brake pedal, the following reaction times (in milliseconds) were obtained. (One millisecond, ms, is one one-thousandth of a second.)

520	420	490	610	590
414	305	685	478	502

Describe this small distribution by computing its mean and standard deviation.

3. Given below are two groups of weight measurements, in pounds:

GROUP I					GROUP II			
27	39	54	29		28	26	28	54
16	21	53	42		24	43	39	22
47	35	38	41		51	34	57	39

Compare the two groups by computing each group's

a. mean   d. variance (MS)
b. range   e. standard deviation
c. variation (SS)

4. Below are the number of years of schooling completed by 20 workers in one department of a factory. Describe the educational level of this group by computing the mean and standard deviation.

8.5	12.0	13.5	12.0	9.0
12.0	10.5	9.5	12.0	13.0
6.5	14.0	11.5	8.0	10.5
8.0	10.0	12.0	11.0	12.0

5. What do you know about a distribution or group if its standard deviation is zero? What do you know about your computations if you produce a value for SS that is a negative number?

6. Fill in each blank with one of the following terms: range, interquartile deviation, mean deviation, standard deviation, variation, variance, standard score.

   a. A _____ indicates directly how many standard deviations above or below the mean an individual score lies.

   b. _____ is the sum of the squared deviation scores.

   c. The _____ is an index of variability determined entirely by two scores.

   d. _____ is a measure of variability less sensitive to skew than the standard deviation.

   e. The _____ is probably the best measure of variability to choose when the measure of central tendency is the mean.

   f. The absolute value sign in the equation for _____ makes this a difficult statistic to manipulate algebraically and thus accounts for its unpopularity.

   g. _____ is also known as the mean square.

7. A class of sixteen students was asked to record the number of miles they traveled from home to attend school. Their responses: 40, 15, 20, 60, 22, 13, 110, 9, 13, 18, 15, 43, 3,001, 75, 50, 18. Compute a mean and standard deviation for this group of sixteen scores, and then compute a mean and standard deviation excluding the extremely high value (3,001). Which mean and standard deviation most accurately characterize this group? How might you deal with this situation in writing a report on "miles traveled to attend school"?

8. From the frequency distribution tables in problems 8 and 9, Chapter 2, compute the grouped-data standard deviation for each distribution. From cumulative graphs, determine each group's interquartile range.

9. Complete the following table. You should be able to complete the row for each group with the information given:

	N	SS	MS	s
a. Group I	10			15
b. Group II	100	14,400		
c. Group III		105	7	
d. Group IV		176,400		70

10. 5,342 people take an aptitude test. The mean score for this group is 69.04 and the standard deviation (s) is 19.00. What z-score is associated with the following scores?

   a. $X = 64$ d. $X = 100$

   b. $X = 69$ e. $X = 2$

   c. $X = 48$

11. John took two comprehensive examinations that were also given to hundreds of other people. The overall results from these exams were

    Exam 1: $\bar{X} = 102$; $s = 13$.
    Exam 2: $\bar{X} = 540$; $s = 90$.

    a. John scored 112 on exam 1 and 610 on exam 2. On the basis of his z-score for each exam, which was the better relative performance?
    b. Mary scored 117 on the first exam and 600 on the second. Compute the mean z-score for her two exams, and compare it with the mean of John's two z-scores. Who had the better overall performance?

12. Convert John's and Mary's test scores from problem 11 to T-scores in a distribution with a mean of 50 and a standard deviation of 10. If the instructor is going to use results of both exams to determine grades for these students, what advantages are there in recording T-scores instead of raw scores?

13. A graduate school entrance exam produces raw scores with a mean of 380 and a standard deviation of 45. What z-scores are associated with raw scores of

    a. $X = 390$ c. $X = 258$ e. $X = 350$ g. $X = 485$
    b. $X = 427$ d. $X = 295$ f. $X = 375$ h. $X = 510$

14. For the exam described in problem 13 above, students scores are converted to transformed standard scores, distributed with a mean of 500 and a standard deviation of 100. Determine the transformed standard score for each raw score given in problem 13.

15. Scores on a manual dexterity test are distributed with a mean of 25 and a standard deviation of 3. Convert the following raw scores on that test into transformed standard scores in a distribution with a mean of 100 and a standard deviation of 20.

    a. 21.5   d. 24.3
    b. 30.0   e. 26.6
    c. 29.4   f. 22.4

5. The Normal Distribution: A Mathematical Function Versus Accidents of Nature

After reading this chapter you should be able to do the following:

1. *Give some examples of naturally occurring frequency distributions that approximate the normal distribution.*

2. *List the essential characteristics of the mathematically defined normal distribution.*

3. *Explain how areas under a frequency distribution graph are related to relative frequencies of observed values.*

4. *Use the unit normal table to find relative frequencies of values in any normal distribution.*

Bell-Shaped Curves in Nature

If we could, on a single afternoon, visit every living American man and measure his height and then describe all of those measurements with a frequency polygon, the graph would be very similar to Figure 5.1.

Since height is a continuous variable, and since this distribution would include many millions of observations, we would have to group the data. Even if we used many intervals (instead of just ten or twenty), there would still be many people in each; the graph would appear very smooth, as Figure 5.1 does. Most of the observations would fall near the mean value (assume it is 68 in.), and fewer and fewer observations would appear as we proceeded further and further in either direction from the mean. It is almost certain that no one would appear with heights greater than 8 ft 11 in. or less than 26 in.[1] This is all well and good but not terribly interesting.

An interesting fact would emerge, however, if we continued the process and measured other physical characteristics besides heights, such as hat sizes, volumes of ear canals, or weights of big toe bones. Frequency polygons of these measurements would have the same shape as Figure

---

[1]Extreme heights are taken from *The Guinness Book of World Records* (New York: Bantam Books, 1976).

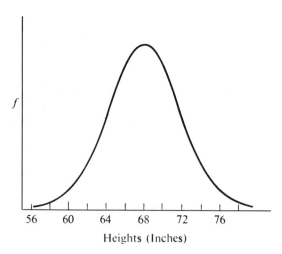

Figure 5.1   Frequency distribution that might result if the height of every living American man were measured. The distribution would be bell-shaped, with a mean of about 68 in.

5.1, even though all would represent distributions of different variables. (The frequency polygons for each variable would, of course, have differently scaled horizontal axes.)

Bell-shaped curves would also appear if we measured people's performances on various tasks. A machinist, for example, might be making airplane engine parts that are supposed to be exactly 1.000 in. in diameter. Human and machine errors, however, would make it impossible for them all to be 1.000 in. exactly, so some would be very slightly larger and some very slightly smaller. If we took all finished parts from the same hypothetical machinist and measured them with an ultraprecise micrometer, the frequency distribution of part diameters would look very similar to Figure 5.2, another bell-shaped curve.

Bell-shaped curves appear quite often when groups of empirical observations are tallied and graphed. Many (but by no means all) of these are nearly identical to each other in shape and also *nearly* identical to a mathematical expression known as the **normal distribution.**

Heights and other physical characteristics are determined by a number of independent random variables.[2] That is, a person's genetic endowment as well as a multitude of environmental variables contribute to the determination of each physical characteristic. When a number of such variables contribute to the value of some other single variable, values of that variable often appear in a normal distribution. Similarly, magnitudes of errors often take on a normal distribution for the same reason. Many factors undoubtedly contribute to determination of the exact size error the machinist will make on each part — tremors in his arm muscles, irregularities in the metal, and so on. When "chaos" contributes something to each observation, the group of observations often take on a very

___

[2]The term *random variable* will be discussed more fully on pp. 231–233.

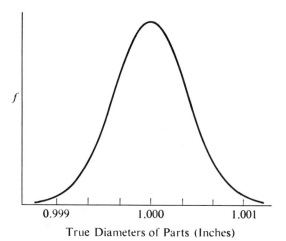

True Diameters of Parts (Inches)

Figure 5.2   Frequency distribution of true diameters of parts made by a hypothetical machinist. Although he intends every part to have a diameter of 1.000 in. exactly, human and machine errors might produce the distribution of part sizes shown. The distribution has a mean of 1.000.

predictable order. For this reason, the normal distribution was once called the "Law of Error," a name that seems almost to be a contradiction in terms. On the matter of order in apparent chaos, Sir Francis Galton commented in 1889:

> I know of scarcely anything so apt to impress the imagination as the wonderful form of cosmic order expressed by the "Law of Frequency of Error." The law would have been personified by the Greeks and deified, if they had known of it. It reigns with serenity and in complete self-effacement amidst the wildest confusion. The hungrier the mob, and the greater the apparent anarchy, the more perfect is its sway. It is the supreme law of Unreason. Whenever a large sample of chaotic elements are taken in hand and marshalled in the order of their magnitude, an unsuspected and most beautiful form of regularity proves to have been latent all along.[3]

When a group of empirical observations can be considered to be normally distributed, the statistician's task is greatly simplified. Statistical properties of the normal distribution have been thoroughly investigated and specified; these properties are then assumed to apply to the real observations as well. Also, as we shall see in Chapter 9, distributions of *statistics* will take on a normal shape, even when the group of potential observations from which they are computed is not normally distributed. Here, too, properties of the mathematical abstraction are assumed to apply to characteristics of real observations. To appreciate the importance of this state of affairs, we must leave the real world for a moment and look at the theoretical curve.

---

[3]Sir Francis Galton. *Natural Inheritance* (London: Macmillan and Co., 1889), p. 66.

The Normal Curve[4]

$$Y = 2X + 1$$

Those who have studied algebra or analytic geometry may remember that expressions like the preceding one specify a relation between two variables. X and Y are symbols that stand for two different variables, and the equation says that there is a mathematical relation between them. When X has a certain value, then the value of Y must be equal to $2X + 1$. Furthermore, such relations can be graphed between two perpendicular axes, with the horizontal axis specifying the X values and the vertical axis specifying the Y values corresponding to each X. The preceding formula and two others are graphed in Figure 5.3 to refresh your memory on the general nature of graphs of equations. These are not frequency distributions but graphs of mathematical relations.

Notice that in each graph there is a Y value for *every* X value — for $X = 2.198374082 \ldots$ and $X = 0.1209834756 \ldots$ as well as for $X = 1.0$ and $X = 2.0$. X and Y are continuous variables, and the points representing each value merge into a continuous line.

We focus here on one mathematical relation that, like the equations graphed in Figure 5.3, specifies a Y value for every X. Don't be frightened by its appearance. Unless you become a professional statistician, or take advanced courses in theoretical statistics, you probably will never have to memorize it or even use it to compute values. We will, however, take a look at some basic characteristics of the curve produced by this monster. The normal curve is generated by the following equation:

$$Y = \frac{1}{\sigma\sqrt{2\pi}} e^{\frac{-(X - \mu)^2}{2\sigma^2}}$$

Two of these symbols are constants, that is, they always have the same value:

$$e = 2.71828 \ldots$$
$$\pi = 3.14159 \ldots$$

If we wanted to produce values of Y for a graph of this equation, we would still need values for μ and σ before computing a Y for each X. These could be any values desired, but once chosen they must not be changed while computing a graph. μ and σ are **parameters** of the curve, and they must be specified before the curve is produced. A parameter is a characteristic of the entire curve, a characteristic whose value helps determine all values in the distribution. It, too, remains constant as long as we are working with the same normal curve; if we change the value of either parameter, we are working then with a different normal curve. If we change the values inserted where e and π go in the equation, then we are no longer working with a normal distribution. Later, in inferential statis-

---

[4]Also known today as the **Gaussian** distribution.

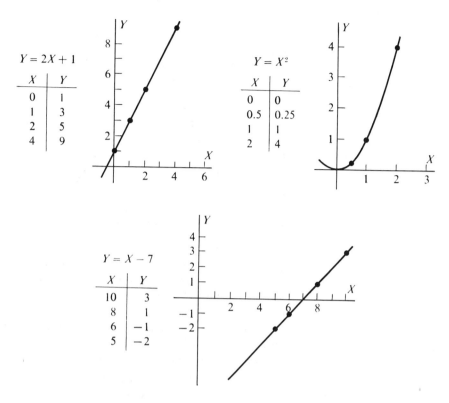

Figure 5.3 Graphs of three mathematical relations. The equations specify a Y value for *every* X value, not just the sample values shown in the table. Hence, the graphs of these equations are continuous lines.

tics (Chapters 9–15) we will be using normal curves to describe populations; so we will be able to use the term parameter to mean a characteristic of an entire population.

The value we use for μ is the "mean" of the distribution; to make an example, assign μ a value of 10. σ is the "standard deviation" of the distribution; say that σ has a value of 2. With the parameter values assigned, we can go ahead and insert all possible values of X into the equation (from $X = -\infty$ (infinity) to $X = +\infty$) and obtain a Y value for each X. The graph of these values is shown in Figure 5.4. You do not have to be able to produce such a curve to interpret the formula. To use the normal curve for solving actual problems, however, you should know the following characteristics of the normal distribution:

1. The normal distribution is a mathematically defined function that is only *approximated* by some frequency distributions.
2. The distribution is symmetrical and centered on μ. The two halves of the curve form mirror images around μ.
3. Points of inflection (where the shape changes from concave to convex)

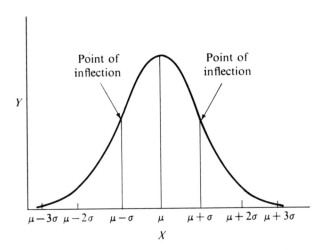

Figure 5.4   The graph generated by the normal-curve equation. The distribution is symmetrical and centered on the value specified for μ. Points of inflection are located at a
distance of σ on either side of μ.

are located at a distance of 1 σ in either direction from the mean (these
are marked on Figure 5.4).

4. The ends of the curve approach but never quite touch the horizontal
   axis. This means that the graph has nonzero Y values for even very
   high and very low X values, but these are very nearly zero.

5. There is actually an infinity of normal curves — a different normal
   curve for every possible value of μ and σ. Some possible normal
   curves are shown in Figure 5.5. Notice that changing the value of μ
   shifts the location of the curve along the horizontal axis; changing the
   value of σ changes the spread of the distribution.

6. Even though some of the curves in Figure 5.5 appear to be different
   from the others, they are all normal curves because they are all generated by the normal-curve equation.

These properties are important because we will soon treat actual frequency distributions as though they were normal curves and had the same
properties as the mathematically defined function.

When Is an Empirical Curve a Normal Curve?

The bell-shaped distribution of heights shown in Figure 5.1 looks very
much like the normal curves in Figure 5.5. But *is* it a normal curve? Could
we specify values of μ and σ that, when inserted in the normal-curve
equation, produce Figure 5.1 *exactly*? Can we apply statistical methods
designed for normal distributions to the group of heights graphed in Figure
5.1?

Simple visual inspection will not provide an answer, since it is not
possible to tell by eye whether or not a certain bell-shaped curve fits the

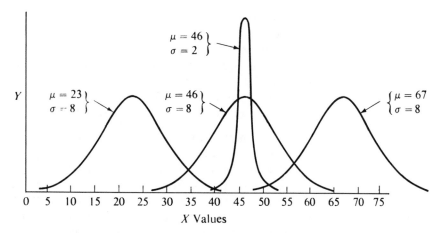

Figure 5.5 Four different normal curves. Notice that changing the value of μ shifts the location of the curve on the horizontal axis. Changing the value σ changes the spread of the distribution.

normal-curve equation. The only definite answer to that question comes from use of some inferential tests, such as the goodness-of-fit test. One version of this test requires use of the chi-square statistic, and we will cover this procedure in Chapter 13. And, sometimes, even that procedure does not provide a definite answer.

Nonetheless, there are some situations where it is appropriate to treat empirical distributions as if they were normal distributions without performing the goodness-of-fit test:

1. *When applying statistics to results from tests that have been designed to produce normally distributed scores.* This is often done by organizations that publish intelligence tests, college and graduate school entrance examinations, or other types of professional, aptitude, and certification exams. Preliminary versions of the tests are tried out on groups of applicants. With information gained from such trial runs, questions are added, deleted, and modified so that frequency distributions of scores are normal when the test is administered to the appropriate people. In any case, the testing organization will (or should) clearly indicate when this has been done and when exam scores can be treated as if they come from a normal distribution.

2. *When differences among observed values are caused by multiple unrelated, uncontrollable factors.* Heights graphed in Figure 5.1 turned out to be normally distributed, since every man's height is the product of many genetic, nutritional, and experiential factors, most of which are unrelated to each other and unknown. Part sizes shown in Figure 5.2 are probably also distributed normally, because the machinist cannot control all sources of error. In fact, the terms "unrelated," and "uncontrollable" should *suggest* the term "error" to you. Frequency distributions of error sizes — for a multitude of errors produced by

machines and human beings — are often normally distributed. Not every "natural" distribution is normal, however; distributions of human reaction times, as well as many specific kinds of error distributions, are skewed. If ceiling effects or floor effects (as defined on p. 54) are operating, it is probably better not to assume that the distribution is normal (reaction times, for instance, cannot be less than zero, but they may be infinitely long).

Incidentally, an examination that produces normally distributed scores is not necessarily superior to an exam that produces a skewed or some other non-normal curve. Some teachers express satisfaction when they give a test and obtain what looks like a normal curve of score values. But the occurrence of a normal score distribution indicates absolutely nothing[5] about the test's ability to discriminate between good and poor students. In fact, if every student in class were to choose responses on a multiple-choice exam by throwing a die, the class's score distribution would closely approximate the normal curve. Remember that a normal curve is also an error curve (see point 2 in the preceding list). The only advantage to using exams that produce normally distributed scores is that the results are easier to analyze statistically.

When actual frequency distributions can be considered normal, their analysis is greatly simplified because the entire distribution is *known*. That is, specifying that a distribution is normal, and specifying a mean (μ) and standard deviation (σ), completely determines the shape of the entire distribution. Furthermore, the exact relative frequencies with which different values appear is known (to the extent that the real distribution is normal). For instance, in every normal frequency distribution

68.26 percent of the observations fall within 1 standard deviation of the mean (above or below);

95.44 percent fall within 2 standard deviations of the mean;

99.74 percent fall within 3 standard deviations of the mean.

If we assume, as in Figure 5.1, that heights of American men really are distributed normally, with $\mu = 68$ in. and $\sigma = 4$ in., then 68.26 percent of American men have heights within 4 in. of 68 in. (that is, between 64 and 72 in.). Similarly, 95.44 percent of these men are between 60 and 76 in. tall (within 2 standard deviations of the mean).

Because these and all other relative frequencies of values in a normal distribution have been determined by statisticians, we can perform two very important types of operations on normally distributed data if μ and σ are known. First, we can find percentiles and percentile ranks from a special table, and second, we can use the same table to calculate the probability that specific values appear. (Probability concepts will be introduced in Chapter 8 and used throughout inferential statistics.) In order

---

[5] "Absolutely nothing" is strong language, but appropriate.

to perform either type of operation, we will have to depart briefly from the real world again and absorb some more mathematical theory.

Areas and Relative Frequencies

In this section we will look at a special property of frequency distribution graphs, and in the next two sections we will apply that principle to normal distributions.

The principle is this: In any frequency polygon[6] the proportion of the observations that fall within an interval on the measurement scale is equal to the proportion of the area under the curve that falls over that interval. This principle is the working basis of inferential statistics and is quite useful in descriptive applications. Fortunately, it can be illustrated graphically much more easily than it can be explained verbally.

Figure 5.6 shows two empirical frequency polygons, clearly non-normal distributions, involving different measurement scales and different Ns. An interval is marked on each scale, and the question is, What proportion of the observations in each group have values in the intervals?

The *total* area under each curve over the horizontal axis is indicated by cross-hatching. The proportion of this area that lies over the specified intervals (shaded) is equal to the proportion of the observations that fall within the interval. If 0.30 of the area under the curve (a) lies over the interval 10–15, then 0.30 (30 percent) of the observations have values between 10 and 15. Similarly, if 0.15 of the area under curve (b) lies above the interval 570–620, then 0.15 (15 percent) of the observations have values between 570 and 620.

Curves (a) and (b) represent observations made on continuous variables, but to understand the relation between areas under curves and relative frequencies, and why it holds with any distribution, it may help to picture a histogram graph for distribution of a discrete variable.

Figure 5.7 is a histogram summarizing the number of children in each of twenty different families. A square representing each observation is placed over the appropriate ages, and the squares stack up to form histogram bars.

Consider, now, the *area* covered by each histogram bar. The bars are each 1 unit wide, and the area of each is thus 1 times the bar height. The area of the bar representing families with three children is 1(3) = 3 sq units, for instance. The total area covered by the entire graph is 20 sq units.

What proportion of the families in this group had one or two children? The values 1–2 designate an interval on the horizontal axis, and the area of the graph over this interval is 10 sq units. This is 10/20, or 0.50, of the total graph area; hence 0.50 of these families had one or two children. Of

---

[6]Any polygon, that is, with linearly scaled axes and a frequency axis that is not cut (see Chapter 2). Remember that frequency polygons can be made only for interval-, ratio-, and absolute-scale data.

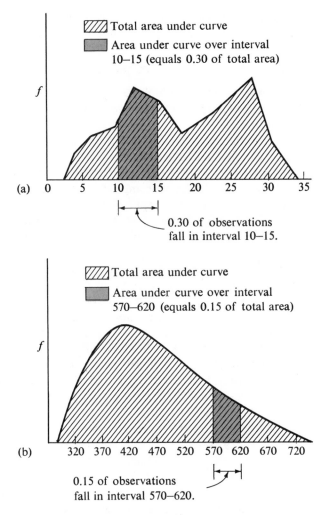

Figure 5.6   The relation between areas under frequency distribution graphs and the relative frequencies of observations in intervals on the measurement scale. The proportion of the total area that lies over an interval is equal to the relative frequency (proportion) of the observations with values in that interval.

course, this example is rather trivial, since we already know N and the number of people in the specific interval. But you should examine closely the correspondence between area and relative frequencies shown here. The same relation holds when the histogram bars are not 1 unit wide; it holds even when the graph represents a continuous distribution, as do Figures 5.1 and 5.2. If we have some way to find areas of parts of graphs, we can specify the proportions of a distribution that fall in various intervals.

Finding percentiles and probabilities associated with observations in normal distributions requires us to know areas under parts of normal

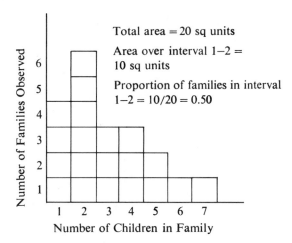

Figure 5.7   Histogram graph showing relation between areas of graphs and relative frequencies. 10/20 of the area lies over the interval 1–2; hence, 10/20 of the observations occur in this interval.

curves. Calculating areas under continuous distributions, however, is a little more difficult than finding areas of simple histograms. In most cases, it requires integral calculus, and in the case of the normal curve, a particularly difficult integral is involved. Fortunately, areas under one special normal curve have been calculated and summarized in table form; we can use this table to find areas under parts of *any* normal curve. We will look first at the special normal curve (pp. 131–136) and then extend its use to all normal curves (pp. 136–139).

The Unit Normal Curve and Its Table

The theoretical curve used in the analysis of *all* normal distributions is the **unit normal curve.** This is just a normal curve (a graph specified by the equation on p. 124) that has a mean of zero and a standard deviation of 1. This curve is graphed in Figure 5.8. By convention the total area under this curve, including the tips of both tails, is assigned a value of 1, regardless of the size to which it is actually drawn. Areas of parts of this distribution will therefore be less than 1.

Now assume that numbers on the horizontal axis of Figure 5.8 represent values of a variable and that the vertical dimension represents frequencies with which various values occur. We can use areas under this curve to answer questions like the following: What proportion of the observations have values greater than *A*? What proportion of the observations have values less than *A*? What proportion of the observations have values between *A* and *B*?[7] *A* and *B* can be any value of the variable

---

[7]Notice that these questions do not include: What proportion of the observations have value *A?* The unit normal table can in most cases only be used to ask questions about the proportion of observations that fall in certain *intervals* on the measurement scale; it cannot

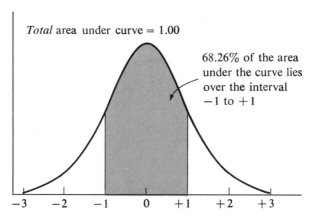

Figure 5.8   The unit normal distribution. Areas under parts of this curve are listed in Table 5.1 and in more complete form in Table A, Appendix 1.

we choose. Notice that each of these asks a question that is about an *interval*, a section of the horizontal axis. All can be answered by finding the appropriate areas in the **unit normal table,** Table A, Appendix 1.[8]

Table A actually specifies areas under the curve for certain intervals above the mean zero. In order to use the table for the entire normal distribution, we will have to keep in mind that the distribution is symmetrical, with 0.5 of the area above zero and 0.5 of the area below zero. Probably the best way to show how Table A is used is to ask each of the preceding questions in some examples. An abbreviated version of Table A is presented here as Table 5.1.

The table has three columns:

z — A point on the horizontal axis of the unit normal graph.

Area between μ and z — The area under the curve over an interval extending from the mean (μ) to the specified z.

Area beyond z — The area under the curve over an interval extending from z to infinity.

be used to find a proportion of cases that fall at a certain *point* on the scale. This is because the normal curve used in this manner is not really a frequency distribution. Strictly speaking, it is a *probability density function* for a continuous variable (more on this in Chapter 8). In such a distribution the relative frequency associated with any one value of the variable is essentially zero, since there are an infinite number of different values possible. When the normal curve is used as a model for an actual variable, that variable is treated as if it had the properties of the model. Hence, the vertical axis is more correctly labeled "probability density," not "frequency."

It will not, however, be harmful to treat the unit normal distribution as if it were a frequency distribution at this point. Probability density is very closely related to the concept of relative frequency; all the procedures discussed for determining frequencies in *intervals* are perfectly correct.

[8]Also called the *standard normal* table.

Table 5.1 *Abbreviated Unit Normal Table*

z	AREA BETWEEN μ AND z	AREA BEYOND z
0	0.0000	0.5000
0.25	0.0987	0.4013
0.50	0.1915	0.3085
0.75	0.2734	0.2266
1.00	0.3413	0.1587
1.25	0.3944	0.1056
1.50	0.4332	0.0668
1.75	0.4599	0.0401
2.00	0.4772	0.0228
2.25	0.4878	0.0122
2.50	0.4938	0.0062
2.75	0.4970	0.0030
3.00	0.4987	0.0013
3.50	0.4998	0.0002
4.00	0.49997	0.00003

These areas are labeled on the unit normal curves in Figure 5.9. Be sure you understand the meaning of the values in each column of the table and what they represent on the unit normal graph.

When the observations at hand are values in the unit normal distribution, these values are listed directly in the column labeled z; we can thus use the table to answer directly questions about relative frequencies and values of the variable.

In this distribution what proportion of the observations have values between 0 and 1.00? We need to find the area under the curve shaded in Figure 5.9(a) that lies over this interval. Since the mean of this distribution is zero, and since values of the variable are listed in the z column, the area is given in the second column, Area Between μ and z. In this case, 0.3413 of the observations have values between 0 and 1.00.

What proportion of the observations have values greater than 1.00? The needed area is shaded in Figure 5.9(b) and given directly in the third column, Area Beyond z: 0.1587 of the observations have values greater than 1.00.

The third question we can ask is really a request for a percentile rank: What proportion of the distribution has values less than 1.00? The shaded area in Figure 5.9(c) suggests that we will have to perform addition. Table 5.1 lists the area between the mean and 1.00 as 0.3413. To this we will have to add the area under the left half of the distribution, 0.5000. Thus, the total area to the left of $1.00 = 0.3413 + 0.5000 = 0.8413$. Slightly more than 84 percent of the observations in this distribution will fall below 1.00 in value.

Now consider the same kinds of questions about negative values of the

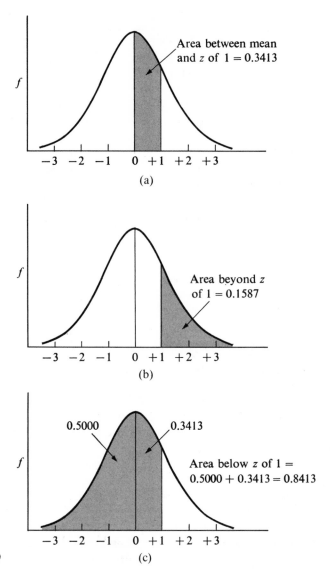

Figure 5.9

variable. What proportion of the observations have values greater than
-1.00? The area is shaded in Figure 5.9(d), but negative z values are not
listed in the table. Remember, however, that the normal distribution is
symmetrical, so that the area under the curve between 0 and $+1.00$ is
exactly equal to the area under the curve between 0 and -1.00. This area
is found to be 0.3413, and to it is added the area of the half of the
distribution above zero, 0.5000. Thus, 0.8413 of the distribution has
values above -1.00. Similarly, 0.1587 of the group's values lie below
-1.00, because the area beyond a z of -1.00 is exactly equal to the area
beyond a z of $+1.00$.

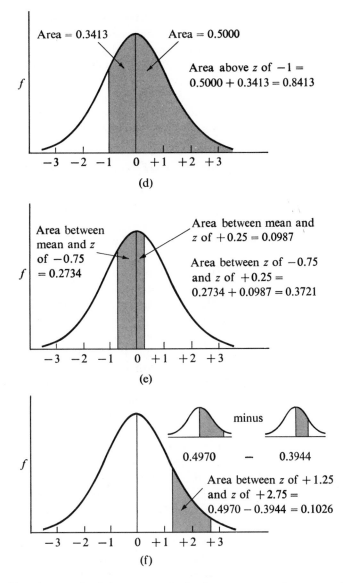

Figure 5.9   (cont.)

Keeping track of the areas required to answer a given question is tricky, and it is always a good idea to sketch the distribution and the required areas. The need for sketching becomes clearer when we ask questions like the following: What proportion of the distribution has values between −0.75 and +0.25? The shaded area in Figure 5.9(e) indicates that we will have to add two values from the table. The area between −0.75 and 0 is the same as the area between 0 and +0.75, that is, 0.2734. The area between 0 and +0.25 is given directly as 0.0987. Adding these together, we find that 0.3721 of the distribution lies in this interval.

Finally, consider the even more intricate question, What proportion of these values fall between 1.25 and 2.75? Figure 5.9(f) shows the needed area, which will have to be found by subtraction. As the figure suggests, we will find the area between the mean and 2.75 and subtract from it the area between the mean and 1.25; the difference will equal the area between 1.25 and 2.75. Thus, 0.4970 − 0.3944 = 0.1026 of the distribution falls in this interval.

If you are ever in doubt about whether to add or subtract, or what values are needed from the table, sketch the normal curve, identify the interval, and then look in the table. This cannot be overemphasized.

Areas Under Other Normal Curves

Questions about actual situations will not, of course, appear initially as questions about intervals on the unit normal curve. What proportion of the population is overweight? What proportion of the nation's school children are mentally handicapped? Will I decrease my chances of getting lung cancer if I smoke half a pack of cigarettes a day instead of a full pack? As suggested in Chapters 1 and 2, these questions can be approached statistically only if they are phrased in terms of questions about observable variables. Sometimes observations on such variables will approximate a normal curve; sometimes *statistics* computed from such observations will form a normal distribution, even though the observations themselves do not come from a normal distribution. In either case, you would be working with areas under parts of the normal curve to find relative frequencies, percentiles, and probabilities.

Such normal distributions will not, however, have a mean of zero and a standard deviation of 1. Figures 5.1, 5.2, and 5.3, for example, are all normal curves but with different means and standard deviations. What proportion of American men are over 6 ft tall? What proportion of the parts made by the hypothetical machinist are within 0.001 in. of being 1.000 in. in diameter? (This seemingly trivial question can actually have great importance; its answer determines how many parts must be discarded after manufacture, and it may even indicate how many will fail in use.)

We can use the unit normal table to answer such questions simply by converting the observations to z-scores. Remember that transforming a group of observations to their respective z-scores does not change the shape of the distribution; if we change a normal-shaped distribution of heights to their respective z-scores, we still have a normal-shaped distribution. Remember also that any distribution of z-scores has a mean of zero and a standard deviation of 1. We can thus change any normal distribution to the unit normal distribution by computing z-scores.

If we want to find the proportion of machine parts falling within 0.001 in. of 1.000 in., we must know that the distribution of such parts is a normal distribution, and we must know the mean and standard deviation of that distribution. Assume that the distribution in Figure 5.2 is normal, that $\mu = 1.000$ in. and $\sigma = 0.0005$ in. The proportion of the distribution that we want is shown in Figure 5.10(a). This will be the proportion of

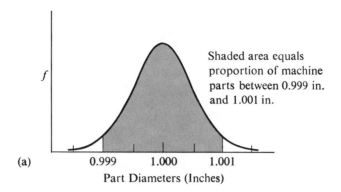

Shaded area equals proportion of machine parts between 0.999 in. and 1.001 in.

(a) 0.999   1.000   1.001

Part Diameters (Inches)

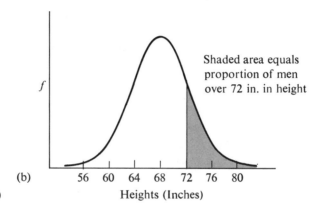

Shaded area equals proportion of men over 72 in. in height

(b)   56   60   64   68   72   76   80

Heights (Inches)

Figure 5.10

machine parts whose true values lie between 0.999 in. and 1.001 in. In order to find the corresponding interval on the unit normal curve, find a z-score for 0.999 and 1.001:

$$\text{For } 0.999, z = \frac{X - \mu}{\sigma} = \frac{0.999 - 1.000}{0.0005} = \frac{-0.001}{0.0005} = -2.00$$

$$\text{For } 1.001, z = \frac{X - \mu}{\sigma} = \frac{1.001 - 1.000}{0.0005} = \frac{0.001}{0.0005} = +2.00$$

The question then becomes one of finding the area between a z of -2.00 and $+2.00$ on the unit normal curve. From Table A, the area between the mean and $+2.00$ is 0.4772. Since the normal curve is symmetrical, this is also the area between a z of -2.00 and 0. Thus, the total area between $z = -2.00$ and 2.00, and the total proportion of machine parts between 0.999 and 1.001 in. is the sum of these two areas: $0.4772 + 0.4772 = 0.9544$. Of the parts this machine turns out, 95.44 percent will be within 0.001 in. of the desired 1.000-in. diameter.

What proportion of the American men are over 6 ft tall? If this height comes from a normal distribution, and we assume it does, the question can be answered by finding the z-score for 6 ft. The mean of the height distribution is given as 68 in. and the standard deviation as 4 in., and 6 ft = 72 in. The z-score is thus:

$$z = \frac{X - \mu}{\sigma} = \frac{72 - 68}{4} = \frac{4}{4} = +1.00$$

On the distribution in Figure 5.10(b) we can see that the question now is What proportion of the area under the unit normal curve lies beyond a z value of 1.00? The answer from Table A is 0.1587.

Finding Percentiles in Normal Distributions

The final operation to be covered in this chapter uses skills you already have, but in a different way than we have used them to this point. Here, we will learn how to answer the question "What score (value of the variable) is associated with the nth percentile?"

Consider, for instance, a normal distribution of human heights with a mean of 65.0 in. and a standard deviation of 3.0 in. What height is associated with the 95th percentile? To answer such questions, we again use Table A, but we do not begin searching in the z column. The question about the 95th percentile has given us information about areas under the curve, and we must use that to determine a z-score and raw score, instead of the other way around. The height at the 95th percentile has 0.9500 of the distribution below it (this is simply the definition of percentile); thus, the score we are looking for must be associated with a z-score that has 0.4500 of the area under the curve between itself and the mean, and 0.0500 of the area under the curve beyond it. That much information was contained in the original question. Now we go to Table A and search the "Area Between μ and z" column until we find the value 0.4500. This turns out to be associated with a z-score of +1.645. We know now that the height is 1.645 standard deviations above the mean, or $(3.0)(1.645) + 65.0$. The height is thus 69.9 in.

In finding values of variables that lie at different percentiles, we must often use the process of interpolation. Suppose, instead, that we had asked instead for the height associated with the 30th percentile. Thus, we need the z-score associated with an area between the mean and z of 0.2000. We see from Table A that a z of 0.52 has 0.1985 of the area between itself and the mean, and a z of 0.53 cuts off an area between itself and the mean of 0.2019. None of the tabled z values cuts off an exact 20 percent of the area, however.

To find the z-score associated with an area between itself and the mean of 0.2000, take the two area values that *are* given, and subtract the smaller from the larger:

$$0.2019 - 0.1985 = 0.0034$$

Now subtract the smaller of these values from the value we are concerned with, 0.2000:

$$0.2000 - 0.1985 = 0.0015$$

Divide the first remainder into the second:

$$\frac{0.0015}{0.0034} = 0.4412$$

This tells us that the z value we are searching for lies 0.4412 of the way from a z of 0.52 to a z of 0.53 (the z-scores associated, respectively, with areas of 0.1985 and 0.2019). These z-scores differ by 0.01. Thus, the z we want is (0.01)(0.4412) above a z of 0.52. The z, associated with a height at the 30th percentile, is calculated as:

$$z = (0.01)(0.4412) + 0.52 = 0.5244$$

Finally, we know that the height at the 30th percentile is 0.5244 standard deviations below the mean. Thus:

$$X = 65 - (0.5244)(3.0)$$
$$= 63.4$$

Thus a height of 63.4 in. has 30 percent of the distribution below it.

SUMMARY

Many frequency distributions of actual observations are characteristically bell shaped. Often these are similar in shape to each other and to a mathematically specified curve called the normal curve. Statistical properties of the mathematical expression have been thoroughly determined; when the empirical distribution is close enough in shape to the normal curve to be considered essentially normal, then statistical properties of the empirical distribution also are known.

There are actually an infinite number of different normal curves that fit the normal curve in question; in order to specify one of these, values for the mean and the standard deviation must be specified. The normal distribution is symmetrical and centered on the value of the mean.

Determining whether or not a given distribution is close enough to the normal distribution to be treated as one requires a statistical test that is beyond the scope of this book. However, it is usually safe to assume that observations are normally distributed when (1) observations represent scores from tests that have been specially constructed to produce normally distributed scores, and (2) when differences among values are produced by error, that is, by multiple unrelated, uncontrollable, and unknown factors.

When a distribution is said to be normal, and when values for the mean and standard deviation are specified, the shape of the entire distribution is specified. This, in turn, means that the relative frequencies with which different values of the variable occur can be determined. Determining relative frequencies of various values in the frequency distribution makes use of a special property of frequency polygons: in any frequency polygon the proportion of the observations falling within an interval on the measurement scale is equal to the proportion of the area under the curve that lies over that interval.

Tables have been prepared that specify the area under the normal curve above specific intervals; these tables are computed for a normal distribution with a mean of zero and a standard deviation of 1, the unit normal distribution.

If the mean and standard deviation of any normal distribution are known, the unit normal tables can be used to find proportions of the distribution that lie in various intervals. To use the tables in this manner, the raw scores must be converted to their equivalent z-scores.

KEY CONCEPTS

normal (Gaussian) distribution

parameters of the normal distribution

relationship between area under the curve and relative frequency

unit normal curve

unit normal table

SPOTLIGHT 5   **The Normal Distribution**

The normal curve that is so familiar in modern statistics was known among early-nineteenth-century mathematicians as the Law of Error — a term that describes its first practical applications in science.[9]

Astronomers of the 1830s were greatly concerned with errors. Mathematical theorems had been developed that predicted the orbits of various heavenly bodies very precisely, but theory could not always be advanced or verified through the telescope because human error entered into every observation. Differences in observers' reaction times and minute variations in spatial judgments from observation to observation added a personal equation to every measurement, one that had to be circumvented if possible. Thus, astronomers turned to mathematicians like Pierre La Place and Karl Gauss for detailed descriptions of errors. These men, perhaps more than any others, laid the foundation for modern statistics by pointing out that errors are distributed as a normal curve (although the term normal curve did not come into use until several decades later) and that the probabilities associated with different error sizes can be predicted from a knowledge of the curve. This made it possible to predict the "most probable orbit" of heavenly bodies from a number of observations.

Sir Francis Galton, however, should receive primary credit for pointing out that the Law of Error also describes a multitude of biological and psychological distributions. Making the bridge between mathematics and biology, Galton wrote in *Natural Inheritance*:

[9]Abraham DeMoivre published the equation for the curve in 1733, but it saw little practical application for nearly 100 years. Those interested in a more complete history of the normal curve — and of many other methods treated in this book — should see Helen M. Walker's classic, *Studies in the History of the Statistical Method* (Baltimore: Williams and Wilkins Co., 1929).

I need hardly remind the reader that the Law of Error upon which these Normal Values are based, was excogitated for the use of astronomers and others who are concerned with extreme accuracy of measurement, and without the slightest idea until the time of Quetelet that they might be applicable to human measures. But Errors, Differences, Deviations, Divergencies, Dispersions, and individual Variations, all spring from the same kind of causes. . . . The Law of Error finds a footing wherever the individual peculiarities are wholly due to the combined influence of a multitude of "accidents". . . . All persons conversant with statistics are aware that this supposition brings Variability within the grasp of the laws of Chance, with the result that the relative frequency of Deviations of different amounts admits of being calculated, when these amounts are measured in terms of any self-contained unit of variability.[10]

Galton was fully aware that using the mathematical abstraction to describe real distributions required an ultimately unprovable assumption, namely, that the underlying characteristic *was* normally distributed; he was also aware that many naturally occurring distributions are *not* normal in shape. He made it clear, however, that he believed the assumption of normality justified, in some cases, even if not provable:

It has been objected to some of my former work, especially in Hereditary Genius, that I pushed the applications of the Law of Frequency of Error somewhat too far. I may have done so, rather by incautious phrases than in reality; but I am sure that with the evidence now before me, the applicability of that law is more than justified. . . . I am satisfied to claim that the Normal Law is a fair average representation of the Observed Curves during nine tenths of their course . . . the agreement of the Curve of Stature with the Normal Curve is very fair, and forms the mainstay of my inquiry into the laws of Natural Inheritance. It has already been said that mathematicians laboured at the Law of Error for one set of purposes, and we are entering into the fruits of their labours for another.[11]

PROBLEMS

1. To the best of your ability, sketch normal curves with the given parameters. Be sure to label your axes clearly.

   a. Scores on a nationally administered college entrance exam, normally distributed with a mean of 500 and standard deviation of 100.

   b. Human weights, distributed normally with a mean of 150 lb and a standard deviation of 25 lb.

   c. Intelligence test scores, distributed normally with a mean of 100 and a standard deviation of 10.

   d. Machine part diameters distributed normally with a mean of 2.510 cm and a mean of 0.010 cm.

2. a. For the distribution described in problem 1a, what z-score is associated with a test score of 750? A score of 445? A score of 620?

[10]Sir Francis Galton, *Natural Inheritance* (London: Macmillan and Co., 1894), pp. 54, 55. Reprinted by permission of Macmillan, London and Basingstroke.
[11]Ibid., pp. 56, 57.

b. For the distribution of weights described in problem 1b, what z-score is associated with a weight of 180 lb? A weight of 162 lb? A weight of 115 lb?

c. For the machine parts described in problem 1d, what z-score is associated with a diameter of 2.532 cm? A diameter of 2.504 cm? A diameter of 2.499 cm?

3. Which distributions below might you expect to be normal?

   a. Salaries of major league baseball players
   b. Weights of adult pigeons in a laboratory pigeon colony
   c. Waist measurements for British men
   d. The amount of time it takes a number of rats to run the length of a laboratory alleyway.

4. Listed below are five z-scores from a normal distribution. For each one, indicate what proportion of the normal curve's area falls below the z value and what proportion falls above the z value (these two values should, of course, add up to 1.0000 for each value).

   a. $z = 1.45$ d. $z = 2.77$
   b. $z = 0.49$ e. $z = -0.99$
   c. $z = -0.01$

5. A normal distribution of aptitude test scores has a mean of 60 and a standard deviation of 7. What is the percentile rank of the following test scores?

   a. 65   d. 44
   b. 72   e. 55
   c. 81   f. 50

6. In the normal distribution described in problem 5 above, what test score is associated with the

   a. 50th percentile   c. 99th percentile
   b. 95th percentile   d. 5th percentile

7. When test scores are normally distributed with a mean of 140 and a standard deviation of 33, approximately what must your score be so that you have a percentile rank of

   a. 10   c. 50   e. 90
   b. 20   d. 75

8. In a normal distribution of measurements having a mean of 500 meters and a standard deviation of 82 meters, what percent of the distribution falls

   a. Between 600 and 700 meters?
   b. Below 400 meters?
   c. Above 620 meters?
   d. Between 300 and 450 meters?

9. A cylindrical automobile part is designed to be exactly 2.000 cm in diameter. However, the machine that produces these parts does not make them all precisely the same size; part diameters are normally distributed with a mean of 2.000 cm and with a standard deviation of 0.001 cm.

   a. If any part that is more than 0.0015 cm from the intended size (2.000) is classified as "defective," what proportion of this machine's parts are defective?

   b. What proportion of this machine's parts are within 0.0001 cm of the intended size?

   c. What proportion of this machine's parts are within 0.00001 cm of the intended size?

10. In a normal distribution, what proportion of the distribution falls

    a. Within 1.5 standard deviations of the mean?

    b. Within 2.5 standard deviations of the mean?

    c. Within 3.5 standard deviations of the mean?

    d. Within 4.0 standard deviations of the mean?

11. In a normal distribution of weights, with a mean of 200 lb and a standard deviation of 40 lb, what proportion of the distribution members have weights

    a. above 250 lb?   d. between 150 and 230 lb?

    b. between 160 and 220 lb? e. below 190 lb?

    c. below 180 lb?

12. For the distribution described in problem 11 above, what weight is associated with the

    a. 60th percentile   c. 45th percentile

    b. 20th percentile   d. 10th percentile

13. Assume that the heights of men are normally distributed, with a mean of 68 in. and a standard deviation of 4 in.

    a. Orville is 71 in. tall. What is his percentile rank on the height variable?

    b. Otto is 65 in. tall. What proportion of the adult male population is taller?

    c. What proportion of the same population falls between Orville and Otto in height?

    d. Oliver claims that he is taller than 75 percent of the men in this population. How tall is he?

    e. Only one man out of every thousand is taller than Oscar. How tall is Oscar?

14. Assume that adult male waist sizes are normally distributed with a mean of 35 inches and a standard deviation of 3 inches. Remember that a size measurement of, say, 36 inches has true limits of 35.5–36.5 inches.

    a. A 36-inch belt fits men with waist sizes 35–37 inches. About what proportion of the male population would wear a 36-inch belt?

    b. If a 38-inch belt fits men with waist sizes 37–39 inches, what proportion of the same population would wear a 38-inch belt?

    c. About what proportion of the population would wear a 40-inch belt, assuming such a belt fits waist sizes 39–41 inches?

    Note: In this problem, be a stickler for detail and take into account the true limits of the specified waist sizes. Thus, a size measurement of, say, 36 inches should include values of 35.5 to 36.5 inches, and so on. Sometimes such details make a great deal of difference in statistical results — as you can verify for yourself by working this problem with and without consideration of true limits. Answers given in Appendix 4 are based on the true limits.

15. A college entrance exam administered every year to thousands of high-school seniors produces a mean score of 550 and a standard deviation of 100. The scores are normally distributed.

    a. A certain university admits only students scoring in the top 25 percent of those taking this exam. What score does a student have to receive in order to be eligible for admission to that university?

    b. The principal of Adams H.S. awards a scholarship each year to any student from his school scoring in the top 1 percent of all applicants. Cecilia scored 785. Did she receive a scholarship?

    c. Roger was informed by the testing organization that he had scored at the 40th percentile. What was his raw score?

    d. What proportion of students scored within 100 points of the mean?

    e. What proportion of students scored lower than 300?

6. Correlation: Measuring Relations

After reading this chapter you should be able to do the following:

1. *Give examples from everyday experience that demonstrate correlation between variables.*

2. *Discuss the essential characteristics of the correlational relation: direction and degree of relation.*

3. *Make a scatter-plot diagram for some paired observations and explain how the chart shows direction and degree of a relation.*

4. *Explain the purpose and fundamental characteristics of correlation coefficients.*

5. *Compute and interpret the Pearson r for a group of paired observations.*

6. *Compute and interpret the Spearman rho.*

7. *Discuss some ways in which correlation coefficients are used to evaluate tests.*

8. *Compute and discuss some uses of the point biserial correlation coefficient.*

Relations Among Variables

There are probably many good ways to describe the goals of science. One might be to say that science attempts to discover relations among natural phenomena, to describe them accurately, and then, ultimately, to explain why they exist. The correlational methods we cover in this chapter and the next address the first two of these goals, discovering and describing relationships.

If you reflect for a minute on any of the science courses you have taken, you should see easily that most of the basic "laws" that are taught as items of scientific knowledge are really statements about relationships among variables. In chemistry, the pressure of a gas is related to its volume and temperature; in economics, the market price of a commodity is related to its supply and the demand for it; in psychology, the amount of information remembered in a memory study is related to the amount of time spent studying it, and so on.

In the natural sciences, many of the lawful relationships among **correlated variables** are known with such precision that they can be stated quantitatively. If one variable has this value, then another variable must have that value. Einstein's famous mass-energy equation, for instance, shows the relationship between the amount of energy contained in a bit of matter (one variable) and its mass (essentially, its weight, the second variable):

$$E = mc^2$$

This relationship is stated quantitatively in that it specifies values for each variable involved: the energy (in ergs) contained in a body is equal to its mass (in grams) multiplied by the squared speed of light (in centimeters per second).

This particular law was stated theoretically many years before it was verified empirically. Very often, however, science progresses in the opposite direction: relationships may be discovered through observation, and only after many repeated observations are the relationships known with precision. It is this latter kind of scientific progress, forming laws after collecting many, many observations, that calls for statistical analyses such as correlational methods. This is particularly true in the behavioral sciences.

Economists may suspect that the rate of inflation is related to the prime lending rate; psychologists may suspect that a person's reaction to a persuasive speech is related to the identity of the speaker. But these relations are much harder to discover than physical laws, much harder to summarize precisely, for several reasons. As the discussion on pages 19–32 indicated, measurement in the behavioral sciences is often prone to larger error than measurement in the physical sciences, and furthermore, we are often less sure that observable indicators (such as test scores) correspond closely to hypothesized constructs such as intelligence, anxiety, attitude strength, personality traits, and so on. These problems contribute to the difficulty of determining relations among variables; they account for the fact that the behavioral sciences cannot yet summarize many relations with neat mathematical laws.

This does not mean that we cannot hope to learn something about relations among behavioral phenomena, even though these relations are complex or obscure. **Correlational** methods are techniques designed specifically for identifying relations among variables; correlational methods give us a means of describing and measuring relations even in situations where the relations are difficult to see.

Kinds of Relations

Here we will confine our discussion to relations between two variables. Two variables are related when changes in the value of one are systematically related to changes in the value of the other. There are two things we will want to know about relations: direction and degree.

When we speak of **direction** we mean that relations between variables are either positive or negative. A **positive relation** means that as values of one variable increase, values of the other tend also to increase. For human beings, the relation between height and weight is positive. People who register high values on the height variable tend also to register high values on the weight variable. There is probably also a positive relation between the number of shots a basketball player takes in practice and his shooting percentage during game play. Those who get lots of practice probably have higher shooting percentages than those who get little practice. There is also a positive relation between your scores on examinations in any class and the grade you ultimately receive; those who receive high exam scores tend to receive high grades.

A **negative relation,** as you might suspect, indicates the opposite relation between variables. High values on one variable tend to be accompanied by low values on the other. As the interest rate for mortgages goes up, the number of new home loans issued in a given time period tends to go down. There are volumes of statistics that suggest a negative relation between the number of cigarettes smoked over a lifetime and age at death.[1]

Relations between variables can also be characterized by **degree. Degree of relation** refers to the extent that observed values adhere to the designated relation. The highest degree of relation is a **perfect relation,** in which knowing the value of one variable determines exactly the value of the other. To give a trivial example, your age in years is perfectly correlated with your age in months. If you register 20.5 on the "years" variable, then you must register 246 on the "months" variable. Physical laws such as $F = MA$, $y = \frac{1}{2}gt^2$ and $E = mc^2$ are further examples of perfect relations. In a perfect relation there are no exceptions to the rule.

At the other extreme is the lowest degree of relation, the **zero relation;** this is just another way of saying that no relation exists between variables. You can probably think of any number of these, such as the relation between the number of coyotes living in Montana during any given year and the number of yellow taxicabs in New York City during the same year.

The vast majority of relations studied in the behavioral sciences range in degree between zero and perfect. The relation between intelligence scores of parents and those of their children is positive but not perfect. That is, parents who score high on intelligence tests tend to have children who score high, and vice versa, but there are many exceptions to this rule.[2]

Correlational methods offer a means of determining whether or not such relations exist, that is, whether variables are "co-related" or **correlated.** They can tell us the direction of the relation and its degree. Quite

[1] See, for example, R. Pearl, "Tobacco smoking and longevity," *Science* 87 (1938): 216–217.

[2] Many studies designed to examine this question have been reviewed by L. Erlenmeyer-Kimling and L. F. Jarvik, "Genetics and Intelligence: A Review," *Science* 142 (1963): 1477–1479.

often we simply cannot ascertain this information by casual observation; quite often the analysis surprises us by showing little or no relation between variables. You might expect, for example, that people who get high grades in college would also tend to have more successful careers and more prestigious professions in later life, but available statistics suggest there is no relation between college grades and later success.[3] The only answer to questions about many such relations comes from empirical evidence; that is, collect the data and apply correlational analysis.

Scatter-Plot Diagrams

When a single variable is under study, the frequency distribution graph functions as a group picture of the data. From visual inspection of the frequency polygon, one gets a rough idea of central tendency and variability. Similarly, when one is interested in the relation between two variables, there is a graphical technique that can give a rough idea of the direction and degree of the relation. This is the **scatter-plot diagram.**

To make a scatter plot we need a number of **paired observations** on two variables. Usually, one variable is arbitrarily designated X and the other Y. We can make paired observations on these if in the group of interest there is one Y for every X and vice versa.

Do people who spend much time studying get better grades than those who spend little time studying? Suppose we collected some paired observations on the two variables mentioned here (grades and time spent studying) and present the results in a scatter plot.

Ten people volunteer to provide us with data, and for each person we record the number of hours per week spent studying (X variable) and the current grade-point average (Y variable). These data are summarized in Table 6.1. Simply looking at Table 6.1 may or may not suggest an answer to the question of whether or not there is a relation between X and Y, but the scatter plot in Figure 6.1 clearly does. Notice the following characteristics of the scatter plot:

1. Each variable is represented by one of the axes; each axis is scaled to include only the range of values observed on one variable. Thus, the X axis covers only the range of hours spent studying, from 6 to 16; the Y axis covers the grade-point averages from 2.1 to 3.9.

2. Both axes are about the same length, making the diagram square in shape.

3. Each point represents one *pair* of observations. The point representing observations on Cleo, for instance, is labeled to indicate that for this pair of values, $X = 8$ and $Y = 2.6$.

4. The dots are *not* connected with a line (as they would be in a frequency polygon).

---

[3]This information will be comforting to some and disturbing to others. You may wish to read more on the subject in D. McClelland, "Testing for Competence Rather than for Intelligence," *American Psychologist* 28 (1973): 1–14.

Table 6.1 *Paired Observations on Grade-point Averages and Hours per Week Spent Studying for Ten People*

	HOURS PER WEEK STUDYING (X)	GRADE-POINT AVERAGE (Y)
Bob	12	3.00
Carol	9	2.30
Ted	6	2.10
Alice	12	3.70
Cleo	8	2.60
Julius	13	3.40
Mark	14	3.80
Anthony	7	2.40
Mickey	10	3.10
Minnie	16	3.90

In order to get a better idea of the message conveyed by this graph, draw a light pencil line around all of the points (this is not part of the scatter plot, just an aid to interpreting it). Notice that the group of dots tends to rise toward the right; this indicates a *positive* relation — values of both variables tend to increase together. Scatter plots indicating positive

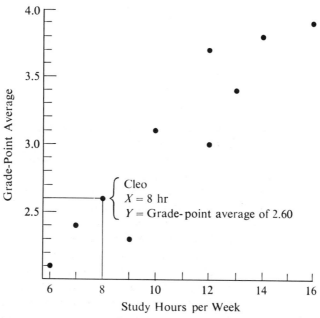

Figure 6.1 Scatter-plot diagram for the paired observations presented in Table 6.1. The body of dots suggest that as hours per week of studying increase, grade-point averages increase.

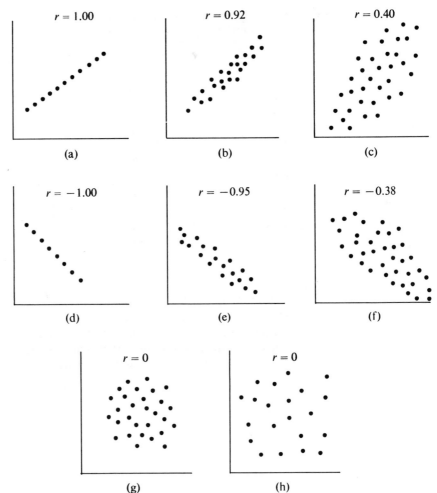

Figure 6.2

relations are also shown in Figures 6.2(a), (b), and (c). (The *r* values shown will be explained shortly.)

The body of dots slopes in the opposite direction when the two variables are related negatively. If the value of one tends to decrease as the value of the other increases, the group of points will form an oval sloping down toward the right, as in Figures 6.2(d), (e), and (f). Thus, the direction of the slope indicates the direction of the relation.

The scatter plot also permits rough estimation of *degree* of relation. If all the dots line up perfectly in a straight line, the relation between X and Y is a perfect relation. Figure 6.2(a) shows a perfect positive relation, and 6.2(d) shows a perfect negative relation. The more the dot pattern differs from a straight line, the lower the degree of relation.

In the grade-point/study hours example [Figure 6.1], the correlation between variables is relatively high; this is reflected in a thin, compact oval. Scatter plots showing other high correlations are pictured in Figures

6.2(b) and (e). The scatter plot indicates a low-degree relation with a relatively fat oval. Low-degree relations are represented by scatter plots 6.2(c) and (f).

Finally, when no relation exists between the variables, the scatter plot will take the general shape of a circle [Figure 6.2(g)] or some other irregular form [for example, Figure 6.2(h)], in which there is no discernible upward or downward slope.

Correlation Coefficients

The scatter plot is a quick and ready method for picturing correlational relations, but it is not the endpoint of correlational analysis. Statistics designed to specify precisely the direction and degree of relations are known as **correlation coefficients.**

There are many available correlation coefficients, and their respective computational formulas are quite different; each is designed for particular kinds of data and particular kinds of variables. However, all of them share some important characteristics.

Most correlation coefficients range in value from -1.00 to $+1.00$. The computational methods that produce them have been designed to produce coefficients with negative values when a negative relation exists and positive values when a positive relation exists. So all you have to do to determine the direction of a relation is to look at the *sign* of the correlation coefficient describing the relation. Coefficients such as -0.90, -0.37, and -0.66 indicate negative relations, and coefficients like $+0.72$, $+0.62$, and $+0.21$ indicate positive relations.

The various correlation coefficients have also been designed so that the *degree* of the relation is specified by the *size* of the coefficient, irrespective of its sign. Perfect relations produce correlation coefficients of $+1.00$ (perfect positive relation) or -1.00 (perfect negative relation). The larger the size, the higher the degree of correlation. Values of $+0.97$, -0.92, $+0.83$, and $+0.92$ appear when the degree of correlation is high; values such as $+0.09$, -0.23, $+0.31$, and -0.14 turn up when computed for variables correlated to a low degree. Correlation coefficients for the relations graphed in Figure 6.2 are shown on the respective scatter plots as *r* values.

We will begin by considering the most useful correlation coefficient: the **Pearson product-moment correlation coefficient.**

Pearson Product-Moment Correlation Coefficient

The Pearson product-moment correlation coefficient has such a long name that we will call it either the Pearson *r*, or simply *r*. In descriptive statistics *r* is the symbol for this statistic.

In order to appreciate the message conveyed by the Pearson *r* and its derivation, consider again the meaning of relation between two variables. In the grade-point/study hours example, observations on one variable range between 2.1 and 3.9, while on the other variable the numbers range from 6 to 16. A measure of correlation must be scale-independent in that the size of the numbers registered on the scale should not be reflected in the correlation statistic. We might even be searching for relations between variables in which observations on one variable are on the order of 0.00003 and observations on the other variable are numbers like 6 million, and so on. When two variables are associated in a positive relation, we would like to find a statistic that says, "Those who score relatively high on X tend to score relatively high on Y *and* those who score relatively low on X tend to score relatively low on Y"; the important concern is the *relative* positions within the two groups. This should suggest to you that z-scores might be useful here.

Figure 6.3 shows how the preceding statement can be rephrased in

In a positive relation

High z-scores in both distributions and

Low z-scores in both distributions come together

In a negative relation

High z-scores on X come with Low z-scores on Y and

Figure 6.3   Low z-scores on X come with High z-scores on Y

terms of z-scores. The distributions suggest that we are searching for a statistic that says, "In a positive relation those with high z-scores in the X distribution receive high z-scores in the Y, and those with low z values in the X distribution obtain low z values in the Y." Conversely, the statistic should also be designed to say that in *negative* relations high z-scores in the X distribution are associated with low z-scores in Y, *and* low z-scores in X are associated with high z-scores in Y. The degree to which individual paired observations conform to these rules — the degree of the relation — should also be reflected in the statistic.

The Pearson r is designed to do all of this. However, before it is applied to a set of paired observations, several conditions must be met. These are sometimes referred to as the *assumptions underlying the Pearson r*. This means that, when you summarize your data with a Pearson r, and report that statistic, it is assumed that the following conditions have been met (otherwise the statistic can present a misleading picture of the relationship between variables):

1. We must have at least interval-scale data on both variables. z-scores and statistics derived from them make use of distances on the measurement scale. These distances have meaning only for interval-, ratio-, or absolute-scale measurements.

2. We assume that the relation between X and Y is approximately **linear.** One meaning of linear relation is that the graph of the relation between X and Y is a *straight line*.[4] Figure 6.4 shows several X–Y relations, but only those in 6.4(a) are linear, since only these are represented by straight lines. Notice that the relation in 6.4(c) is a perfect relation — specifying a value of X determines exactly what the value of Y will be. Graph 6.4(c) is not, however, a linear relation. Although the correlation is perfect, the relation will not be accurately measured or detected by the Pearson r. Figure 6.4(c) is a *curvilinear* relation; the Pearson r is a measure of *linearity* in a relation.

Incidentally, there are some instances in the behavioral sciences where *curvilinear* relations appear. Figure 6.4(c) might represent, for instance, the relation between people's hand strength and chronological age. If the vertical axis represents strength and the horizontal axis represents age, Figure 6.4(c) would suggest that hand strength increases up to a certain age and then decreases.[5]

Most scatter plots, as you know, do not appear as straight lines. But the Pearson r is appropriate if it is assumed that the underlying relation between X and Y is essentially linear. The relation is probably linear if the scatter plot resembles those in Figures 6.5(a), (b), or (c). The bends in the ovals of scatter plots 6.5(d), (e), and (f) suggest nonlinear relations.

---

[4] More formally, the relation between X and Y is linear if it is of the form $Y = a + bX$, where a and b are constants.

[5] For an introduction to correlation with nonlinear relations, see J. P. Guilford and B. Fruchter, *Fundamental Statistics in Psychology and Education*, 5th ed. (New York: McGraw-Hill, 1973), pp. 285–293.

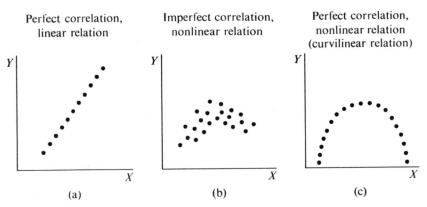

Figure 6.4   Three kinds of *X–Y* relations: Plot (a) appears to show a perfect linear relation, since all the points fall on a straight line. Plot (b) shows an imperfect correlation, since specifying a value of *X* does not determine exactly the value of *Y*. Plot (c) appears to represent a perfect correlation, since each value of *X* is associated with one value of *Y*. Plot (c) does not show a linear relation, however, since the scatter plot points do not fall on a straight line; it represents a typical *curvilinear* relation.

In most situations there is probably little harm in computing a Pearson *r* for data that depart somewhat from the linear relation; the consequences of doing so are a correlation coefficient with a spuriously low size. The Pearson *r* measures only linear relations; relations that exist but that are nonlinear do not show up in the value of *r*.

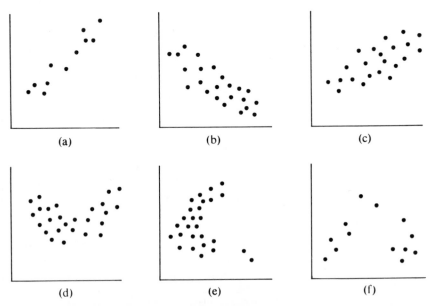

Figure 6.5   Linear and nonlinear *X–Y* relations. Although none of the scatter plots show a perfect linear relation, it is probably safe to assume that plots (a), (b), and (c) represent essentially linear relations. The shapes of scatter plots (d), (e), and (f) suggest that the underlying relations are nonlinear.

3. When computing the Pearson r, we assume **homoscedasticity** of variances. This means essentially that the width of the scatter-plot oval is relatively uniform throughout its length. Figure 6.6 shows one scatter plot with homoscedasticity and two without it. In a homoscedastic relation the variance of Y values above one interval on the X axis is about the same as the variance of Y values above any other interval on the X axis, such as in Figure 6.6(a). Naturally, no real scatter plot meets this condition exactly, but scatter plots such as 6.6(b) and 6.6(c) depart sufficiently from homoscedasticity to make the use of the Pearson r inappropriate. If one has at least interval-scale data on both variables, and if inspection of the scatter plot suggests that conditions of linearity and homoscedasticity are met, the Pearson r will yield useful information about the relationship between the variables.

Just as we did with the standard deviation in Chapter 4, we will first compute the Pearson r by means of its definition formula. The definition formula is more cumbersome to work with than the (optional) computational formula, but it shows clearly how z-scores are related to the value of r. Defined verbally, the Pearson r is the mean of the "cross-products" of the z-scores. We take each pair of observations, multiply the z-score in the X distribution by the z-score in the Y distribution, add these cross-products, and divide by the number of pairs of observations. The statistic obtained has all the characteristics of the correlation coefficient described on page 151. In symbols:

$$r_{XY} = \frac{\sum\limits_{i=1}^{N} z_{X_i} z_{Y_i}}{N}$$

Do not let all of the subscripts in the above expression confuse you — they are there to indicate what each symbol represents. It may be useful for some people to take this formula part by part:

(a)	(b)	(c)

Figure 6.6 Illustration of homoscedasticity and nonhomoscedasticity of variances. Scatter plot (a) has about the same width all along its length. Scatter plots (b) and (c) are much wider in some places than others. Thus, (a) exhibits homoscedasticity of variances, while (b) and (c) do not.

r_{XY} is the correlation coefficient r between two variables.

z_{X_i} represents the z-score associated with individual i's score on the X variable.

z_{Y_i} represents the z-score associated with individual i's score on the Y variable.

N represents the number of paired observations.

Obtaining a value of r does not require any computational feats beyond those already covered in Chapter 4; it is an exercise in careful bookkeeping, however, since we will have to obtain many z-scores and keep track of several other statistics as well.

The data from Table 6.1 have been reproduced in Table 6.2, together with some other values needed to find the value of r. Using the preceding formula, compute r in steps as follows:

1. First find the mean and standard deviations of the X and Y distributions (mean $X = \bar{X}$, mean $Y = \bar{Y}$; standard deviation for Xs $= s_X$ and for Ys $= s_Y$). These are computed exactly as described in Chapters 3 and 4. Of course, Y values do not enter into computation of the X statistics, and vice versa. The computed values for these statistics are shown at the bottom of Table 6.2.

2. Use the means and standard deviations to find a z-score for each X and for each Y score within their respective distributions. The z-score for $X = 12$, for instance, is:

$$z_{X_i} = \frac{X_i - \bar{X}}{s_X} = \frac{12 - 10.70}{3.07} = \frac{1.30}{3.07} = +0.423$$

3. For each pair of observations multiply the z-score for the X value (z_X) by the z-score for the Y value (z_Y) to obtain the cross-product; these are listed in the last column of Table 6.2. For example, the first person listed registered a z-score of $+0.423$ on the X variable and a z-score of -0.048 on the Y. The product of these two values is -0.0203.

4. Find the mean cross-product. This is done by adding the cross-products and dividing by the number of paired observations. The result is the Pearson r, here equal to $+0.93$ (r is usually specified to two decimal places).

For the data collected, the correlation between hours per week studying and grade-point averages turned out to be $+0.93$. But what does this mean? Now that we have the correlation coefficient, what do we do with it?

Interpreting the Pearson r

Interpreting the sign of the correlation coefficient is easy. The plus sign indicates a positive relation — those who study more tend to have higher grade-point averages.

Table 6.2 *Computation of the Pearson r*

i	X_i	$X_i - \bar{X}$	z_{X_i}	Y_i	$Y_i - \bar{Y}$	z_{Y_i}	$z_{X_i} z_{Y_i}$
1	12	+1.3	+0.423	3.00	−0.03	−0.048	−0.0203
2	9	−1.7	−0.554	2.30	−0.73	−1.166	+0.6460
3	6	−4.7	−1.531	2.10	−0.93	−1.485	+2.2735
4	12	+1.3	+0.423	3.70	+0.67	+1.070	+0.4526
5	8	−2.7	−0.879	2.60	−0.43	−0.687	+0.6039
6	13	+2.3	+0.749	3.40	+0.37	+0.591	+0.4426
7	14	+3.3	+1.075	3.80	+0.77	+1.230	+1.3222
8	7	−3.7	−1.205	2.40	−0.63	−1.006	+1.2122
9	10	−0.7	−0.228	3.10	+0.07	+0.112	−0.0255
10	16	+5.3	+1.726	3.90	+0.87	+1.390	+2.3991

$$\sum X_i = 107 \qquad \sum z_{X_i} = 0.0 \quad \sum Y_i = 30.30 \qquad \sum z_{Y_i} = 0.000 \quad \sum z_{X_i} z_{Y_i} = 9.3063$$

$$\bar{X} = 10.70 \qquad \bar{Y} = 3.03$$
$$s_X = 3.07 \qquad s_Y = 0.626 \qquad r_{XY} = \frac{\sum z_{X_i} z_{Y_i}}{N} = \frac{9.3063}{10} = +0.93$$

N = Number of paired observations = 10

Interpreting the size of the coefficient is a more subtle matter. Because correlation coefficients look like proportions, many people respond to them as if they *were* proportions, thinking that a correlation of 0.70 is twice as high as a correlation of 0.35. This is unfortunate, because r itself is not a proportion. Interpretation of the magnitude of r depends somewhat on the number of observations,[6] but more important for our immediate purposes, on the value of r^2. r^2 *is* a proportion, specifically the proportion of variation in the scatter plot accounted for by a linear equation, but we will have to wait until the next chapter to appreciate that definition completely. For the moment, we will do best simply to characterize some sizes (absolute values) of r as being high and others as being low. These labels are presented with values of r^2 in Table 6.3.

The most common misuse of correlation coefficients occurs when people assume that a high correlation between variables proves that a cause-and-effect relationship also exists between these variables. For instance, there have been several recent medical studies showing a positive correlation between coffee consumption and heart attacks. But this evidence, by itself, does not prove that coffee drinking *causes* heart problems any more than it proves that heart problems cause coffee drinking. It is entirely possible that a third factor causes changes in both variables. Perhaps people who drink excessive amounts of coffee also

---

[6]The size of N and the magnitude of r allow us to make a decision about the *significance* of r. Significance is a technical term used in inferential statistics (Chapters 8–15), and it need not concern us as long as we restrict ourselves to descriptive statistics.

Table 6.3 *A Rough Guide to Degree of Correlation Indicated by Different Sizes of r's Absolute Value and r²*

$|r|$	r^2	DEGREE OF CORRELATION (ROUGH GUIDE)
1.00	1.00	Perfect
0.98	0.96	
0.95	0.90	
0.90	0.81	High
0.85	0.72	
0.80	0.64	
0.70	0.49	
0.60	0.36	Moderate
0.50	0.25	
0.40	0.16	Low
0.30	0.09	
0.20	0.04	Negligible
0.10	0.01	

tend to be people with sedentary jobs, which would mean that many of them do not get enough proper exercise. At any rate, the discovery of a high correlation coefficient in no way establishes that changes in one variable cause changes in another variable. However, correlational methods are very useful for *predicting* the value of one variable if the value of another correlated variable is known. In fact, all of Chapter 7 will deal with the subject of making predictions from correlational information.

C **Obtaining *r* Through Covariation**

Computing *r* with the definition formula and methods shown above is a rather laborious process because one must compute means, standard deviations, and then *z*-scores for each *X* and *Y* score. The many computational steps in that definition formula require an excessive amount of time for both human and electronic computers to execute and they increase the likelihood of rounding error and other mistakes. So, now that you know how *z*-scores are involved in the value of *r*, we will cover a simpler definition formula. Of course, many people will consider this section optional.

This alternative approach makes use of a statistic called **covariation.** The covariation for a set of *X* and *Y* scores is obtained by taking the *X* deviation score for each pair, $(X_i - \bar{X})$, multiplying it by the *Y* deviation score for that pair, $(Y_i - \bar{Y})$, and then adding up all such cross-products for the entire group. In symbols:

$$XY \text{ covariation} = \sum_{i=1}^{N} (X_i - \bar{X})(Y_i - \bar{Y})$$

The Pearson r is then obtained as follows:

$$r_{XY} = \frac{XY \text{ covariation}}{\sqrt{X \text{ variation}} \sqrt{Y \text{ variation}}}$$

$$= \frac{\sum (X_i - \bar{X})(Y_i - \bar{Y})}{\sqrt{\sum (X_i - \bar{X})^2} \sqrt{\sum (Y_i - \bar{Y})^2}}$$

To show how this approach works, Table 6.4 extracts some of the information from Table 6.2 and shows the necessary computational steps. Notice that here, unlike the example in Table 6.2, the cross-products in the last column are cross-products of deviation scores rather than cross-products of z-scores.

The steps in computing r with the covariation approach may be summarized:

1. Obtain the values of $\bar{X}$ and $\bar{Y}$.

$$\bar{X} = \frac{\sum X_i}{N} = \frac{107}{10} = 10.7 \quad \text{and} \quad \bar{Y} = \frac{\sum Y_i}{N} = \frac{30.30}{10} = 3.03$$

2. Obtain a deviation score for each X_i value within the X distribution and each Y_i within the Y distribution. For the first X and Y pair:

$$X_1 - \bar{X} = 12.0 - 10.7 = 1.3$$

Table 6.4 *Steps in Computing* r *with the Covariation Approach*

i	X_i	$X_i - \bar{X}$	$(X_i - \bar{X})^2$	Y_i	$Y_i - \bar{Y}$	$(Y_i - \bar{Y})^2$	$(X_i - \bar{X})(Y_i - \bar{Y})$
1	12	+1.3	1.69	3.00	−0.03	0.0009	−0.0390
2	9	−1.7	2.89	2.30	−0.73	0.5329	+1.2410
3	6	−4.7	22.09	2.10	−0.93	0.8649	+4.3710
4	12	+1.3	1.69	3.70	+0.67	0.4489	+0.8710
5	8	−2.7	7.29	2.60	−0.43	0.1849	+1.1610
6	13	+2.3	5.29	3.40	+0.37	0.1369	+0.8510
7	14	3.3	10.89	3.80	+0.77	0.5929	+2.5410
8	7	−3.7	13.69	2.40	−0.63	0.3969	+2.3310
9	10	−0.7	0.49	3.10	+0.07	0.0049	−0.0490
10	16	+5.3	28.09	3.90	+0.87	0.7569	+4.6110

$\sum X_i = 107 \quad \sum (X_i - \bar{X})^2 = 94.10 \quad \sum Y_i = 30.30 \quad \sum (Y_i - \bar{Y})^2 = 3.9210$

$\sum (X_i - \bar{X})(Y_i - \bar{Y}) = 17.8900$

$$\bar{X} = \frac{\sum X_i}{N} = \frac{107}{10} \qquad \bar{Y} = \frac{\sum Y_i}{N} = \frac{30.30}{10} = 3.03$$

$$r = \frac{\sum (X_i - \bar{X})(Y_i - \bar{Y})}{\sqrt{\sum (X_i - \bar{X})^2} \sqrt{\sum (Y_i - \bar{Y})^2}} = \frac{17.8900}{\sqrt{94.10} \sqrt{3.9210}} = +0.93$$

and
$$Y_1 - \bar{Y} = 3.00 - 3.03 = -0.03$$

3. Square the deviation scores:

For $i = 1$, $(X_1 - \bar{X})^2 = (1.3)^2 = 1.69$ and
$$(Y_1 - \bar{Y})^2 = (-0.03)^2 = 0.0009$$

4. Obtain the separate X variation and Y variation as the sums of the squared deviation scores:

$$SS_X = \sum (X_i - \bar{X})^2 = 94.10 \quad \text{and} \quad SS_Y = \sum (Y_i - \bar{Y})^2 = 3.9210$$

5. Obtain the cross-product of $(X_i - \bar{X})$ and $(Y_i - \bar{Y})$ deviation scores for each pair of deviation scores. For $i = 1$,

$$(X_1 - \bar{X})(Y_1 - \bar{Y}) = (1.3)(-0.03) = -0.0390$$

6. The sum of these cross products is the *covariation*.

$$\sum (X_i - \bar{X})(Y_i - \bar{Y}) = 17.8900$$

7. The Pearson r is then obtained as

$$r = \frac{XY \text{ covariation}}{\sqrt{X} \text{ variation} \sqrt{Y} \text{ variation}} = \frac{\sum (X_i - \bar{X})(Y_i - \bar{Y})}{\sqrt{\sum (X_i - \bar{X})^2} \sqrt{\sum (Y_i - \bar{Y})^2}}$$

$$= \frac{17.8900}{\sqrt{94.10} \sqrt{3.9210}} = +0.93$$

The value of r obtained in this way must equal the r obtained from the z-score formula in the preceding section.

[C]   The Computational Formula for r

The simplest and most direct method of computing r from raw scores has the most complex formula (the i subscripts are omitted because it is clear they are implied):

$$r = \frac{\sum XY - \dfrac{(\sum X)(\sum Y)}{N}}{\sqrt{\sum X^2 - \dfrac{(\sum X)^2}{N}} \sqrt{\sum Y^2 - \dfrac{(\sum Y)^2}{N}}}$$

This is the formula commonly used when computers are programmed to compute correlations because it is faster than the previous methods shown. You will also find it fastest and easiest in hand or small calculator computations because only raw scores are used in the process and it actually has the fewest number of steps involved.

Again, a table will be helpful in showing how to use the formula. We

Table 6.5 *Values Needed for the Computational Formula Method of Finding r.*

FROM TABLE 6.2				
X_i	Y_i	X_i^2	Y_i^2	X_iY_i
12	3.00	$(12)^2 = 144$	$(3.00)^2 = 9.00$	$(12)(3.00) = 36.00$
9	2.30	81	5.29	20.70
6	2.10	36	4.41	12.60
12	3.70	144	13.69	44.40
8	2.60	64	6.76	20.80
13	3.40	169	11.56	44.20
14	3.80	196	14.44	53.20
7	2.40	49	5.76	16.80
10	3.10	100	9.61	31.00
16	3.90	256	15.21	62.40
$\sum X_i = 107$	$\sum Y_i = 30.30$	$\sum X_i^2 = 1239$	$\sum Y_i^2 = 95.73$	$\sum X_iY_i = 342.10$

will use the X_i and Y_i values from the preceding examples, their squared values, and the cross products of the *raw* scores. These are shown in Table 6.5.

Notice first that each radical sign ($\sqrt{}$) in the denominator contains a variation (SS) for one of the variables. That is:

$$\sum (X_i - \bar{X})^2 = \sum X^2 - \frac{(\sum X)^2}{N} \quad \text{and}$$

$$\sum (Y_i - \bar{Y})^2 = \sum Y^2 - \frac{(\sum Y)^2}{N}$$

We have already met these parts of the computational formula in Chapter 4, as part of the computational formula for s. From Table 6.5, these variations are easily found:

$$\sum (X_i - \bar{X})^2 = 1239 - \frac{(107)^2}{10} = 94.10$$

and

$$\sum (Y_i - \bar{Y})^2 = 95.73 - \frac{(30.30)^2}{10} = 3.921$$

The numerator of the computational formula for r is simply the computational formula for the covariation statistic introduced in the preceding section:

$$\sum (X_i - \bar{X})(Y_i - \bar{Y}) = \sum XY - \frac{(\sum X)(\sum Y)}{N}$$

$$= 342.10 - \frac{(107)(30.30)}{10}$$

When these elements, X-variation, Y-variation, and XY covariation, are arranged as shown, we obtain the same Pearson r:

$$\frac{\sum XY - \dfrac{(\sum X)(\sum Y)}{N}}{\sqrt{\sum X^2 - \dfrac{(\sum X)^2}{N}}\ \sqrt{\sum Y^2 - \dfrac{(\sum Y)^2}{N}}} = \frac{17.89}{\sqrt{94.10}\ \sqrt{3.92}} = 0.93$$

Pitfalls in the Use of r

A recurring theme in this book is that statistics can be dangerous if used blindly. In order to use and interpret statistics correctly, you must be able to do more than apply formulas to data. This is as true with r as it is with any other statistic.

Consider two kinds of situations where r can mislead you about the nature of a relationship. The first of these derives from the scatter plot shown in Figure 6.7. Here, an instructor gave a "screening test" to all 20

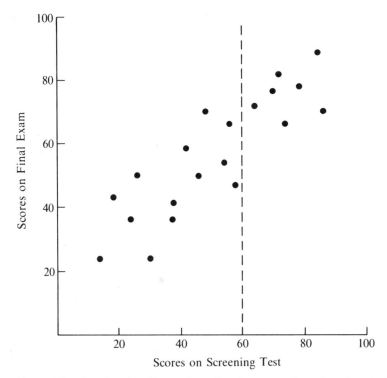

Figure 6.7   Scatter plot showing the effects of restricted range on the value of r. The Pearson r between variables is +0.90 for the whole scatter plot, but only +0.64 for the values to the right of the dotted line.

members of his statistics class at the very first session. The final exam scores for these same people, taken at the end of the term, were then paired with the screening test scores, with the resulting scatter plot shown in Figure 6.7. The Pearson r between screening test scores and final exam scores is +0.90, and one could therefore conclude that there is a high positive relationship between these variables. However, suppose now that he had admitted only the top seven students to class — the people with screening test scores to the right of the dotted line in Figure 6.7. This process *restricts the range* of the first variable (screening test scores), and we now have a much smaller scatter plot made up of only 7 points.[7] It may be instructive for you to draw a pencil line around the complete scatter plot, and then the smaller scatter plot; the smaller scatter plot's line is less elongated. Not surprisingly, the r for these seven people is lower than the overall r: +0.64. This shows how arbitrarily restricting the range of one of the variables in a correlational relationship can produce a spuriously low r.

There are also times when including all of the data produces a spuriously high r. One of these is shown in Figure 6.8. This, again, represents a screening test/final exam situation like the one just described. The r for these data is +0.90, and from this statistic one could again conclude that there is a strong positive relationship between the two variables. Notice here, though, that two people, marked A and B, by themselves account for the slope and apparent narrowness of the scatter plot. Draw a pencil line around the scatter plot with all points included. You may recognize this situation as a violation of the homoscedasticity assumption. Now draw another line excluding A and B. Without A and B in the picture, the value of r between variables for the remaining eight people is only +0.28.

What do you report if your data appear in a form such as that shown in Figure 6.8? One approach would be to compute r without A and B, and then be sure to mention in your report that you are excluding two sets of paired observations because they are anomalous. In any case, the decision to include or exclude those people is based on information other than that obtained from the scatter plot; it is simply a situation calling for judgment on the part of the statistician.

[C]   **Spearman Rank-Difference Correlation Coefficient**

There are many times, of course, when you may want to compute correlation statistics but do not have data that meet the assumptions necessary for the Pearson r — particularly with respect to the assumption about interval-level data or better. Scores on essay exams, statements of attitude strength, and confidence ratings, for instance, are often more ap-

---

[7]It should be noted that a different result would probably have appeared if the instructor *randomly selected* seven students (rather than picking the seven highest). Random selection is covered in Chapter 9, but we can say here that it means selecting seven people in such a way as to ensure that each of the original twenty has an equal chance of being selected. With random selection, one is not arbitrarily restricting one variable's range.

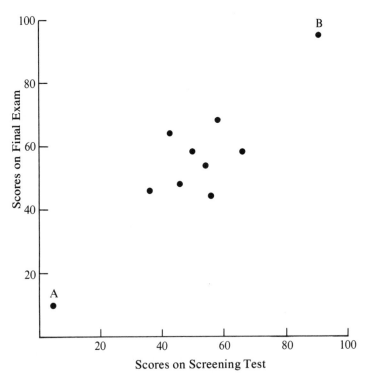

Figure 6.8 Scatter plot illustrating how a spuriously high value of *r* can appear when only a few points account for most of the linearity in a relationship. With all individuals, the Pearson *r* between variables is +0.90; without *A* and *B*, the value of *r* is only +0.28.

propriately treated as ordinal-level data rather than as interval- or ratio-level data. In this situation, you may wish to compute the Spearman rho rather than the Pearson *r*.

The **Spearman rho** (or more simply, rho) is also applicable when there are very many pairs of observations and you must compute a coefficient without the help of an electronic computer; rho is faster and easier to compute than *r* for most data. And, rho is often the correlation coefficient of choice when there are one or two anomalous observations, perhaps in a situation such as that illustrated in Figure 6.8.

Rho does everything a correlation coefficient should do, as discussed on page 151, and it can be interpreted almost in the same way that *r* is. The difference, however, is that rho makes use of *ranks* rather than actual values of measurements.

Return to the question of whether the variables "hours per week spent studying" and "grade-point averages" are related. We might want to compute a Spearman rho instead of a Pearson *r* for the data in Table 6.1 (1) because we are in a hurry or because an electronic calculator isn't available; (2) we do not have at least interval-level observations for all the people involved. Ted, for instance, might not want to tell us what his grade-point average is, except to say that it is below everyone else's. In this case, we can still rank-order all observations on the Y variable from high to low, even though we do not know Ted's exact Y value.

The first step in computing the Spearman rho is to rank-order the subjects on both variables. They may be ranked from high to low or from low to high, but both variables must be ranked in the same direction. For the people listed in Table 6.1, the rank orders on each variable, together with the X and Y values, are listed in Table 6.6. There are several points to consider in this listing:

1. People are ranked separately on X and Y variables.
2. The rank orders on X differ slightly from the rank orders on Y.
3. Two people are tied for the fourth and fifth ranks on the X variable. When ties occur, both people receive the *mean* of the two tied ranks. Here, Bob and Alice are tied for fourth and fifth ranks, and the mean rank is $(4 + 5)/2 = 4.5$. If more than two people were tied for a given rank, the mean of all tied ranks would be used for all of them. Notice that the rank after Alice is 6.

Spearman rho specifies the correlation between *ranks* on variable X and variable Y for the people observed. In order to compute the statistic we will have to determine the *difference* between ranks (D) for each person. The correlation coefficient itself is computed as follows: rho is equal to 1, minus six times the sum of the squared differences between ranks divided by $N(N^2 - 1)$, where N is the number of paired observations. In symbols:

$$\text{rho} = 1 - \frac{6 \sum D^2}{N(N^2 - 1)}$$

This formula looks a little strange perhaps, but it has been designed to produce a statistic that obeys all the requirements for a correlation coefficient. The number 6 is a constant, always present regardless of the situation in which rho is applied.

Table 6.6 *Data Prepared for Computation of the Spearman rho*

RANK ON X	PERSON	X VALUE	RANK ON Y	PERSON	Y VALUE
1	Minnie	16	1	Minnie	3.90
2	Mark	14	2	Mark	3.80
3	Julius	13	3	Alice	3.70
4.5	Bob	12	4	Julius	3.40
4.5	Alice	12	5	Mickey	3.10
6	Mickey	10	6	Bob	3.00
7	Carol	9	7	Cleo	2.60
8	Cleo	8	8	Anthony	2.40
9	Anthony	7	9	Carol	2.30
10	Ted	6	10	Ted	2.10

Table 6.7 *Computation of the Spearman rho*

	RANK ON X	RANK ON Y	DIFFERENCE BETWEEN RANKS (D)	D^2
Minnie	1	1	0.0	0.00
Mark	2	2	0.0	0.00
Julius	3	4	−1.0	1.00
Bob	4.5	6	−1.5	2.25
Alice	4.5	3	+1.5	2.25
Mickey	6	5	+1.0	1.00
Carol	7	9	−2.0	4.00
Cleo	8	7	+1.0	1.00
Anthony	9	8	+1.0	1.00
Ted	10	10	0.0	0.00

$$\sum D = 0 \qquad \sum D^2 = 12.50$$

$$\text{rho} = 1 - \frac{6 \sum D^2}{N(N^2 - 1)} \qquad (N = \text{Number of paired observations})$$

$$= 1 - \frac{6(12.50)}{10(100 - 1)}$$

$$= 1 - \frac{75}{990}$$

$$= 1 - 0.076$$

$$= 0.92$$

Table 6.7 summarizes the steps in computing rho for these data.

1. List the paired observations. Here, each pair of observations represents one person, so the people are listed in the first column.

2. List the ranks each person obtained on variable X and variable Y. *Do not list the X values or the Y values.* (Using the X and Y values at this point is the most frequent mistake students make in computing the Spearman rho. This statistic is concerned only with *ranks*.)

3. Obtain the difference between ranks, D, for each person. Subtract the rank on Y from the rank on X to obtain each person's D. The D values are presented in the fourth column. (As a check on your computations, the sum of the D values should be equal to zero.)

4. Square the D values. The fifth column of Table 6.7 lists D^2 for each person.

5. Add the D^2 values and insert them into the formula.

The obtained value for the Spearman rho is +0.92, very close to the obtained value of +0.93 for the Pearson r for the same data (except when there are many tied ranks). The difference between rho and r is due to the fact that rho ignores distances on the measurement scales and deals only with the rank orders of observations; thus it utilizes less information in the data than does the Pearson r. However, the difference between rho and r is usually small, and rho offers a quick means of computing correlation that is very close to the value of the product-moment correlation coefficient. Of course, if the data are not interval level or above, rho is definitely better than r because statements about r and underlying variables are not correct (see Chapter 1).

⦿ The Point Biserial Correlation Coefficient

One point mentioned in Chapter 1 was that very often in the behavioral sciences, the variables we are interested in can be observed, at best, at the nominal or ordinal levels. The **point biserial correlation coefficient** (or r_{pb}) is one statistic that can be used when data on one variable are nominal or ordinal — *if that variable is truly dichotomous*. That is, the nominal or ordinal variable must be able to take only two values. We will use r_{pb} for finding the correlation between a continuous variable and one that is really dichotomous.

There are many ready examples of dichotomous variables: sex (male/females); performance on a single test item (correct/incorrect); ownership of an automobile (yes/no). There are times when we will want to know if the status of the dichotomous variable is correlated with the values of a continuous variable. We might, for instance, want to know if car ownership is correlated (positively or negatively) with grades that college students receive. To show how r_{pb} can be used, we will resurrect our ten students again for an illustration of test-item analysis.

Consider the problems facing a teacher in making up a multiple-choice test. The test should consist of good items, but what can be used as criteria for "goodness"? One characteristic of a good test item is that it discriminates among students who know the material and those who do not. The teacher would hope that those who do well on the overall test would tend to get the item right, and those who do poorly should tend to get that specific item wrong. Our correlation coefficient gives us a measure of how well an item does this.

Table 6.8 presents our ten students again, their scores on a final examination in English, and indicates whether or not they got multiple-choice item #13 correct. These were last year's scores. The teacher wants to determine if the same item should be used again on this year's test. A quick look at the table suggests that some of the students who scored highest overall missed the item. We can see if there really is a negative correlation between overall scores and performance on #13 by computing r_{pb}:

$$r_{pb} = \frac{\bar{X}_p - \bar{X}_q}{s_X} \sqrt{pq}$$

Table 6.8   *Data for Computation of the Point Biserial r Coefficient*

STUDENT	SCORE ON ENGLISH FINAL EXAM	PERFORMANCE ON TEST ITEM #13
Bob	86	Wrong
Carol	71	Wrong
Ted	63	Correct
Alice	79	Wrong
Cleo	76	Correct
Julius	82	Wrong
Mark	69	Correct
Anthony	62	Wrong
Mickey	70	Correct
Minnie	97	Wrong

where $\bar{X}_p$ = the mean exam score for those passing the item

$$= \frac{63 + 76 + 69 + 70}{4} = 69.50$$

$\bar{X}_q$ = mean exam score for those failing the item

$$= \frac{86 + 71 + 79 + 82 + 62 + 97}{6} = 79.50$$

s_X = the standard deviation of the exam scores

$$= 10.29$$

p = the proportion of people in the group passing the item

$$= \frac{4}{10} = 0.40$$

q = the proportion of people in the group failing the item

$$= \frac{6}{10} = 0.60$$

From these data, r_{pb} is easily computed as:

$$r_{pb} = \frac{69.50 - 79.50}{10.29} \sqrt{(0.60)(0.40)}$$

$$= (-0.97)(0.4899)$$

$$= -0.48$$

Thus, we have a moderate negative correlation between overall performance (the continuous variable) and performance on item 13.* This indicates that the better students tended to miss the item, and the poorer

---

*Some statisticians argue that another correlation coefficient, the biserial r is more appropriate than the point biserial r for evaluating test items this way. Their view is that the dichotomous correct/incorrect variable really represents a continuous underlying variable. (See, for instance, Q. McNemar, *Psychological Statistics*, 4th ed. [New York: John Wiley & Sons, 1969].)

students tended to get it right. Such evidence would be grounds for not using the item again, or else modifying it to be a better discriminator.

Incidentally, r_{pb} does not behave exactly like r and rho, in that it is not always possible to get values of $+1.0$ or -1.0, even when, say, a test item is a perfect discriminator. The values $+1.0$ and -1.0 can appear only when p and q each equal 0.50.

Evaluating Tests with Correlation Coefficients

Anyone who has received a low grade in school, failed a driver's license exam, or not obtained a desired job may have spent a few moments reflecting on the quality of the tests or examinations that led to those consequences. How do we know that tests measure what they are supposed to measure? that the test scores do not reflect errors in measurement?

These questions should interest everyone who lives in a society as test-dependent as ours. Tests determine who gets into college, into graduate school, into the army; who receives professional licenses and certifications; who gets certain jobs; and so on. It seems that for every important event in life there is a test lurking nearby.

Test construction is a very subtle and by no means certain business. In the course of an hour, a teacher cannot ask the student everything he or she is supposed to have learned during the year and so must select only a few questions from a very long list of potential questions. Similarly, an employer may want to know if a job applicant will perform successfully on the job, but the only way to determine this with certainty is to hire the applicant for a period of time. Instead, the employer administers a test that requires the applicant to demonstrate only a few skills. In either case, simply making up a list of questions does not ensure that the test will accomplish its purpose — measuring academic achievement or predicting job performance. Tests must be evaluated before they can be trusted, and correlational methods are especially useful for doing this.

There are many aspects of a test that should be evaluated if it is to be used in important situations or given to many people. Two of these are the test's **reliability** and its **validity.**

A test is reliable if it is *consistent*. Although we take many tests only once, we want a test that would produce the same score every time if we took it more than once. We do not want a test score that is influenced by the weather, by what the person taking the test had for breakfast, or by the fact that the test taker is not feeling well at the time he takes the test. In short, we do not want a test score to reflect fluctuating, extraneous variables.

The most direct approach to assessing the reliability of a test is to administer the test to a group of people, wait a period of time, and then readminister the same test to the same people. The correlation coefficient between scores from the first and second test sessions is known as the **reliability coefficient,** and the method described is known as the **test-retest method** of measuring reliability. If the test is highly reliable, that is, if scores do not reflect chance factors to a very high degree, then the obtained r will be high.

Obviously, the test-retest method of measuring reliability will not be suitable for some forms of tests, such as, for example, intelligence tests. IQ is known to change with age, so if the waiting period between the first test and the retest is very long, it is likely that the characteristic being measured will change and that the obtained reliability of the test will be spuriously low. On the other hand, if the waiting period between test sessions is short, and if exactly the same questions are asked each time, then it is possible that answers in the retest will be contaminated by the practice received on the same questions a short time ago. Thus, the test-retest method of measuring reliability can be used only where it is certain that both test sessions produce unbiased measures of the tested characteristic.

Two commonly used methods of getting around the problems inherent in the test-retest method are the **split-half method** and the **comparable-forms method.** These techniques also use correlation coefficients as coefficients of reliability.

In the split-half method the test is administered to a group of people only once. Unbeknownst to the people taking the test, the questions are divided into two groups (say, one group consists of even-numbered questions and the other group consists of odd-numbered questions). The correlation obtained between the two groups of test items is taken as a measure of the test's reliability.[8]

When the test has been administered to many people, and when the test constructor has good statistics on people's responses to many different questions, it may be possible to construct two comparable forms of the same test. The two forms may contain different questions, but they must be constructed so that any person will achieve the same score using either form. This requires large amounts of statistical information on the questions used and the performances of different kinds of people, and thus comparable forms are more likely to be issued by large commercial testing organizations than by an individual classroom teacher. Nonetheless, when comparable forms are available, a measure of reliability can be obtained by administering *both* forms to the same group of individuals. Each person will thus have a score on both forms, and the correlation

---

[8] If one simply computes the value of r between scores obtained on two halves of the test, r will be spuriously low. (Its value is influenced by the number of observations. The number of observations in each group of items is only half that of the full test.) Using the split-half technique, the Spearman-Brown prophecy formula is applied:

Reliability coefficient $= \dfrac{2r}{1 + r}$

where r = Pearson r between scores on two halves of test.

between obtained scores on both halves will constitute a measure of reliability.

Many commercially made tests are furnished with an instruction book that contains reliability information; usually this includes a description of the method used to obtain reliability coefficients and of individuals involved as subjects. In any case, the coefficients provide a precise answer to the questions, How precise is this test? How large an error is typical for this test?

An issue completely unrelated to the reliability of the test is expressed by the question, Does the test measure what it is supposed to measure? A test can be perfectly reliable — that is, a person could receive the same score every time he takes it — but still be invalid. It might not measure what it was intended to measure at all. This is known as the problem of construct validity or, more simply, **validity.** A test is valid if test scores reflect the underlying characteristic that they are supposed to. If the test is supposed to predict performance on a specific job, then it is valid only to the extent that it actually predicts that performance.

Correlation can also be used to measure **validity** if some independent criterion or measure is also available. Suppose, for example, that an aptitude test is supposed to measure ability of people to perform on a factory job. The extent to which the test is valid, that is, the extent to which it really measures that ability, can be assessed by correlating aptitude-test scores with some independent criterion of job performance. Similarly, college entrance exams are supposed to measure a student's aptitude for college-level work. But do they? A way to test the validity of entrance exams would be to compute the correlation coefficient between entrance-exam scores and grade-point averages obtained 4 years later. Correlation coefficients used in this way are called **validity coefficients.** Correlation coefficients are thus highly useful in evaluating tests.[9] Look for them on booklets that come with the tests or in the articles written about various tests.

SUMMARY

Every person or object can be measured on an almost unlimited number of variables. Some variables seem to be related so that knowing an individual's status on one allows us to predict his status on another. This relation is called correlation, and the methods discussed in this chapter are designed to detect correlation and measure it.

Correlation may have two directions: positive or negative. When two variables are positively correlated, an increase in the value of one tends to be accompanied by an increase in the value of the other, and vice versa. A negative relation exists when an increase in the value of one tends to be accompanied by a decrease in the value of the other, and vice versa. Thus, people's heights and weights are positively correlated, since taller

[9]Reliability, validity, and other statistical concepts in testing are covered in A. Anastasi, *Psychological Testing*, 4th ed. (New York: Macmillan, 1976).

people tend to have higher weights than shorter people. A negative correlation might be represented by the relation between prevailing mortgage interest rates and the number of new homes built in a given year; when the value of one variable is high, the other tends to be low.

Correlation may also be characterized by degree. This refers to the extent to which observed values adhere to the designated relation. The highest degree of relation is a perfect relation, in which the value of one variable determines exactly the value of the other. At the other extreme is the lowest degree of correlation, a zero relation, in which values on one variable are completely unrelated to values on the other. Most relations studied in the behavioral sciences fall between perfect and zero in degree.

Some feeling for the direction and degree of a relation may be obtained by visually inspecting the scatter-plot diagram. To make a scatter plot, paired observations on two variables are graphed. Usually one variable is arbitrarily designated X and the other Y. Such paired observations can be represented as points in a two-dimensional graph if there is a Y value corresponding to every X value. The direction of the slope of scatter-plot points indicates the direction of the relation; the extent to which scatter-plot points fall into a narrow line or deviate from such a line indicates the degree of the relation.

Correlation coefficients are statistics designed to specify quantitatively the direction and degree of a relation. Many of these have been developed, but all share some characteristics. The value of most correlation coefficients will be between -1.00 and $+1.00$. A positive value indicates a positive relation, a negative value indicates a negative relation. The size or absolute value of the coefficients describes the degree of relation.

The Pearson product-moment correlation coefficient, or the Pearson r, is one of the most frequently used correlation coefficients. It is defined as the mean cross-product of the z-scores for pairs of observations. One should not compute the Pearson r, however, unless conditions of homoscedasticity of variances and a linear relation between variables are met. Also, the Pearson r is appropriate only for interval-level data or higher.

When observations on one or both variables are at the ordinal level, the Spearman rank-difference correlation coefficient, or the Spearman rho, may be used. It is computed from the differences in each individual's ranks on the variables and is easier to calculate than the Pearson r.

Precise evaluation of the obtained correlation coefficients requires more sophisticated procedures than can be covered here. However, we can roughly categorize different values as "high," "moderate," "low," or "negligible." The key to rough evaluation is to square the coefficient. The squared r indicates the proportion of one variable's variation accounted for by the other variable's variation.

The final correlation coefficient introduced in this chapter is the point biserial correlation coefficient, or r_{pb}. This statistic gives you an index of the relationship between a continuous variable and a dichotomous variable; for instance, between the overall scores people obtain on an examination and their performance (right or wrong) on a single test item.

A high correlation between two variables in no way proves that a cause-and-effect relation exists between them. It is entirely possible that a third factor causes both variables to vary together.

One of the most useful applications of correlation coefficients is the area of test evaluation. Correlation measures can be used to indicate whether a test is consistent, that is, reliable, and whether it really measures what it is supposed to measure, that is, whether or not it is valid.

KEY CONCEPTS

correlated variables

correlational analysis

direction of relation — positive or negative

degree of correlation

perfect relation

scatter-plot diagram

paired observations

correlation coefficients

Pearson product-moment correlation coefficient (Pearson r)

linear relation

homoscedasticity

covariation

correlation versus cause and effect

Spearman rank-difference correlation coefficient (Spearman rho)

split-half method

comparable forms method

point biserial correlation coefficient

reliability

validity

test-retest method

SPOTLIGHT 6-a Correlation Versus Cause and Effect

Darrell Huff, in *How to Lie with Statistics*,[10] reports there is a positive correlation between the ages of women and the angles of their feet in walking. Younger women tend to point their toes straight ahead; older women tend to walk with toes pointed out. Is it possible that the aging process causes women's feet to turn out? When one assumes, because two variables are correlated, that changes in one *cause* changes in the other, one has a post hoc error. The above relationship no more proves that aging causes feet to turn out than it proves the opposite — turning the feet out causes an increase in age.[11]

There are times, of course, when high correlations appear and no one is tempted to make the post hoc error. Reportedly, there are high correlations between (1) the sizes of schoolboys' feet and the quality of their handwriting;[12] (2) the number of storks' nests and the number of human

[10]Darrell Huff, *How to Lie with Statistics* (New York: W. W. Norton, 1954), Chapter 8.

[11]Huff suggests that older women were raised during a time when toeing out was encouraged; young women today are encouraged to walk with a different posture.

[12]W. A. Wallis and H. V. Roberts, *The Nature of Statistics* (New York: Free Press, 1962), p. 108.

births in northwest Europe;[13] and (3) the salaries of Presbyterian ministers in Massachusetts and the price of rum in Havana.[14] In some cases we can *suggest* a possible third factor that might account for the relation, but these can never be any more than suggestions. Children with bigger feet tend to be older than children with smaller feet, and handwriting probably improves with age. The number of storks' nests in Europe is related to the number of chimneys, which is related to the number of houses, which is related to the number of human births. The reason for relation 3 is anybody's guess.

There are other times, however, when we suspect that correlated variables might indeed represent a cause-and-effect relationship, but we cannot jump to that conclusion strictly on the basis of our correlational information. If, for instance, a medical study found that people who drink more coffee tend to have more coronary heart disease problems than people who drink less coffee, could we conclude that coffee drinking causes heart problems? It is possible that people who drink more coffee have, on the average, more sedentary jobs and thus get less daily exercise. Not until we can rule out all other possible causes can we begin to make cause-and-effect arguments based on correlational data. This is the problem faced by the research specialist in epidemiology (a science that attempts to understand and control disease in populations). Dr. Paul Milvey, a biophysicist who specializes in epidemiological problems, discusses some of the problems involved in demonstrating a causal link between physical exercise and risk of coronary heart disease (CHD):

> Most of the several hundred studies that deal with the relationship of physical activity to CHD or mortality from all diseases show the general picture of those people who are physically active tending to have a lower incidence of disease at any particular age than those who are not.
>
> The scientific and medical field pursuing these studies is called epidemiology. Epidemiologists seek to discover through statistical studies of man and his environment the cause or causes of pathology and disease. Whether the problem is cholera in 19th century England (solved), the Legionnaires' disease in Philadelphia last year (solved), the primary cause of lung cancer (solved), bladder cancer (unsolved) or CHD (unsolved), epidemiologists seek to demonstrate a cause-and-effect relationship between two entities.
>
> The criteria for demonstrating this relationship are quite specific and are difficult to satisfy. Sophisticated mathematical techniques often are employed so the epidemiologist can achieve valid conclusions. It's a difficult discipline because, while it's very easy to prove "association," it is infinitely more difficult to demonstrate "cause and effect."
>
> For example, in the 1960s, [Michael] Yudkin studied the relationship between dietary sugar and CHD and its final manifestation, the heart attack. He took very careful dietary histories of three groups of hospital patients: (1) patients hospitalized for any one of a large variety of reasons (broken legs to

---

[13]Ibid., p. 108.

[14]Huff, *How to Lie with Statistics*, p. 90.

appendectomies); (2) patients suffering from CHD; and (3) patients who had had myocardial infarcts (heart attacks).

The amount of sugar each group consumed over the months and years prior to hospitalization was shown to be least for the "control" patients with the variety of diseases unrelated to their hearts and arteries, intermediate for the CHD patients and highest for the heart attack patients.

This study seemed strongly to indicate (nothing is ever *quite* proven in science) that sugar consumption caused or perhaps was one of several causing factors in the development of CHD and its final manifestation, the heart attack. But several investigators questioned this.

In the 1970s, two groups, working independently, showed that there was a better association or correlation between the development of this disease and the amount of the patients' smoking. And there was an even better correlation — an excellent correlation — between smoking and sugar consumption. (Those who smoked more also consumed more sugar.)

From a scientific and medical point of view, there is no obvious reason why sugar and CHD should have a causal relationship (fats, especially saturated fats, are quite probably related to CHD, but not sugar). But one can suggest any of a number of good, medically sound explanations or mechanisms by which smoking might cause CHD. And there is statistical evidence that also shows this. So we were fooled for a number of years by the earlier study which showed a correlation or association between CHD and sugar consumption, when in fact no *causal* relationship existed.

In this case, the epidemiologist calls sugar a "confounding variable." It goes along with the real causal variable. It's absent when the real cause is absent, it's there when the real cause is there, and it's a damned confusing problem to handle in all epidemiological studies.

Why are sugar and smoking consumption correlated? We can only speculate. We do know that the lower socio-economic groups in our country eat more sugar (junk food) than the higher socio-economic classes. Although I haven't checked it out, they almost certainly smoke more as well. So both smoking and sugar consumption may, in this light, be seen to be "caused" by one's socio-economic status in life.

We also know that CHD is *higher* in the less affluent classes. So perhaps the cause of CHD should be looked for in some common aspect of the environment or life style of the less-advantaged socio-economic classes: Do they go for checkups less frequently, do they smoke and consume fats or salt in larger amounts?

It's very difficult for the epidemiologist to pinpoint the environmental causes. . . .[15]

SPOTLIGHT 6 – b **Surprising Correlations**

Sometimes correlation coefficients only tell us what we already know by confirming the obvious. Just as often, however, unsuspected relations emerge from the data.

Take the relation between the effectiveness of teachers and the subjective ratings students give them. It is only reasonable to suspect that

[15]From P. Milvey, "Getting to the Heart," *Runner's World*, 12 (April 1977): 27–31. Reprinted with permission from *Runner's World*, Mountain View, California.

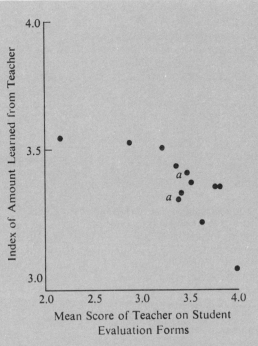

Figure 6.9   Relation between objective and subjective criteria of good teaching ($r = -0.75$). The points labeled *a* are for two sections taught by the same instructor.

Source: From M. J. Rodin and B. Rodin, "Student Evaluation of Teachers," *Science*, 177 (September 1972): 1164–1166. Reprinted by permission from Dr. Miriam J. Rodin and Dr. Burton Rodin and the American Association for the Advancement of Science. Copyright 1972 by the American Association for the Advancement of Science.

college students know a good teacher when they see one and, when asked to grade the performances of their instructors, assign the best grades to the best teachers. One recent study suggests, however, that just the opposite may be true. Rodin and Rodin approached this issue by developing a highly objective measure of the amount of material learned from the instructor in an undergraduate calculus course.[16] They calculated the mean amount learned by each of twelve classes and then determined the mean rating each class assigned its teacher. The scatter plot in Figure 6.9 shows the surprising results: A *negative* correlation between subjective ratings of teachers and amount learned is clearly indicated by the slope of scatter-plot points, and the computed correlation coefficient for these data was $r = -0.75$. Rodin and Rodin appropriately point out that any explanation for the negative correlation offered at this point would be speculative, but they do interpret the results to indicate that "students are

---

[16]M. Rodin and B. Rodin, "Student Evaluation of Teachers," *Science* 177 (1972): 1164–1166.

less than perfect judges of teaching effectiveness if the latter is measured by how much they have learned."[17]

Or, consider the relation that surely must exist between students' backgrounds in mathematics and the grades they receive in statistics courses. Undoubtedly, the student entering a college sophomore-level statistics course having already taken calculus and advanced algebra is better off than a student who has had only elementary algebra. But are the highest statistics grades received by those with the more extensive math backgrounds? The answer is, "That's an empirical question." This means that, strong as our intuitions on the matter may be, the only way to find out for sure is to examine the evidence.

Dr. Leonard Giambra addressed this question by examining the relation among mathematics backgrounds, overall grade-point averages, and grades received by 201 students in his statistics classes.[18] Surprisingly, he found no apparent relation between statistics grades and math backgrounds — students who had taken calculus did not, for example, receive a disproportionate percentage of the A's and B's. There was, however, some correlation (0.25) between statistics grades and overall grade-point averages, suggesting to Dr. Giambra that a student's grade in statistics depends more on his overall ability than on his math background.

The results from these two studies may or may not apply to situations beyond the immediate ones in which they were obtained. But they effectively illustrate the utility of correlational analysis in distinguishing between objective reality and cherished suppositions.

PROBLEMS

1. Indicate whether the following pairs of variables would most likely exhibit positive correlations or negative correlations:

   a. Outdoor temperature and the number of people at the beach.

   b. Body weights of adult men and the amount of time it takes each man to run 1,500 meters.

   c. IQ scores of high-school students and their grade-point averages.

   d. Room temperature and the amount of time it takes an ice cube to melt.

   e. Home prices and the annual incomes of the people who buy them.

2. Workers in a factory are rated annually by their supervisors in two categories. Quality of workmanship is rated on a 5-point scale, with a score of 5 being the highest possible, and a score of 0 being the lowest

---

[17]Ibid., p. 1166.

[18]L. Giambra, "Mathematical Background and Grade-Point Average as Predictors of Course Grade in an Undergraduate Behavioral Statistics Course," *American Psychologist* 25 (1970): 366–367.

possible. Productivity is rated on a point scale with possible scores ranging from 0 (no productivity) to 1,000 (highest possible productivity). Scores for eight workers are given below:

WORKER	QUALITY RATING	PRODUCTIVITY RATING
1	2.1	411
2	3.7	510
3	3.6	430
4	1.0	250
5	1.9	240
6	3.7	637
7	3.9	637
8	3.9	712

a. Prepare and interpret a scatter plot diagram for these data.

b. Compute and interpret the Pearson r between scores on the two rating scales.

3. Twenty-five students from a small high school's graduating class take a college entrance exam, and all are admitted to the state university. After two years, the university registrar prepares a list of their current grade-point averages (GPA) and their entrance exam scores:

STUDENT	EXAM SCORE	GPA	STUDENT	EXAM SCORE	GPA
1	515	2.49	14	200	2.51
2	640	3.45	15	760	3.68
3	315	2.95	16	480	3.22
4	300	2.41	17	521	2.50
5	812	3.42	18	740	3.39
6	241	2.73	19	620	2.62
7	665	3.18	20	157	1.78
8	158	2.05	21	463	2.94
9	760	3.13	22	189	2.26
10	680	2.90	23	550	3.08
11	300	2.02	24	405	2.76
12	580	2.85	25	420	2.48
13	395	2.50			

Prepare a scatter plot for these data and, from it, indicate whether or not you think entrance exam scores and college GPAs are correlated for these students. Indicate the direction of the relationship and characterize the degree as either zero, slight, moderate, or extremely high.

4. What is the value of the Pearson r if

   a. $\sum z_X z_Y = -230$ and $N = 1,000$

   b. $\sum z_X z_Y = 92$ and $N = 110$

   c. The covariation is 421, the X-variation is 371, and the Y-variation is 483.

   d. The covariation is -7.31, the X-variation is 9.20 and the Y-variation is 8.37.

5. Compute the Pearson r for the data in problem 3 above using each of the three methods presented in this chapter. (It might be interesting to record the amount of time required for each of the three calculations; the computational formula method should be faster than the other methods.)

6. Compute and interpret the Spearman rho for the data given in problem 2.

7. Demonstrate how restricting the range can change the value of r. Do this by computing r (use any one computational procedure you prefer) only for the people who scored between 400 and 600 on the exam data given in problem 3 above. If one is concerned only with this middle group, do the exam scores tell very much about what a person's GPA is likely to be?

8. Prepare a scatter plot for each of the following sets of paired observations:

a. X	Y		b. X	Y		c. X	Y
107	3.94		12	650		120	93
109	5.68		7	460		156	75
120	9.75		9	640		102	122
87	6.12		3	350		119	76
116	7.40		1	300		137	138
75	3.71		4	520		104	104

   d. From your scatter plots, which of these shows the lowest *degree* of correlation?

9. Compute the Pearson r for each set of paired observations in problem 8 above.

10. Listed below are three items of financial information about ten people: their annual incomes, the value of the automobiles they own, and the amount of money each spends on entertainment each week:

INDIVIDUAL	ANNUAL INCOME	WEEKLY ENTERTAINMENT EXPENSES	AUTOMOBILE VALUE
1	$22,000	$59	$8,000
2	14,000	30	5,500
3	21,000	50	7,000
4	26,500	68	10,000
5	16,000	36	6,800
6	22,500	56	9,800
7	19,000	42	5,800
8	25,700	70	8,000
9	22,000	53	5,600
10	19,200	45	8,700

   a. Prepare two scatter-plot diagrams, one to show the relationship between annual incomes and weekly entertainment expenses, and the other to show the relationship between annual incomes and automobile values. From your scatter plots, make a verbal comparison of the two scatter plots.

   b. Compute the Pearson r associated with each of the scatter plots you prepared for part a. Interpret the two obtained rs in your own words.

11. Are students' performances in algebra related to their overall grade-point averages? Are students' grade-point averages related to their art grades? Six students happened to be taking both algebra and art one semester. Their averages and grades in these courses are presented in the table. Compute the Spearman rho to examine the relation between grade-point averages and algebra grades; compute also the rho between grade-point averages and art grades. Use the letter grades to rank-order each student's performance in the two courses. Interpret each correlation coefficient in your own words.

	OVERALL GRADE-POINT AVERAGE	GRADE IN ALGEBRA	GRADE IN ART
Richard	3.65	B+	A
Elizabeth	3.70	C−	A
Natalie	2.75	A	C
Robert	2.50	C	C+
Jennifer	3.00	B	B
Elliot	2.10	A+	D

12. Some competitive athletic events, such as diving and gymnastics, are scored by judges who arbitrarily assign numerical ratings to individual performances. Listed in the table are ratings assigned to seven divers by three judges. Consider these ratings to be ordinal-level measure-

ment and compute the Spearman rho between each pair of judges (that is, 1 and 2, 2 and 3, 1 and 3). Which pair of judges are closest in agreement? Which pair is least in agreement?

DIVER	JUDGE 1	RATINGS JUDGE 2	JUDGE 3
Dan	7.9	8.2	9.8
Steve	8.9	9.8	7.8
Gordon	9.8	8.7	8.1
Marty	6.1	7.7	7.7
Terry	7.5	8.1	7.9
Jim	7.4	7.9	8.0
Randy	7.5	7.8	9.1

13. Do college students who own cars get different grades from students who don't? Answer this question after computing the point biserial correlation coefficient for the following data:

STUDENT	GPA	OWN CAR?	STUDENT	GPA	OWN CAR?
1	3.50	Yes	6	2.10	Yes
2	3.40	Yes	7	2.60	Yes
3	3.80	No	8	3.00	No
4	2.70	Yes	9	3.10	No
5	1.90	No	10	2.50	Yes

14. The final exam in a German class included a multiple-choice section. The table below shows the overall exam scores for ten students and their individual performances on three of the multiple-choice items. In order to evaluate the degree to which each of these items discriminated between high scorers and low scorers, compute r_{pb} between overall scores and each item. Which item discriminates best? Which item has a negative r_{pb} and should probably not be used the next time the test is given?

STUDENT	OVERALL SCORE	PERFORMANCE ON		
		ITEM 7	ITEM 9	ITEM 14
Bruce	84	Right	Right	Wrong
Gail	81	Right	Wrong	Wrong
Sandy	73	Wrong	Wrong	Right
Ken	58	Wrong	Wrong	Right
Eva	68	Right	Right	Wrong
Jean	98	Right	Right	Right
Rody	93	Right	Wrong	Wrong
David	75	Wrong	Right	Wrong
Cynthia	62	Wrong	Wrong	Right
Mary Ann	47	Wrong	Right	Right

15. Explain in your own words why a high degree of correlation between two variables cannot, by itself, be taken as evidence that changes in one variable cause (or produce) changes in the other.

7. Regression: Predicting Future Performance from Past Performance

After reading this chapter you should be able to do the following:

1. *Explain the role of correlation in making predictions.*

2. *Describe the nature of a linear relation; discuss the functions of slope and intercept in determining a straight-line graph.*

3. *Explain how the regression equations produce the best possible line through the scatter plot. Use the equation for the line to make statistical predictions.*

4. *Discuss the relation between the two regression lines in any scatter plot and the way this relation depends on the value of r.*

5. *Evaluate the precision of statistical predictions by computing the standard error of estimate.*

6. *Interpret the Pearson r in terms of explained and unexplained variation.*

Correlation and Prediction

From experience we know that a sky full of dark clouds in the morning usually signals rain, so we take an umbrella to work. When we arrive at work, the boss greets us, but the tone of his voice is a little gruff; and we notice that he's smoking a cigar, even though it's only eight in the morning. Experience has taught us that these subtle cues signal a foul disposition, so we decide that today is not a good day to ask for a raise. Later that day, listening to the radio during lunch hour, we hear that banks are going to lower the interest rate on new home loans. That prompts a call to the stockbroker with an order to buy lumber stock, since lower interest rates are usually associated with an increase in home building. These are all examples of an informal kind of predicting we do when we know that two variables are correlated; changes in one variable (appearance of the sky, the boss's tone of voice, interest rates) are usually associated in predictable ways with changes in other variables. The prediction is still something of an educated guess, since it is not absolutely certain that it will rain or that lumber stock will rise in value. Still, the predictions are better than those made on the basis of no information at all.

All of these examples have some important elements in common that illustrate the prediction process to follow. In each prediction situation we already have some *past information*; suppose, for example, that since we have worked for this boss, there have been twenty-three occasions on which he was smoking a cigar early in the morning. Let's call cigar smoking behavior the *X* variable. On most of these occasions he has been irritable; let's call his disposition the *Y* variable. The past information thus indicates that there is a correlation between the *X* and *Y* variables, that is, when the boss is smoking a cigar, he is typically difficult to please all day. Now, with the past information established, some *new information* appears: it is 8 o'clock and the boss is smoking a cigar — we are given a *new* observation on the *X* variable. We can use the past information about the correlation to predict the status of the boss's disposition, that is, the value of the *Y* variable.

Regression analysis is the process of making the same kind of predictions quantitatively, according to a formal procedure. The methods are used, for example, by college admissions officers, who know from years of record keeping that students' high-school grades and scores on college entrance exams are correlated with later academic performance in college. These officers routinely apply regression methods to such grades and scores in order to make entrance selections. Similarly, personnel officers in industry and the military apply regression analysis to scores on various kinds of aptitude tests to determine who will receive certain kinds of training and job assignments. If experience shows a high correlation between the screening test scores and subsequent performance levels, future performances of new applicants can be predicted with a high degree of precision.[1]

Regression methods discussed here are based on the Pearson *r*. We will obtain a number of paired observations on an *X* variable and a *Y* variable, then determine the correlation coefficient describing the relation between them. If *r* is not zero, we can use it to predict *Y* values given a value of the *X* variable.

Strictly speaking, we are going to discuss *linear* regression analysis, and this can be performed only if the requirements for the Pearson *r* are met. Both variables must represent at least interval-level data, the condi-

---

[1] Recently many people have raised serious objections to the use of such predictor tests as selection devices, since the tests do not predict with certainty whether or not any individual will succeed or fail. It is well to keep in mind that regression methods specify how other people with similar predictor scores have subsequently performed, on the average. In some cases, it is better to use this information than to make a guess based on no information. In all cases, however, those who evaluate such predictions and those who are affected by them should understand the meaning of regression analysis and its limitations.

tion of homoscedasticity of variances must exist, and the relation must be essentially a linear one.

The linear relation is, in fact, the fundamental concept in this chapter. At several points we have already suggested the meaning of the linear equation and the linear relation; now we must consider it in more detail before going on.

The Linear Relation

Ultimately, the regression methods will produce a prediction equation that will be a **linear equation.** That is, the predicted value of Y will be specified by an equation in the following form:

$$Y' = a + bX \qquad (Y' \text{ is ``}Y\text{-predicted''})$$

The equation says that the value of Y can be predicted from the known value of X by multiplying X by some value b and adding some other value a to that product. We can thus have a prediction equation if we know the values of a and b.

In discussing the requirements for the Pearson r (pp. 153–155) we mentioned that the graph of a linear equation is a straight line. (Graphing mathematical relations is covered in Chapter 5, together with a sample graph of a linear equation: $Y = 2X + 1$.) Any equation of the above form will produce a straight-line graph if the values of a and b are constant for all values of X. Let's take two linear equations and examine the roles of a and b in the graph.

Table 7.1 shows X and Y values for two linear equations: $Y = 3 + 2X$ and $Y = 10 + (-1)X$. Below the equations are some arbitrarily selected values of X and the corresponding Y values generated by the equations. In Figure 7.1 these two equations are graphed by running a straight line through the tabled points. (It is permissible to represent these equations with a continuous line, since all points on the lines are specified by the equation.)

On graphs 7.1(a) and (b) it is very easy to see the roles played by a and b. a is called the **intercept,** and its value determines where the line crosses the vertical axis. In the first equation a equals 3, and the graph of that equation crosses the vertical axis at the point $Y = 3$. In the other equation

Table 7.1 *Selected X and Y Values for Two Linear Equations*

$Y = 3 + 2X$		$Y = 10 + (-1)X = 10 - X$	
X	Y	X	Y
1	5	1	9
2	7	2	8
3.5	10	4	6
5	13	7.5	2.5

the intercept is 10, and the graph shows the line crossing the vertical axis at the appropriate point.

In a linear relation the value of b is called the **slope** of the line. Just as its name implies, the slope of the line is the *change* in Y values that occurs for an increment of 1.0 on the X variable. In Figure 7.1(a) a change of 1.0 on X brings a change of 2 on Y; hence the slope is 2. Small triangles are sketched on Figures 7.1(a), (b), and (c) to suggest the slope in each case. Figure 7.1(c) is a schematic representation of the line, intercept, and slope.

The point of the preceding discussion is that *any* straight line that can be

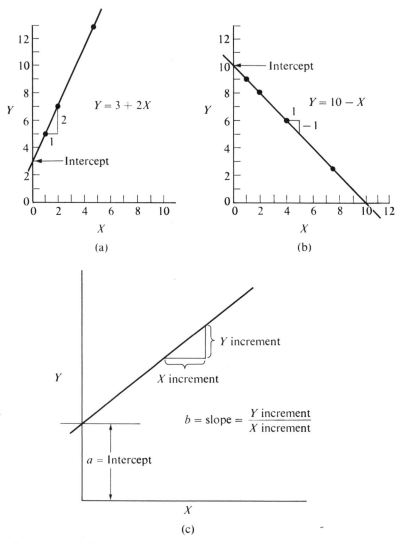

Figure 7.1 Linear X–Y relations. (a) and (b) are graphs of the relations specified in Table 7.1. The graph of *any* straight line is determined by indicating the slope and intercept, as suggested in (c).

drawn through a graph can be specified exactly by finding the appropriate values a and b for the linear equation. Once these constants are determined, the equation specifies a Y_i' for every X_i.

Now, in order to see what all this has to do with correlation and statistical prediction, let's return to the scatter-plot diagram.

The Regression Line

Remember that we are going to develop a linear equation for predicting Y_i values from values of X_i and that the equation will be developed on the basis of previously collected data. Just as with correlation, we need a number of paired observations on two variables, which are recorded in table form and which can be displayed on a scatter plot.

As an example, let's develop an equation for predicting a person's grade-point average (GPA) from the number of hours per week spent studying. Assume that the data in Table 6.1 were collected for this purpose. These data will serve the same function as the *past information* about the boss's cigar-smoking behavior and disposition discussed on page 184. They summarize the relation between X and Y as it existed in the past. In the future, given only the value of X_i, we will use that summary to predict the value of Y_i.

The goal is to run a straight line through the scatter plot of these data (reproduced here as Figure 7.2). We will call it the **regression line**,[2] and it will serve as a basis for the predictions. Actually, we will use the regression methods to find a regression equation that will be both the prediction equation and the graph of a line through the scatter plot. Every point on the regression line will represent both an obtained X_i value and a predicted Y_i' value. Now the problem is to draw the line.

There are, of course, an infinite number of possible lines through the scatter plot, and some of these are suggested as likely candidates for the regression line in Figure 7.2. The one chosen as the regression line is the one that meets the **least-squares criterion.** It will be accepted as the best linear equation available to describe the relation between X and Y.

The meaning of the least-squares concept is suggested in Figure 7.3. This is the same scatter plot that we have seen before, with *the* regression line for Y_i' (Y_i-predicted) running through it. This is the regression line, not some other line, because it minimizes a certain quantity on the graph. The quantity we are interested in is the sum of the squared distances between the obtained points and the regression line. These distances are shown as short vertical lines in Figure 7.3. Notice that the distance between the points and the line is measured parallel to the vertical axis, in

---

[2] The term regression originated with Sir Francis Galton in the late nineteenth century. As part of his studies of heredity, he constructed scatter-plot diagrams showing the relation between heights of parents and heights of their offspring. The diagrams showed that the correlation between parents' heights and children's heights was not perfect; furthermore, the mean heights of the children tended to "fall away" from the parents' heights toward the general population mean height. At first he called this phenomenon *reversion*, and later *regression*, toward the mean.

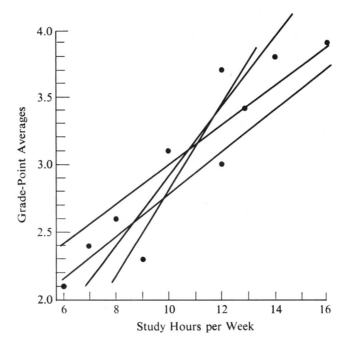

Figure 7.2   Possible regression lines. It is difficult to tell by eye which is the best line.

the Y dimension. Algebraically, we can refer to these distances as $Y_i - Y_i'$, that is, the difference between the predicted Y_i' value (a point on the regression line) and the actual obtained Y_i value. The regression line is the one line that produces the smallest value for the sum of the *squared* $Y_i - Y_i'$ values. Any other line produces a larger $\sum (Y_i - Y_i')^2$; therefore the term "least squares" is used to describe the unique aspect of the chosen line.

Fortunately, we don't have to compute $\sum (Y_i - Y_i')^2$ for a number of lines to find out which one is the least-squares solution. Applications of calculus have yielded some relatively simple equations for determining values of a and b that specify the regression line. Since the development of these equations is beyond the scope of this book, we will simply present them.

Before computing the regression line constants, however, we need some other statistics from our data:

   $\bar{X}$ — Mean of the observed X values.

   $\bar{Y}$ — Mean of the observed Y values.

   s_X — Standard deviation of the observed X values.

   s_Y — Standard deviation of the observed Y values.

   r — Pearson product-moment correlation coefficient between X and Y.

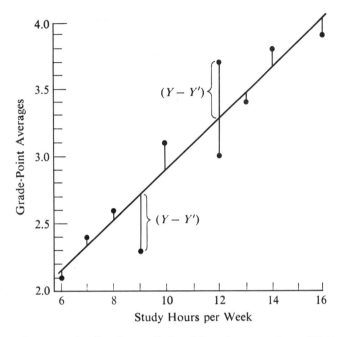

Figure 7.3 The regression line for predicting Y from known values of X. Equation for the line is $Y' = 1 + (0.190)X$, determined by the regression equations for finding intercept (here, $a = 1.0$) and slope (here, $b = 0.190$). This is the line for which the sum of the squared distances between obtained Y values and the regression line is smallest, that is, the line that minimizes $\sum (Y - Y')^2$.

Once these have been determined, the regression constants a and b are found as follows:

Slope: $\qquad b = r \dfrac{s_Y}{s_X}$

Intercept: $\quad a = \left(\bar{Y} - r \dfrac{s_Y}{s_X} \bar{X} \right)$

Fitting these expressions into the linear equation $Y' = a + bX$:

$$Y_i' = \underbrace{\left(\bar{Y} - r \frac{s_Y}{s_X} \bar{X} \right)}_{(a)} + \underbrace{r \frac{s_Y}{s_X} X_i}_{(b)}$$

For the example at hand, the values of the preliminary statistics were computed in Table 6.2. These were:

$$\bar{X} = 10.70 \qquad \bar{Y} = 3.03$$
$$s_X = 3.07 \qquad s_Y = 0.626$$
$$r = 0.93$$

Inserting these values in the regression equation, we can now predict any student's grade-point average if we know how many hours a week he or she spends studying:

$$Y_i' = \left(\bar{Y} - r \frac{s_Y}{s_X} \bar{X}\right) + r \frac{s_Y}{s_X} X_i$$

$$= \left[3.03 - (0.93) \frac{0.626}{3.07} (10.70)\right] + (0.93) \frac{0.626}{3.07} X_i$$

$$= 3.03 - 2.03 + 0.190X_i$$

$$= 1.00 + 0.190X_i$$

All of the preceding parts of this section were included to show you how a slope and intercept are involved in creating predicted values. Now that you know this, however, we can simplify the expression (formula) for Y_i' by factoring out a common element in both terms:

$$Y_i' = \bar{Y} + r \frac{s_Y}{s_X} (X_i - \bar{X})$$

Of course this formula produces exactly the same values for Y_i' as does the longer version:

$$Y_i' = 3.03 + (0.93) \left(\frac{0.626}{3.07}\right) (X_i - 10.70)$$

$$= 1.00 + 0.190X_i$$

It goes without saying that one should not perform regression analysis unless the data requirements for the Pearson r are met. Both variables should represent at least interval-scale data, the relation between X and Y should approximate a linear one, and the condition of homoscedasticity of variances should not be violated.

Making and Evaluating Predictions

Suppose, now, that we learn a student's schedule allows him or her only 10 hr/week study time. What grade-point average do we predict for the student? From our summary of past information (the regression equation) and the new information ($X_i = 10$), we can predict what the value of Y_i will be:

$$Y_i' = 1.00 + (0.190)(10)$$

$$= 1.00 + 1.90$$

$$= 2.90$$

The same prediction can be made from the graph. In Figure 7.3 locate the X_i value (10 hr) on the X scale; proceed straight up until you meet the regression line. From the point on the regression line directly over 10, proceed horizontally, to the left, until you come in contact with the vertical axis. You ought to come in contact with this axis right at the appropriate Y' value.

If, instead, the student manages to find 15 hr/week to study, what is his or her predicted grade-point average?

$$Y_i' = 1.00 + (0.190)(15)$$
$$= 1.00 + 2.85$$
$$= 3.85$$

These are the best guesses we can make based on the known information.

To appreciate the value of such predictions, imagine the situation that exists when regression analysis is not used. Suppose, for instance, that you are a student at Ficticia University. Walking across campus one day, a stranger approaches you and says he is considering coming to your college. He asks, "What is the typical GPA at this school?" You reply that the mean GPA is 3.03. What can he tell from this about GPAs of *individuals*? With no other information to go on, the best guess he can make about anybody's GPA, say, yours, is this value: 3.03. However, if he knows the correlation between study hours and GPA and if he also knows you spend 15 hr/week studying, he can make a better estimate. Now the question is, *How much* better is the estimate made with the regression line than the estimate made without it?

One way to evaluate the quality of estimates is in terms of *precision*. *Without* the regression equation, everyone obtains the same predicted Y_i value, which is $\bar{Y}$. The precision of these estimates is indicated by the standard deviation of obtained scores around the predicted scores — the extent to which observed values differ from the predicted ones. In this instance, that is simply s_Y. Figure 7.4 includes a line showing the size of s_Y to suggest the precision of these estimates.

Now consider the situation that exists *with* the regression equation. Not everyone receives the same Y_i'; only people who have the same X_i value obtain the same Y_i'. Everyone who studies 10 hr/week, for example, receives a Y_i' of 2.90. In the future, not everyone who receives $Y_i' = 2.90$ will actually score that value, but based on the scatter plot we can assume that these people will obtain Y_i values both above and below 2.90.

Similarly, everyone who has 15 hr as the X_i variable will receive an estimated GPA of 3.85 — but, of course, some of these people will actually have GPAs greater than 3.85, some less than 3.85.

In either case, we can specify the precision of the estimate as the standard deviation of obtained values around the regression line. This is equivalent to saying that precision is indicated by the typical difference between predicted and observed Y_i values. This entity is known as the **standard error of estimate** and *could* be determined as follows:

$$\begin{array}{c} \text{Standard deviation} \\ \text{of obtained values} \\ \text{from predicted values} \end{array} = \begin{array}{c} \text{Standard error} \\ \text{of estimate} \end{array} = s_{\text{est}_Y} = \sqrt{\dfrac{\sum (Y_i - Y_i')^2}{N}}$$

In this form, s_{est_Y} looks much like the standard deviation that it is. However, computing it by means of the preceding formula would be

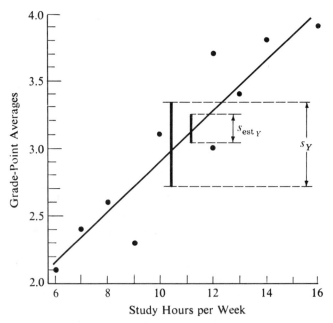

Figure 7.4   Without using the regression equation, the best prediction of the value of Y is $\bar{Y}$. The standard deviation of obtained values around this predicted value is s_Y, whose magnitude is suggested on the graph. When the regression equation is used to predict Y values, all Ys lie on the regression line and the standard deviation of obtained values around these predicted values is given by s_{est_Y}, whose magnitude is also suggested on the graph. Thus, the precision of estimate is improved by using the regression equation.

rather laborious, and we are fortunate that the standard error of estimate can also be found in the following way:

$$s_{\text{est}_Y} = s_Y \sqrt{1 - r^2}$$

For the example developed here, the precision of prediction is indicated as:

$$
\begin{aligned}
s_{\text{est}_Y} &= (0.626) \sqrt{1 - (0.93)^2} \\
&= (0.626) \sqrt{0.135} \\
&= (0.626)(0.367) \\
&= 0.230
\end{aligned}
$$

This can be compared with the precision of estimate obtained in the situation *without* the regression equation, as indicated by s_Y, which is 0.626. Thus, estimates made with the regression line are more precise. For comparison purposes, the size of s_{est_Y} is also sketched in Figure 7.4.

The Other Regression Line

The preceding section treated the Y' regression line as if it were the *only* regression line in the scatter plot — and it is, when predicting Y_i from X_i.

Suppose, however, that a student received a GPA of 2.50. What is the best estimate (or prediction) of the number of hours per week he spent studying? Unfortunately, we cannot use the Y' regression line to answer that question; we will have to use the *other* regression line.

Remember that the regression line for predicting Y from values of X was the least-squares solution *in the Y dimension*. The squared deviations of Y values from the line, as well as s_{est_Y}, were aligned parallel to the vertical axis. When proceeding in the opposite direction and predicting X from Y, the Y' line is no longer the least-squares solution. What is needed is the one line that minimizes $\sum (X_i - X_i')^2$ in the X dimension.

The procedure for developing the X' line is essentially the same as that used for the Y' line, except that the Xs and Ys are reversed. The same set of preliminary statistics are used in the following way:

$$X_i' = \left(\bar{X} - r \frac{s_X}{s_Y} \bar{Y}\right) + r \frac{s_X}{s_Y} (Y_i)$$
$$\underbrace{\phantom{\left(\bar{X} - r \frac{s_X}{s_Y} \bar{Y}\right)}}_{\text{Intercept}} \quad \underbrace{\phantom{r \frac{s_X}{s_Y}}}_{\text{Slope}}$$

For the data in the example:

$$X_i' = 10.70 - (0.93)\left(\frac{3.07}{0.626}\right)(3.03) + (0.93)\left(\frac{3.07}{0.626}\right)(Y_i)$$
$$= 10.70 - (4.56)(3.03) + (4.56)Y_i$$
$$= 10.70 - 13.82 + 4.56Y_i$$
$$= -3.12 + 4.56Y_i$$

The X_i' regression line is graphed in Figure 7.5, together with Y_i' and the important $X_i - X_i'$ values in the X dimension.

With the preceding equation we can estimate the number of hours the student spent studying each week in order to obtain a GPA of 2.50:

$$X_i' = -3.12 + (4.56)(2.50)$$
$$= -3.12 + 11.40$$
$$= 8.28 \text{ hr/week}$$

This prediction can also be made graphically using the X' line in Figure 7.5. Notice that an obtained X of 8.28 does not predict a Y of 2.50, even though the reverse is true. This is because the two regression lines do not usually coincide.

As you might expect, it is possible to evaluate the precision of X_i' estimates by computing the standard error of estimate for X_i':

$$s_{est_X} = s_X \sqrt{1 - r^2}$$
$$= (3.07)(0.367)$$
$$= 1.13 \text{ hr/week}$$

The relation between the two regression lines is particularly interesting, since it indicates something of the degree of correlation between X and Y. Figure 7.6 shows how the relation between regression lines changes with

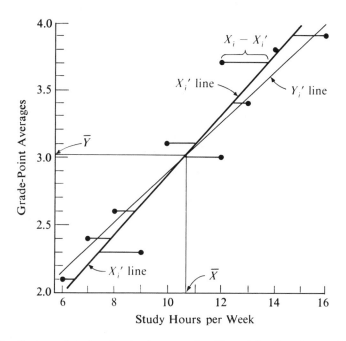

Figure 7.5   Scatter plot showing both regression lines. The line for predicting values of X given values of Y is slightly different from the Y' line. The X' line minimizes the sum of squared deviations from the line *in the X dimension.* Notice that the two lines cross at the point representing $\bar{X}$ and $\bar{Y}$.

different values of r. When the two variables are perfectly correlated, the two regression lines are superimposed, as in Figure 7.6(a). In this case, s_{est_Y} and s_{est_X} are zero — another way of saying that the standard deviation of obtained points around the regression line(s) is zero. There is no error in prediction because the correlation is perfect.

At the opposite extreme, a correlation of zero produces two perpendicular regression lines, as in Figure 7.6(b). For all obtained values of X, the predicted value Y' is equal to the mean $\bar{Y}$. It is the best guess you can make. Similarly, for all obtained values of Y, the predicted value X' is equal to the mean $\bar{X}$, because that, too, is the best prediction to be made in the absence of correlation.

When the correlation falls between these extremes, the two regression lines are neither superimposed nor perpendicular. The larger the value of r, the closer the two lines [see Figures 7.6(c) and 7.6(d)].

A Final Word on Interpreting r

In the literature of the behavioral sciences the Pearson r turns up about as frequently as any other statistic. We may read that the correlation between IQs of parents and children is about 0.50, while the correlation

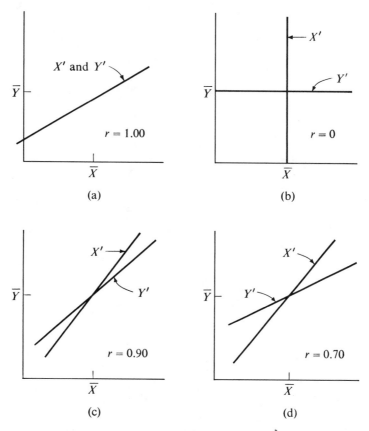

Figure 7.6 Relation between X' and Y' regression lines under different values of r. Notice that regardless of the value of r the lines cross at a point representing $\bar{X}$ and $\bar{Y}$.

between IQs of identical twins raised together is about 0.87.[3] The sample problem considered in this and the previous chapter produced a correlation of 0.93 between hours spent studying per week and GPAs. How should these values of r be interpreted? With understanding of the regression line concept, we can more fully appreciate the meaning of the Pearson r.

Consider once more a scatter plot like that in Figure 7.3. Among the obtained Y_i values there is a certain quantity of variation,[4] namely the sum of the squared deviations about $\bar{Y}$. It may be represented as:

$$SS_Y = \sum (Y_i - \bar{Y})^2$$

The total variation among Y values, when computed, will be a number

---

[3]See, for example, L. Erlenmeyer-Kimling and L. F. Jarvik, "Genetics and Intelligence: A Review," *Science* 142 (1963): 1477–1479.

[4]It may be useful to go back to Chapter 4 and review the relations among variation (SS), variance (MS), and the standard deviation.

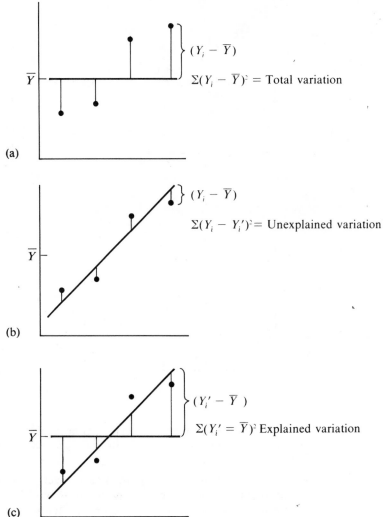

Figure 7.7 (c)

— quite meaningless by itself, but very useful when compared to some other variations in the scatter plot. The deviations involved are shown in Figure 7.7(a). It so happens that this number can be broken into two components that, when added together, produce SS_Y. One of these components is the variation of obtained values around the regression line, a quantity called the **unexplained variation,** or $SS_{\text{unexplained}}$. The deviations to be squared and added here are sketched in Figure 7.7(b); this quantity could be calculated by the formula:

$$SS_{\text{unexplained}} = \sum (Y_i - Y_i')^2$$

The other source of variation in the scatter plot is the variation of points on the regression line around the mean Y_i value, suggested in Figure

7.7(c). The sum of these squared distances is the **explained variation,** or $SS_{explained}$. In symbols:

$$SS_{explained} = \sum (Y'_i - \bar{Y})^2$$

These three variations are interesting only because they obey the following relation:

$$SS_Y = SS_{unexplained} + SS_{explained}$$

This, in turn, is interesting because the relative magnitudes of the explained and unexplained variations at the right are an aid in the interpretation of r.

Where the correlation between X and Y is high, $SS_{explained}$ will be large relative to SS_Y. Most of the variation among Y scores will arise from differences among points on the regression line; thus most of the variation will be explained by the linear relation. On the other hand, when the correlation is low, $SS_{unexplained}$ will be large relative to SS_Y, and most of the variation among Y scores will arise from differences between individual points and the regression line. In the latter situation, most of the variation is unexplained by the regression equation.

Fortunately, we do not have to compute the sums of squares involved in order to compare explained and unexplained variation. The Pearson r is related to these variations in the following way:

$$r^2 = \text{Proportion of variation explained} = \frac{SS_{explained}}{SS_Y}$$

Thus, in the sample scatter plot showing relations between study hours and grades, $(0.93)^2$, or 0.865, of the variation among Y scores is explained by the linear equation. Since r^2 is useful in its own right, it is sometimes given the name **coefficient of determination.**

Conversely, the proportion of variation in a scatter plot not explained by the regression equation is sometimes given the name **coefficient of nondetermination,** and it equals $1 - r^2$. That is:

$$1 - r^2 = \text{Proportion of variation unexplained} = \frac{SS_{unexplained}}{SS_Y}$$

So, in the same sample problem, the proportion of variation among Y scores not accounted for by the linear relation is equal to $1 - (0.93)^2$, that is, 0.135.

This is why we suggested in Chapter 6 that you compute r^2 if you want to obtain an idea of the magnitude of r. r^2 is a proportion that can be interpreted graphically in terms of the scatter plot.

SUMMARY

When two variables are uncorrelated, knowing the value of one does not help us to predict the value of the other. When the variables are correlated, however, we can make an educated guess about the value of Y_i

given the value of X_i, and such guesses are more precise than estimates made on the basis of no information.

Regression, as discussed here, is based on the concept of the linear relation. In fact, regression methods produce a linear equation that (1) is used to make predictions, and (2) specifies the best possible straight line through the scatter plot. The line is found by computing its slope and intercept; the regression formulas produce the slope and intercept of the one line through the scatter plot that satisfies the least-squares criterion.

When predicting Y_i values from given X_i values, the predicted Y_i values (Y_i') all fall on the Y_i' regression line. If one is predicting X_i values from known Y_i values, however, the X_i' regression line must be used. Thus, there are two regression lines through the scatter plot — one that satisfies the least-squares criterion in the Y dimension and one that satisfies the criterion in the X dimension. If $r = 1.0$, the two lines will be superimposed.

The regression line and scatter plot aid in the interpretation of the Pearson r, or more accurately, r^2. r^2 is the proportion of the total variation among Y values that is explained by the linear equation.

It should be kept in mind that regression methods do not specify with certainty what any individual will do or how he will perform. The predictions are limited in the sense that they indicate only what other people with similar X scores have subsequently scored on the Y variable, on the average.

KEY CONCEPTS

regression analysis	least-squares criterion	unexplained variation
the linear relation	standard error of estimate	explained variation
intercept	relation between r and the two regression lines	coefficient of determination
slope		coefficient of nondetermination
regression line		

SPOTLIGHT 7 – a **It Tastes Good, But Is It Good for You?**

Sometimes regression lines are useful for purposes other than for predicting future events from past events. The coefficients that define the line — slope and intercept — are sometimes valuable in their own right for analyzing characteristics of a linear relation between two variables.

Consider, for instance, a problem faced by two General Foods Corporation researchers, Elisabeth Street and Mavis Carroll. They were involved in an attempt to develop an "easy-to-prepare, nutritious, on-the-run meal," code named H.[5] In the laboratory it is quite easy to ensure that such concoctions *contain* certain nutrients, but there is little a priori insurance that humans will metabolize those nutrients efficiently; there is even less assurance that they will find them palatable. Such factors must be determined empirically by testing the finished product on living subjects.

A rigorously controlled test using human subjects indicated that, indeed, H was as tasty as another product C (so called because it has a casein base), which was already on the market. The researchers were also interested, however, in finding whether or not the protein content of H would be metabolized efficiently under conditions of actual use. They approached this question in another carefully controlled study using rats as subjects. Rats — rather than humans — were used in this study because the metabolic processes involved in protein utilization are fairly similar in rats and humans and because the relative efficiency of that utilization is more clearly and quickly reflected in the animals' body weights.

Thirty rats were randomly divided into three groups of ten each; each group contained animals of comparable weights, that is, all groups had equal-age light, medium, and heavy rats.[6] All animals were individually weighed before the experiment, and then each group was given a different diet for 28 days. One group was fed only on a liquid version of H, another group received only a solid version of H, and the third (control) group was fed on casein. Each of the three diets contained about 9 percent protein by weight. During this experimental period the animals were allowed to eat as much as they wished of the designated food, and the food intake of each animal was carefully monitored.

At the end of the 28-day feeding period, the researchers had two measures on each animal: body weight gain (in grams) over the 28 days, and the 28-day protein intake. These data constitute paired observations on two variables for each rat and are displayed in the scatter-plot diagram shown in Figure 7.8. Each dot or cross thus represents data from one animal.

Figure 7.8 shows clearly that animals on liquid and solid H took in more protein *and* gained more weight than animals on the casein diet. The researchers reasoned, however, that more analysis of these data was needed in order to answer the question of whether or not H protein was used more efficiently than casein protein. It is possible, after all, that H was simply more tasty than casein. Casual inspection of Figure 7.8 does not indicate whether or not a gram of H protein can be expected to

[5] E. Street and M. B. Carroll, "Preliminary Evaluation of a New Food Product." In J. Tanur et al. (eds.), *Statistics: A Guide to the Unknown* (San Francisco: Holden-Day, 1972), pp. 220–228.

[6] The procedure used here was similar to the matched pairs experimental design described in Chapter 11. Here, matched *trios* of rats were arranged before the experiment; one member of each trio was randomly assigned to each group.

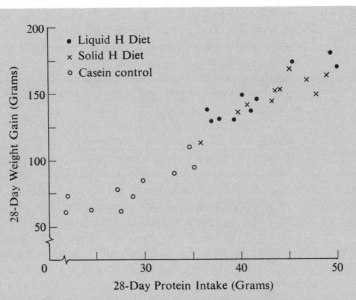

Figure 7.8 Relationship of 28-day protein intake and weight gain in young male rats.

Source: Figures 7.8 and 7.9 are taken from E. Street and M. B. Carroll, "Preliminary Evaluation of a New Food Product," in J. Tanur et al. (eds.), *Statistics: A Guide to the Unknown* (San Francisco: Holden-Day, 1972), p. 223. Reprinted by permission.

produce a greater weight gain than a gram of casein protein. So, a separate regression line was computed for each group's paired observations. These are shown (without data points) in Figure 7.9.

Notice that the H diets produced *steeper* regression lines than did the

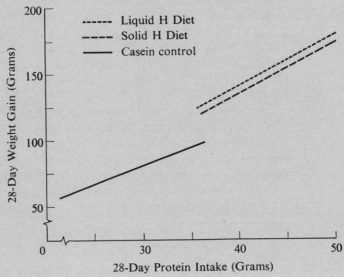

Figure 7.9 Estimated regression of 28-day weight gain on protein intake for young male rats.

casein diet. The calculated slopes, b, of the liquid and solid H groups were respectively, 3.72 and 3.66, while b for the casein group was 2.91. So, the researchers concluded, differences in weight gain between H and casein groups were not due entirely to differences in protein intake. The differences in b values indicate that "For a given increase in protein intake, the H diets resulted in a greater increase in weight gain than did the casein diet."[7]

SPOTLIGHT 7–b Predicting Military Achievement

Regression analysis is used heavily by college admissions offices and personnel offices in industry and the military. When deciding who should be assigned to which job, or who should be admitted to school and who should not, it is vital to have some way to predict future performance. A recent study by Dr. M. Reeb illustrates how regression may be used in this way, and it also shows some problems arising from the use of a single predictor variable.

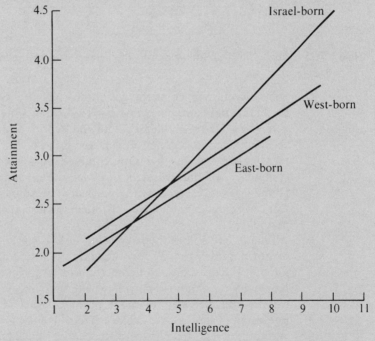

Figure 7.10 Regressions of attainment on intelligence for three ethnic groups.

Source: Figures 7.10, 7.11, and 7.12 are taken from M. Reeb, "Differential Test Validity for Ethnic Groups in the Israel Army and the Effects of Educational Level," *Journal of Applied Psychology*, 61 (1976): 253–261. Copyright © 1961 by the American Psychological Association. Reprinted by permission.

[7]Street and Carroll, op. cit., p. 224.

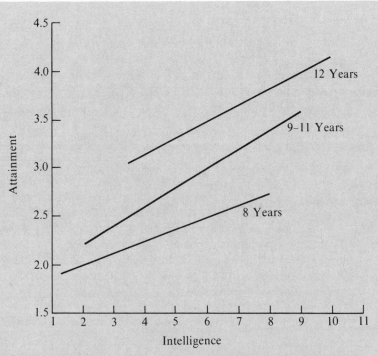

Figure 7.11   Regressions of attainment on intelligence for three educational levels.

Dr. Reeb was concerned with the validity of intelligence tests for predicting military achievement in the Israeli army.[8] Intelligence tests are given to all Israeli soldiers, and the scores often determine personnel selection decisions. Achievement, the criterion variable in this study, was measured simply as the rank attained by a soldier after 2.5 years of compulsory military service; this was scored on a scale ranging from 0 to 5, and it is represented by the vertical axis in Figures 7.10, 7.11 and 7.12. Data were collected from 7,000 male soldiers at the end of one 2.5 year period.

Figure 7.10 shows the regression lines for predicting military achievement from intelligence scores for three different ethnic groups within the army: (a) soldiers born in Eastern countries, primarily the middle east and North Africa; (b) those born in Western countries, primarily Eastern and Central Europe; and (c) those born in Israel. The Pearson r between intelligence scores and achievement ("attainment") for Easterners, Westerners, and Israelis were, respectively, 0.28, 0.28, and 0.43. Clearly the scores predict attainment with more validity for Israelis than for the other ethnic groups. If the mean regression line for all three groups is used as a predictor for all soldiers, then this differential validity is problematic;

[8] M. Reeb, "Differential Test Validity for Ethnic Groups in the Israel Army and the Effects of Educational Level," *Journal of Applied Psychology*, 61 (1976): 253–261.

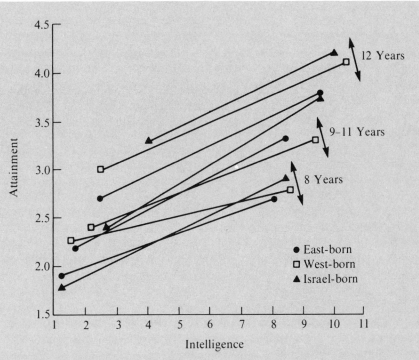

Figure 7.12 Regressions of attainment on intelligence within three educational levels for three ethnic groups.

achievement for Israelis would be underpredicted in the higher score ranges and overpredicted in the lower intelligence ranges. The reverse kinds of errors would be made for the other ethnic groups.

As a possible solution to this dilemma, Dr. Reeb suggested that another predictor, such as educational level, be examined. Figure 7.11 shows regression lines for the same data, except that here soldiers are divided into three groups based on education (number of years of school completed). Figure 7.12 also shows regression lines for the entire data set, with a different line for each educational level within each ethnic group. In Figure 7.12, notice that within each educational level, the regression line slopes for each ethnic group are not very different, but that the intercepts for each ethnic group are different. This means that, within each educational level, equal differences in intelligence scores predict equal differences in achievement, but that the levels of achievement for a given intelligence score are different for the different ethnic groups. Thus the impression created by Figure 7.10, that intelligence scores predict differently for different ethnic groups, is supported only if an overall average regression line is used. If level of education is incorporated in the prediction, ethnic group membership makes little difference.

PROBLEMS

1. Which of the following expressions represent linear relationships?
   a. $Y = X$ c. $Y = (0.207)(X^2)$ e. $Y = X^4 - 4X$
   b. $X = \dfrac{Y}{2}$ d. $X = Y - 23 + Y^2$ f. $Y^2 = X^2 + X^3$

2. For each of the regression lines below, indicate the value of the slope and intercept. Then graph each line. (The simplest way to make each graph is to set up a short table of some X and Y values. When points in the table are plotted, all should fall in a straight line, the line should cross the appropriate axis at the intercept, and should have the indicated slope.)
   a. $Y' = 2 + 3X$ d. $X' = 1 + Y$
   b. $Y' = 4.5 - X$ e. $X' = 4 + \dfrac{Y}{2}$
   c. $Y' = 3X - 2$

3. Determine the slope of the Y' regression line and the X' line for each set of statistics:
   a. $r = 0.50$, $s_Y = 297.1$, $s_X = 14.2$
   b. $r = 0.98$, $s_Y = 20$, $s_X = 10$
   c. $r = -0.78$, $s_Y = 0.071$, $s_X = 110.75$
   d. $r = -0.20$, $s_Y = 50.8$, $s_X = 1072.1$

4. What do you know about the relationship between X and Y if
   a. X' and Y' lines are exactly the same.
   b. X' and Y' lines cross at a right angle.
   c. $s_{est_Y} = s_Y$
   d. $s_{est_Y} = 0$

5. Explain why s_{est_Y} can never be greater than s_Y.

6. A company that breeds laboratory rats has discovered that the amount of food animals eat each day (in grams) is correlated positively with the animals' body weights (also in grams); the Pearson r between these variables is $+0.80$. Suppose that in a particular strain of rats, the mean body weight is 300 grams, and the standard deviation of weights is 30 grams. Suppose also that these animals eat a mean of 40 grams of food per day ($s = 10.0$ grams).
   a. How much food would you expect a 400-gram rat to consume in a day?
   b. How much food would you expect a 270-gram rat to eat a day?
   c. What would you predict as the daily consumption of a 330-gram rat?
   d. Suppose that the caretaker finds a cage door open and one animal gone. If the missing animal ate 50 grams of food for the 24 hours before he left, what size animal do you predict is on the loose?

e. What would your answer for part d be if the animal had eaten only 22 grams on the day before leaving?

7. Listed below are intelligence test scores and performance ratings for ten people:

PERSON	INTELLIGENCE TEST SCORE	PERFORMANCE RATING
1	110	5.1
2	105	4.2
3	95	3.1
4	130	6.8
5	141	6.7
6	102	3.7
7	101	3.2
8	90	3.0
9	87	1.8
10	110	4.4

Assume that in the future you will want to predict performance ratings from intelligence scores. What ratings do you predict for intelligence scores of

a. 100   b. 125   c. 119   d. 89

8. Use the data given in problem 7 to predict intelligence test scores for people with performance ratings of

a. 4.0   b. 2.1   c. 3.7   d. 3.3

9. Over the years a certain aptitude test has been given to thousands of applicants for a highly technical job; the test scores have a mean of 140 and a standard deviation of 30. The employers have developed a rating scale for measuring employee efficiency and productivity; for the past few years the employees have obtained a mean score of 500 on this test, with a standard deviation of 50. The correlation between aptitude tests and job performance scores is 0.75.

a. The criterion for satisfactory performance is a score of 425 on the employee rating scale. Peter scored 120 on the aptitude test. Do you predict satisfactory job performance?

b. During an emergency period twenty employees were hired without taking the aptitude test. What is the best estimate of the rating each will obtain?

c. Wayne's employee rating is 600, but the personnel manager lost records of his aptitude test results. What is the best estimate of his aptitude test score?

10. In order to determine whether high-school grades or college entrance

exam scores are better predictors of college performance, a college registrar tabulated the following information from twelve students:

STUDENT	HIGH-SCHOOL GPA	ENTRANCE EXAM SCORE	COLLEGE GPA
1	2.33	340	2.21
2	2.98	500	3.00
3	3.85	720	3.14
4	3.65	630	3.47
5	2.31	360	2.50
6	2.74	420	2.89
7	4.00	700	3.20
8	3.43	570	3.82
9	2.87	430	2.71
10	3.72	610	3.97
11	3.12	520	3.50
12	2.51	380	2.10

a. A student has a high-school GPA of 3.10. What is your best estimate of her college GPA?

b. What college GPA do you predict for a student with a high-school GPA of 2.60?

c. For a student who scored 650 on the entrance exam, what college GPA do you predict?

d. What is the expected college GPA for a student who scored 450 on the entrance exam?

e. Which of these two predictors — high-school GPA or entrance exam scores — produces the smallest standard error of estimate in predicting college grades?

11. Listed below are heights and weights of 10 people:

PERSON	HT.	WT.
1	68 in.	140 lb
2	62	120
3	71	190
4	67	140
5	64	135
6	68	125
7	61	160
8	74	210
9	59	110
10	69	195

Use these data to predict the
a. height of a 150 lb person
b. height of a 200 lb person
c. height of a 120 lb person
d. weight of a 64 in. person
e. weight of a 73 in. person
f. weight of a 60 in. person

12. Determine the standard error of estimate (s_{est_y}) when
   a. $r = +0.30$ and $s_Y = 3.7$
   b. $r = +0.95$ and $s_Y = 3.7$
   c. $r = -0.92$ and $s_Y = 55.6$
   d. $r = +0.48$ and $s_Y = 0.017$

13. Compute the coefficient of determination and the coefficient of non-determination for the data given in problem 11 above. Interpret the coefficient of determination in your own words.

14. What proportion of either variable's variation is accounted for by the regression line when $r = $
   a. -0.75 b. $+0.32$ c. -0.43 d. $+0.98$

15. What value of Pearson r must we have in order to account for _____ of the variation of either variable with the regression line?
   a. 80%   b. 50%   c. 90%   d. 10%

2 | Inferential Statistics

**Methods that make possible
generalizations beyond the data at hand**

8. Probability: Measuring Uncertainty

After reading this chapter you should be able to do the following:

1. *Interpret probability statements in terms of implied relative frequency.*

2. *Describe two methods for determining probability of outcomes in a simple experiment (relative-frequency observations and a priori calculations).*

3. *Use the counting rule for permutations to determine how many orders a group of things may be arranged in, and use the counting rule for combinations to determine how many different smaller groups may be selected from a larger group.*

4. *Compute the answer to questions like: What is the*

probability that A and B occur? What is the probability that either A or B occurs?

5. *Explain how a probability distribution can be used to find probabilities associated with a discrete random variable.*

6. *Compute probabilities associated with different outcomes in a binomial experiment; construct a binomial distribution given its parameters.*

7. *Use the unit normal table to find the probability that observations occur in different intervals of a normally distributed continuous random variable.*

Measuring Uncertainty

An element of uncertainty is involved whenever inferential statistics are applied. Remember that inferential statistics are essentially methods for selecting a sample from a population and then drawing conclusions about the population based on what is learned from the sample. A polling organization may select 1,000 voters from across the state, ascertain their preferences among candidates in November's election, and on the basis of these sample observations estimate that 70 percent of the state's electorate favors Smith over Jones. The medical researcher may conclude that one component of a food additive is linked to liver cancer; his conclusion may apply to all mammals, but it is based on observations of a few laboratory animals. Both situations involve inferential statistics, and both include the chance that estimates or conclusions are in error. The possibility for error exists because not every member of the population of interest is included in the sample; what is true for the sample may not be true for the population. The beauty of inferential statistics lies not only in the making of such estimates and conclusions, but also in the fact that one can specify the chances that the estimates and conclusions are in error. The techniques for determining the likelihood of error involve probability concepts.

Understanding probability is particularly important if one wants to understand what is meant by "scientific proof." The news media make us all aware, for instance, of current cancer research to determine if various substances either cause cancer or cure it. In such situations it is rarely possible to prove absolutely that substance A increases the risk of cancer or that substance B definitely cures it. Instead, "proof" from experiments usually takes the form of a probability statement. The researcher can say that, based on his or her results, substance A increases the risk of cancer, and this conclusion has a certain probability of being in error.

Determining what constitutes this probability of error requires the use of probability concepts that we will introduce here and in the next few chapters.

All of us are already somewhat familiar with probability, and this is reflected in our language. Everyday expressions such as "chances are," "maybe," and, of course, "probably" suggest an intuitive understanding of the concept. However, in order to deal with probability on a more formal level, we will have to use numbers and precisely defined terms in place of loosely defined expressions.

First of all, probability will appear in **probability statements.** These specify the degree of uncertainty associated with *particular events*. Each probability statement contains two parts: (1) the identification of a specific event, and (2) the probability that the event will occur.

When we hear that the chances of rain tomorrow are 90 percent, or that Seabiscuit is a 10 to 1 long shot in the fifth race, or it's a "fat chance" I passed the chemistry exam, a specific event is clearly identifiable in each case: it rains tomorrow, Seabiscuit wins, I pass the exam. There is also an expression of the likelihood that the event occurs: 90 percent, 10 to 1, and "fat" (that is, tiny) chance.

Thus, in order to use probability statements, we must have an event in mind; we must also be able to recognize its occurrence or nonoccurrence.

Statisticians speak of events as being either **certain** or **uncertain.** A **certain event** is something that either absolutely will happen or absolutely will not happen. Certain events are thus either inevitable or impossible events. Strictly speaking, events in this category are extremely rare; perhaps they include "I walk to the moon this afternoon" or "death" (but not "taxes"). Much more common are **uncertain events,** which either may or may not occur, and these are the events that receive most of our attention in statistics. In particular, we wish to specify the *degree* of uncertainty associated with uncertain events.

Specifying Probability

Probability statements appearing in this book will often take the following form:

$$p(A) = 0.75$$

This means that the probability that event A will occur is equal to 0.75. Notice that probability is specified as a proportion, represented by the symbol p, and that the actual probability value on the right side of the equation is a value between 0 and 1. The meaning of these values will be taken up in the next section.

The event is placed in parentheses next to p and may be a symbol or an expression. The following probability statements, for instance, are in acceptable form:

$$p(\text{A penny flipped in the air lands heads up}) = 0.50$$

$$p(\text{It rains tomorrow}) = 0.70$$

$$p(\text{The ace of hearts is drawn from a full deck}) = 0.019$$

In each case, the event of interest is clearly identifiable. Since writing such expressions takes up considerable space, we might designate each event with an alphabetic symbol:

A = A penny flipped in the air lands heads up

B = It rains tomorrow

C = The ace of hearts is drawn from a full deck

Then we can refer to the probabilities associated with each of these events as:

$$p(A) = 0.50$$
$$p(B) = 0.70$$
$$p(C) = 0.019$$

In statistics, the events we are interested in consist of variables taking on specific values. If X represents the amount of rainfall received tonight, we can use the following symbols to indicate the probability that 2 or more inches of rain will fall:

$$p(X \geq 2)$$

Or let the symbol Y stand for students' scores on a chemistry exam. If we select one student from the class by drawing his name from a hat, the probability of choosing someone with an exam score of 60 or less is indicated by the expression:

$$p(Y \leq 60)$$

With respect to the numerical probability values assigned to these events, *proportions* are almost always used in statistical applications.

Table 8.1 *Various Ways of Expressing Equivalent Probabilities*

PROPORTION	FRACTION	PERCENTAGE	"ODDS"
0.50	$\frac{1}{2}$	50%	50–50
0.01	$\frac{1}{100}$	1%	99 to 1 against
0.80	$\frac{8}{10}$	80%	4 to 1 for

Proportions, of course, range in value between 0 and 1, the extreme values of this range being the probabilities assigned to *certain* events. An event that is certain to occur has a probability of 1; an event certain not to occur has a probability of zero. All other probabilities, that is, probabilities associated with uncertain events, have values between these extremes. Thus, you will never see probabilities greater than 1 or less than zero. The closer an event's probability is to 1, the more likely it is to occur; the closer to zero, the less likely it is to occur.

It should be pointed out that probability is expressed in several ways besides proportions in everyday usage. Proportions are nothing more than fractions, and the ordinary fraction notation may be used instead of the decimal proportion. Thus, a probability of 0.50 equals a probability of $\frac{1}{2}$. Weathermen prefer to express probability in terms of percentages. Thus, a 70 percent chance of rain means that $p(\text{Rain}) = 0.70$.

Probabilities can also be expressed as odds, such as 6 to 5, or 3 to 1. When the odds are said to be 5 to 1 against throwing a three spot with a single die, this means that there is one way to get a three spot but five ways to get something else. Gamblers are especially fond of this approach to probability; however, they also use the term *odds* to indicate the relative amount of money one can win or lose on a specific game, and this can be quite different from the *probability* that one actually wins or loses.

Don't let the various probability notations confuse you. They are presented here simply to show that probability appears in several forms in everyday life. Table 8.1 shows equivalent probability values variously expressed. In this book only proportions are used to express probability.

Probability as Relative Frequency

It is time to give precise meaning to those proportions. Historically, probability could not be defined adequately until the eighteenth century, when James Bernoulli introduced the idea of infinity into the definition of probability. We will do so as well, but without making a formal presentation of Bernoulli's theorem.

Consider the event "flipping a coin and obtaining heads." For the moment, accept on faith that this event is associated with a probability of 0.50. Using the symbol H to represent this event:

$$p(H) = 0.50$$

Table 8.2 *Convergence of Relative Frequency and Probability with Increasing Trials*

FLIP NO.	HEADS	TAILS	RELATIVE FREQUENCY OF EVENT "HEADS"
1	1	0	1.0000 (1/1)
2	2	0	1.0000 (2/2)
3	2	1	0.6667 (2/3)
4	2	2	0.5000 (2/4)
5	3	2	0.6000 (3/5)
6	3	3	0.5000 (3/6)
7	3	4	0.4286 (3/7)
8	4	4	0.5000 (4/8)
9	4	5	0.4444 (4/9)
⋮	⋮	⋮	⋮
100	49	51	0.4900 (49/100)
101	49	52	0.4851 (49/101)
102	50	52	0.4901 (50/102)
⋮	⋮	⋮	⋮
1000	502	498	0.5020 (502/1000)
1001	502	499	0.5015 (502/1001)
⋮	⋮	⋮	⋮

On any single flip, either heads or tails will appear (for our purposes, there is no intermediate event that can occur). The 0.50 probability does not strictly apply to a single flip, a dozen flips, or even a trillion flips. The proportion is the value approached as the number of flips approaches infinity.[1]

To see how $p(H)$ approaches the value 0.50 as the number of flips increases, we could keep a chart showing the relative frequency of heads over an increasing number of flips. It might be similar to Table 8.2. The first flip turned out to be heads, so we had at that point one head out of one flip, for a relative frequency of 1.00. After four flips the tally showed two heads and two tails, so the relative frequency was 2/4, or 0.50. On successive flips the relative frequency fluctuated above and below 0.50, but as the number of flips increase, the fluctuations from 0.50 become smaller and smaller. Bernoulli proved mathematically that when such a process is repeated many, many times it is very likely that the difference between obtained relative frequency and the true probability is very small. So, if $p(H) = 0.50$, we are uncertain about the outcome of any

---

[1]In mathematical terminology, the value 0.50 is the *limit* approached as the number of flips goes to infinity.

single flip but very sure that in the long run heads will turn up half of the time.

Probability in the Simple Experiment

The idea of probability as relative frequency can be extended beyond the coin-flipping situation if we define some more basic terms. Many activities can be described as **simple experiments,** that is, processes or actions that lead to a single, well-defined **outcome.** Many outcomes may be possible, but only one must occur on each **trial** (conducting the experiment once constitutes a trial). Tossing a coin leads either to heads or tails. Measuring the shoe size of the first adult who enters the library also produces a single, well-defined outcome. If a process can be described, if it leads to one outcome on each trial, and if occurrence or nonoccurrence of the various possible outcomes is clearly recognizable, then it is a simple experiment. Table 8.3 lists some experiments and the group of possible outcomes associated with each.

Notice that the list of outcomes for each experiment is **exhaustive.** This means that everything that can happen on a trial fits into one of the outcome categories. In order to make an exhaustive list for some experiments, it is necessary to include an "other" category, as in the last experiment in the table. Also, for each experiment, there are several possible ways to create exhaustive lists. The outcomes of "measuring an individual's height" might include only two categories instead of the three shown in the table — "under 6 ft" and "6 ft or over." Other outcome lists could similarly be specified for other experiments, but each list must be exhaustive.

All outcomes listed must also be **mutually exclusive events.** This is to say that the results of each trial must fit into one, and only one, category. For instance, it would not be appropriate to list the outcomes for the card-drawing experiment as (1) heart, (2) spade, (3) red card, (4) club, since some events (for example, "3 of hearts") fit into more than one outcome category.

Table 8.3 *Experiments and Possible Outcomes*

EXPERIMENT	POSSIBLE OUTCOMES
Selecting an individual and identifying him or her by sex	(1) female, (2) male
Measuring an individual's height	(1) 5 ft or under, (2) between 5 ft and 6 ft, (3) 6 ft or over
Drawing a card from a deck	(1) heart, (2) spade, (3) diamond, (4) club
Selecting a car from a parking lot and determining its country of origin	(1) U.S., (2) Germany, (3) Japan, (4) Italy, (5) England, (6) other

Thus, to analyze probabilities associated with outcomes in a simple experiment, we must list everything that can happen on each trial (make an exhaustive list) and classify outcomes so that any event falls into only one outcome class (list mutually exclusive events).

Having defined an experiment, we can now look at the probabilities involved. We might be interested simply in finding the probabilities associated with each outcome. These can be estimated by repeating the experiment many times and determining the relative frequency with which each outcome appears.

Many of the questions that interest experimental psychologists are approached in this way. A researcher may, for instance, be studying the ability of pilots to understand radio communications through earphones when the message is accompanied by static and other extraneous noises. As part of this study it might be useful to know the probability that a human being can detect a specific earphone stimulus, and this must be determined empirically. A human subject is fitted with earphones and seated in front of a small panel with a light bulb and two push buttons. He is told that when the light goes on, he is to listen for a tone in the next three seconds. If he hears it, he must push one button; if not, he must push the other. The experimenter may conduct 1,000 trials in which a tone follows the light. Mixed in among these trials are 1,000 other trials when the light flashes but no tone is presented. The subject never knows whether a tone will follow the light or not. If the tone is presented on 1,000 trials and the subject detects it on 480 of them, an estimate of the probability of detecting that stimulus is taken as the relative frequency of detection, 480/1,000 or 0.48. Since there were 520 trials on which the tone was given but not detected, the probability of not detecting the stimulus is taken as 520/1,000 or 0.52. The trials on which the tone was presented fit the definition of a simple experiment with the outcomes (1) detection, and (2) no detection.

The obtained relative frequencies illustrate an important characteristic of probabilities in the simple experiment: The sum of the probabilities associated with all outcomes is equal to 1. Here:

$$p(\text{Detection}) + p(\text{No detection}) = 0.48 + 0.52 = 1.00$$

(The trials on which no tone was presented are called catch trials, and they form a different simple experiment.)

In some experiments the only way to find probabilities is simply to run many trials and determine the relative frequencies empirically. However, there are some situations where these can be calculated before the first trial is run. Such probabilities are produced **a priori** — by reason prior to observation rather than by experience. Some statisticians, in fact, trace the development of statistics back to the late seventeenth and early eighteenth centuries, when gentlemen gamblers sought the advice of Pascal, De Moivre, and other prominent mathematicians with respect to the odds, or a priori probabilities, in games of chance.

Consider the situation where a single card is drawn from a well-shuffled deck. What is the probability of drawing a heart? This can be calculated a priori if we know how many cards there are in the deck and how many are hearts. The probability of the outcome "heart" is its relative frequency in the deck. For one card chosen:

$$p(\text{Heart}) = \frac{\text{Number of heart cards in deck}}{\text{Total cards in deck}} = \frac{13}{52} = 0.25$$

The probability of drawing an ace is similarly obtained:

$$p(\text{Ace}) = \frac{\text{Number of aces in deck}}{\text{Total cards in deck}} = \frac{4}{52} = 0.077$$

At the state university there are 5,500 men and 4,500 women enrolled. One name is drawn from a (very large) hat. What is the probability that a woman is picked?

$$p(\text{Woman}) = \frac{\text{Number of women at university}}{\text{Total students at university}} = \frac{4,500}{4,500 + 5,500} = 0.45$$

When calculating a priori probabilities, each student or each card is considered an **elementary event.** If we can count the number of elementary events in each outcome class, and if each elementary event is equally likely, then the probability of outcome class A can be computed:

$$p(A) = \frac{\text{Number of elementary events that are } A}{\text{Total number of elementary events}}$$

It should be mentioned here that a priori calculations apply only if the individual person or card is selected **randomly.** "Random" is a term that appears several times in this book in connection with probability and sampling. Here, a randomly chosen individual is one selected such that no bias or special preference is given to any outcome category. Hence, each person is *equally* likely to be selected. The a priori probability must not be altered by using a poorly shuffled deck, or by picking students from the university in a way that favors one outcome over others, for instance, by picking the first person to leave the women's gym.

Thus, there are two ways of finding the probabilities of various outcomes in the simple experiment. The first is to repeat many trials and obtain the relative frequencies. The other is to compute the probabilities a priori from the number of equally likely elementary events in each outcome category. In the long run, as the number of trials approaches infinity, the relative frequencies will approach (come to be approximately equal to) the a priori probabilities. This was demonstrated in Table 8.2 for coin flipping. In this case, the a priori probability of tossing a head, given only two equally likely outcomes, is 0.50. Over many trials the relative frequency of that event approached 0.50. Sometimes, of course, we cannot compute a priori probabilities and must simply run the experiment many times to estimate probabilities.

Probability has thus been defined as relative frequency. What, then, does it mean when the weatherman says the chances of rain tomorrow are

70 percent? Tomorrow is only going to occur once, and it is not so easy to see where relative frequency is involved. Such statements mean that if the current weather conditions were repeated on an infinite number of days, it would rain the next day 0.70 of the time.[2]

Two Counting Rules

The a priori probability model just presented is simple in concept: simply place the total number of possible outcomes on the bottom of a fraction, and place the number of those outcomes comprising the event of interest on top. As students in more advanced probability courses quickly learn, however, it is a simple concept that can be extremely difficult to apply. The problem is that both of these numbers may involve some tedious counting. The purpose of this section is to give you two counting rules that can alleviate some of the work in finding a priori probabilities.

Consider one example — again involving a game of chance. If 5 cards are drawn randomly from a well-shuffled deck of 52, what is the probability that they constitute the poker hand "royal flush"?[3] Think for a moment how you will have to go about calculating that event's a priori probability. You will need two numbers arranged as follows:

$$p(\text{royal flush}) = \frac{\text{Number of 5-card hands that are royal flushes}}{\text{Number of equally likely 5-card hands}}$$

Now, the numerator of that fraction is easy: there are only four possible royal flushes, one for each suit. But what about the denominator? How many different 5-card hands can come from a 52-card deck?

One approach would be to list all of the different hands, perhaps with the help of a chart to keep track of which hands had been counted and which hadn't. Since there are over two and a half million possible hands, however, it is not likely that anyone would do so. Instead, a counting rule must be used. It is for probability situations such as this that we should learn at least two counting rules. Let us temporarily put aside the royal flush problem and take the simplest counting rule first.

Suppose that we have five people milling around and we ask them to form a single line. What is the probability that they line up in alphabetical order? We know that there are two possible ways to arrange a line in alphabetical order — left to right and right to left — so the numerator of our a priori fraction must be 2. For the denominator, however, we must know the total number of different ways five people can line up. Rather than counting the ways individually, we can determine the number of ways N objects or people can be arranged in order by calculating the

---

[2]Probability statements about the weather, such as the one discussed here, may reflect the analysis of dozens, or even hundreds, of weather variables. For an introduction to this kind of forecasting, see R. G. Miller, "The Probability of Rain," in J. M. Tanur et al. (eds.) *Statistics: A Guide to the Unknown*. (San Francisco: Holden-Day, Inc., 1972).

[3]A royal flush consists of the Ace, King, Queen, Jack, and ten, all of which must be of the same suit. It is the rarest poker hand and it beats all other poker hands.

number of **permutations** possible. The number of permutations possible when $N = 5$ is given by the expression:

Possible permutations of N objects $= N!$

The symbol $N!$ is not a way of saying that we are enthusiastic about N. Rather, it is the symbol for an operation (a computation) called a **factorial**. N objects have N-factorial permutations. Some illustrations should make the idea of factorials clear:

$$6! = (6)(5)(4)(3)(2)(1) = 720$$
$$4! = (4)(3)(2)(1) = 24$$
$$11! = (11)(10)(9)(8)(7)(6)(5)(4)(3)(2)(1) = 39,916,800$$

and, in general,

$$N! = (N)(N - 1)(N - 2) \cdot \cdot \cdot (3)(2)(1)$$

By convention, $0! = 1$, and factorials are undefined (they don't exist) for fractions, mixed numbers (such as 1.3, 2¼, etc.), or negative numbers.

In how many ways, then, can five people line up? The number of permutations possible with five people is $5!$, or $(5)(4)(3)(2)(1)$, that is, 120. Now we can answer the question about people lining up in alphabetical order:

$$p \begin{pmatrix} \text{5 people line up} \\ \text{in a randomly} \\ \text{determined order that} \\ \text{also turns out to be} \\ \text{alphabetical order} \end{pmatrix} = \frac{\text{Number of ways 5 people can line up in alphabetical order}}{\text{Number of different ways 5 people can line up}}$$

$$= \frac{2}{5!} = \frac{2}{120} = 0.0167$$

Permutations are thus a very simple counting rule to use in setting up a priori probability computations. Three people can be arranged in $(3)(2)(1)$ different orders, that is, six ways; seven items can be arranged in $7!$, or 5,040 different ways. That is our first counting rule.

An even more useful counting rule, built on factorials, is the rule for determining the number of possible combinations involved in taking r objects[4] from a group of N objects. This is the computation we need for our royal flush problem and many other more important probabilities.

Combinations, too, are perhaps best described by giving an example. If, for instance, we have ten people available to make up a four-person bridge team, how many different four-person teams are possible? In symbols we want to know how many different groups of size r (here $r = 4$) can be made when selecting from a larger group of size N (here $N = 10$). It does not matter what order these r selected objects or people appear in; we pay

---

[4]Regrettably, the symbol r used in this process is the same symbol used for the Pearson product-moment correlation coefficient. Do not confuse them.

attention only to *which* individuals are selected. If we have one group of four selected, and then exchange one of the selected persons for another who was previously not selected, we have a new combination.

The symbol for the combinations we are after is $\binom{10}{4}$, or, in its general form, $\binom{N}{r}$. This is not a fraction, but rather a symbol indicating that the following operation is to be performed:

$$\binom{N}{r} = \frac{N!}{r!\,(N-r)!}$$

For our question about bridge teams, this is:

$$\binom{10}{4} = \frac{10!}{4!\,6!} = \frac{(10)(9)(8)(7)(6)(5)(4)(3)(2)(1)}{(4)(3)(2)(1)\quad(6)(5)(4)(3)(2)(1)}$$

$$\text{where } r = 4,\ N = 10,\ (N-r) = 6$$

Before computing, notice how many numbers appear both above and below the line. This means that substantial cancelling can be done:

$$\binom{10}{4} = \frac{(10)(9)(8)(7)\cancel{(6)(5)(4)(3)(2)(1)}}{(4)(3)(2)(1)\quad\cancel{(6)(5)(4)(3)(2)(1)}}$$

$$= \frac{(10)\cancel{(9)}^{3}\cancel{(8)}(7)}{\cancel{(4)(3)(2)}(1)} = \frac{(10)(3)(7)}{1} = 210$$

Thus, there are 210 different bridge teams that one could make given the original 10 people. Notice especially the use of cancellation; without it, we would have to compute the entire factorials of very large numbers. One such example occurs in the royal flush problem with which we started this section. Remember that we need to know how many different groups of five cards (five-card hands) can be assembled using the cards in a 52-card deck. This is given by:

$$\binom{52}{5} = \frac{52!}{5!\,47!} = \frac{(52)(51)(50)(49)(48)\cancel{(47)\cdots(2)(1)}}{(5)(4)(3)(2)(1)\quad\cancel{(47)(46)\cdots(2)(1)}}$$

(Notice that 47! cancels out of both numerator and denominator)

$$= \frac{(52)(51)(50)(49)(48)}{(5)(4)(3)(2)(1)}$$

$$= 2,598,960$$

Thus, there are 2,598,960 five-card hands that can be dealt. If we assume that all of them are equally likely, and that there are only four of these that constitute a royal flush, then the probability of getting that hand is:

$$p(\text{royal flush}) = \frac{4}{2,598,960} = 0.0000015$$

Probability computations using permutations, combinations, and other counting rules can become quite complex. The intent here is to show you how two counting rules can be used, rather than to explore their complex uses. Since permutations and combinations will turn up again in this book, perhaps one more example would be useful before leaving the subject.

A baseball manager has nine players to arrange in a batting order. How many different batting orders can be had using these nine? This calls for *permutations*:

$$N! = (9)(8)(7)(6)(5)(4)(3)(2)(1) = 362,880$$

Now, if three players are to be picked to bat in the first inning, how many different groups of three can he choose? Remember that we are not paying attention here to which order that they bat in the first inning. This is a question calling for calculation of *combinations*:

$$\frac{N!}{r!\,(N-r)!} = \frac{9!}{3!\,6!} = \frac{(9)(8)(7)(6)(5)(4)(3)(2)(1)}{(3)(2)(1)\quad(6)(5)(4)(3)(2)(1)}$$

$$= \frac{(3)(4)(7)}{1} = 84$$

So, there are 362,880 different batting orders possible with nine players; 84 different groups of three may be selected from them.

Probability Operations

Probabilities from simple experiments can be used in probability operations to answer questions that include an "or" or "and" condition, like What is the probability that a child contracts the measles *and* the mumps in a given year? What is the probability that a machinist turns out a part that is either too large *or* too small? What is the probability of selecting a student by chance who is both an *A* student *and* male? To answer such questions we must subject the simple experiment outcome probability to some calculations.

Consider an experiment listed in Table 8.3, "Selecting a car from a parking lot and determining its country of origin." Each car in the lot is an elementary event, and if we use a system of choosing that ensures that every elementary event is equally likely, then the probability of each outcome class is its relative frequency in the lot. Suppose that out of 1,000 cars, 90 were German and 100 were Japanese. Thus:

$$p(\text{Picking a German car}) = \frac{90}{1,000} = 0.09$$

$$p(\text{Picking a Japanese car}) = \frac{100}{1,000} = 0.10$$

What is the probability that a car selected as described is *either* German *or* Japanese? When outcome categories represent mutually exclusive events:

$$p(A \text{ or } B) = p(A) + p(B)$$

We have defined a new event in parentheses, (A or B). If an elementary event in either category occurs, it is an instance of the "or" event in parentheses. Here:

p(Picking a German car or a Japanese car) $= 0.09 + 0.10 = 0.19$

When an "or" joins two or more mutually exclusive events, the probability that any one of them occurs is equal to the sum of the individual probabilities. The "or" situation for mutually exclusive events is illustrated in Figure 8.1. The case of "or" when events are not mutually exclusive will be discussed later.

Let's look now at another kind of question — one involving an "and." In the simple experiment, by definition, only one outcome can occur on each trial. But suppose we conduct more than one trial of the experiment, or even two different experiments. In these situations, we may be interested in particular combinations of outcomes, that is, in **joint events.** As an illustration, two simple experiments, their individual outcomes, and the possible joint events are listed in Table 8.4. In this example, a joint event occurs when we conduct trials of both experiments. One joint event might be "heads *and* 5." For this pair of experiments, there are twelve possible joint events.

When the simple experiments in no way interact, that is, when conducting a trial of one has no influence on the outcome of the other (an important consideration, as we will see shortly), probabilities can be found as follows:

$p(A \text{ and } B) = p(A)p(B)$

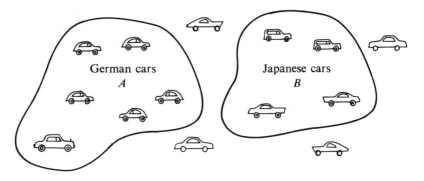

Figure 8.1 *A* and *B* are mutually exclusive events (one car cannot be both German and Japanese). For such events, $p(A \text{ or } B) = p(A) + p(B)$. Here, a ring is drawn around all outcomes in event *A* (German car), and another ring is drawn around all outcomes in event *B* (Japanese car). If a single car is selected at random from this group, the probability that it belongs to *A* or *B* is simply the total number of cars in both rings, divided by the total number of cars in the lot. The total number of cars in both rings is found by adding the number in each.

Table 8.4   *Individual Outcomes and Joint Events for Two Simple Experiments*

Experiment	Flipping a coin	Throwing a die
Individual outcomes	Heads, tails	1, 2, 3, 4, 5, 6
Probabilities	$p(H) = 0.50$ $p(T) = 0.50$ $\overline{\sum p = 1.00}$	$p(1) = 0.1667$ $p(2) = 0.1667$ $p(3) = 0.1667$ $p(4) = 0.1667$ $p(5) = 0.1667$ $p(6) = 0.1667$ $\overline{\sum p = 1.0000}$
Joint events		$(H$ and $1), (H$ and $2),$ $(H$ and $3), (H$ and $4),$ $(H$ and $5), (H$ and $6),$ $(T$ and $1), (T$ and $2),$ $(T$ and $3), (T$ and $4),$ $(T$ and $5), (T$ and $6)$

For the die and coin experiments listed in Table 8.4, the probability for one joint event is:

$$p(H \text{ and } 5) = p(H)p(5) = (0.50)(0.1667) = 0.0833$$

The calculation of probabilities associated with joint events is summarized in Table 8.5. In the preceding example, we considerd only joint events in which the individual events were *independent* (see left side of Table 8.5). The concept of independence and joint events composed of nonindependent events will be discussed later.

In probability operations, remember that "and" means multiply and "or" means add. Operations can be combined as in the following example, which specifies the probability of the joint event "heads on the coin and an even number on the die."

What is $p[(H \text{ and } 2) \text{ or } (H \text{ and } 4) \text{ or } (H \text{ and } 6)]$? Solve the inner parentheses first:

$$p(H \text{ and } 2) = p(H)p(2) = (0.50)(0.167) = 0.0833$$

$$p(H \text{ and } 4) = (0.50)(0.167) = 0.0833$$

$$p(H \text{ and } 6) = (0.50)(0.167) = 0.0833$$

Now, all of these joint events are connected with "or," so the probability of getting one of them is:

$$0.0833 + 0.0833 + 0.0833 = 0.25$$

Or consider the throwing of two dice as two unrelated simple experiments. What is the probability of getting a 6 on the first and a 1 on the second?

Table 8.5   *Calculation of Probabilities for Joint Events (Independent and Nonindependent)*

INDEPENDENT EVENTS	NONINDEPENDENT EVENTS
$p(A \text{ and } B) = p(A)p(B)$	$p(A \text{ and } B) = p(A)p(B \mid A)$
Example	$p(A \text{ and } B) \neq p(A)p(B)$
A = Heads on coin $p(A) = 0.50$	*Example*
B = 6 on die $p(B) = 0.17$	A = Freshman $p(A) = 0.50$
Coin is flipped and die thrown.	B = Science major $p(B) = 0.72$
$p(A \text{ and } B) = (0.50)(0.17) = 0.085$	$p(B \mid A) = 0.80$
Conditional probability:	A student is selected at random.
$p(A \mid B) = p(A)$	$p(A \text{ and } B) = (0.50)(0.80) = 0.40$
The probability that heads occurs given that a 6 turned up on the die is exactly equal to the simple probability of heads occurring.	$0.40 \neq (0.50)(0.72)$
	Conditional probability:
	$p(A \mid B) \neq p(B)$

This is given as:

$$p(6)p(1) = (0.167)(0.167) = 0.0278$$

Notice that this question is different from the question, What is the probability of throwing a 6 and a 1 regardless of order? Here we must consider two joint events:

Joint Event 1: (6 on 1st die, and 1 on 2nd die)

Joint Event 2: (1 on 1st die and 6 on 2nd die)

These two joint events are joined by an "or" when we ask for the probability of getting a 6 and a 1 in one order *or* the other; in this case, we are asking for the following probability:

$p[(6$ on 1st die and 1 on 2nd die) or (1 on 1st die and 6 on 2nd die)]

The probability of the event in brackets is:

$$(0.167)(0.167) + (0.167)(0.167) = 0.055$$

Perhaps these examples suggest to you that computing probabilities can become a complex process; this can be readily confirmed by examining other applied statistics books. Some texts begin with a long chapter on probability; some even devote a majority of their pages to probability topics. However, the intent of this book is not to study probability in its own right but to present what is essential for understanding basic techniques in inferential statistics. Thus, the most difficult applications of probability we will have to consider are general applications of the "and" and "or" rules described in the preceding pages.

Table 8.6 represents some information extracted from the catalog of Ficticia University, a school that, conveniently, has only two colleges. The frequencies in the table can be used to calculate some a priori probabilities for two different simple experiments. First consider the

Table 8.6   *Enrollment by Class and College at Ficticia University*

	LIBERAL ARTS	SCIENCE	TOTALS
Freshmen	1,000	4,000	5,000
Sophomores	750	1,750	2,500
Juniors	550	950	1,500
Seniors	500	500	1,000
Totals	2,800	7,200	10,000

process of selecting one student and determining his class standing. If the selection process gives every student an equal chance of being chosen, the probability of picking a senior is 1,000/10,000, or 0.10. Similarly, the probability of selecting a junior is 0.15. Or we might treat the same data as part of a different simple experiment. What is the probability that the chosen student will be a Liberal Arts major? The column total for "Liberal Arts" is 2,800, so the probability in question is 2,800/10,000, or 0.28. When each experiment is handled alone, the probability of each outcome is easily found.

If, however, we ask "and" or "or" questions, we run into a problem. Outcomes in the two experiments are not unrelated, that is, each student is an elementary event in *both* experiments. We can therefore make joint events of outcomes that are not necessarily mutually exclusive. This possibility has consequences for the calculations involved in "or" and "and" probability operations.

The general rule for the "or" situation in the case of events that are not mutually exclusive is:

$$p(A \text{ or } B) = p(A) + p(B) - p(A \text{ and } B)$$

Consider a simple experiment in which a student is selected from Ficticia University. What is the probability that he is either a senior or a science major? If A is the event "senior" and B is the event "science major," we can use Figure 8.2 to illustrate that A and B are not mutually exclusive. Drawing one ring around all seniors and another ring around all science majors, we find that the two rings overlap; 500 students fall in both categories. The probability that a selected student is either a senior or a science major is:

$$p(\text{Senior or Science}) = p(\text{Senior}) + p(\text{Science}) - p(\text{Senior and Science})$$
$$= 0.10 + 0.72 - 0.05$$
$$= 0.77$$

(The probability 0.05 for the joint event was determined directly from Table 8.6: 500/10,000 of the students are both seniors and science majors.) The subtraction must be performed, because the outcome categories "senior" and "science" are not mutually exclusive. Simply adding the probabilities associated with "senior" and "science" would count some students twice. This can be seen by going back to Table 8.5 and computing the a priori probabilities.

The probability of selecting a senior is:

$$p(\text{Senior}) = \frac{\text{Number of seniors}}{\text{Total number of students}} = \frac{500 + \textcircled{500}}{10,000} = 0.10$$

The probability of selecting a science major is:

$$p(\text{Science major}) = \frac{\text{Number of science majors}}{\text{Total number of students}}$$

$$= \frac{4,000 + 1,750 + 950 + \textcircled{500}}{10,000} = 0.72$$

Notice the circled 500 in each equation; these are the same 500 students. Therefore, adding p(Senior) to p(Science) counts these students twice and they must be subtracted once.

Similarly, what is the probability of selecting one student who is either a Liberal Arts major or a freshman?

$$p(\text{Liberal Arts major or Freshman}) = p(\text{Liberal Arts}) + p(\text{Freshman})$$
$$- p(\text{Liberal Arts and Freshman})$$
$$= 0.28 + 0.50 - 0.10$$
$$= 0.68$$

When the "or" joins two mutually exclusive events, the subtraction is not required, because there is no "and" event possible (as when the "or"

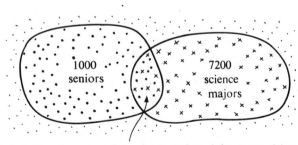

500 students who are senior science majors, joint event, (A and B)

Figure 8.2 A and B are not mutually exclusive events (one student can be both a senior and a science major). For such events, $p(A$ or $B) = p(A) + p(B) - p(A$ and $B)$. Here, one ring is drawn around all seniors (represented with dots) and another ring is drawn around all science majors. When a student is selected at random, the probability that he is a member of either ring is equal to the total number of students in both rings divided by the number of students in the university. However, we cannot simply add the number of students in one ring to the number in the other ring without adding some students twice; these must therefore be subtracted once.

connects two outcomes of the same simple experiment). Thus, the probability of picking a student who is either a freshman or a sophomore is just:

$$p(\text{Freshman or Sophomore}) = p(\text{Freshman}) + p(\text{Sophomore})$$
$$= 0.50 + 0.25 = 0.75$$

By definition, two outcomes of the same simple experiment are mutually exclusive; no student can be both sophomore and freshman.

We can also use these data to illustrate the general rule for the joint probability "and" operation. In the coin and die example (Table 8.4) the outcome of the coin flip had no effect on the outcome of the die throw. Under that condition, $p(H \text{ and } 1) = p(H)p(1)$. For the two experiments derived from Table 8.6, however, a different situation exists. From Table 8.6 we can determine the a priori probabilities of three events directly from the appropriate frequencies:

$$p(\text{Freshman}) = \frac{1,000 + 4,000}{10,000} = 0.50$$

$$p(\text{Science}) = \frac{4,000 + 1,750 + 500}{10,000} = 0.72$$

$$p(\text{Freshman and Science}) = \frac{4,000}{10,000} = 0.40$$

Pay particular attention to the third of these events, (Freshman and Science). This refers to the process (simple experiment) of drawing *one* student from the student body, and noting both the person's class standing and college. Here, even though only one person is involved, we have a joint event if we pay attention to both classes of outcome. There are eight joint events possible, and we have focused on one of them: finding a student who is both a freshman and a Science student. We can see directly from the table that 4,000 of the total 10,000 students are members of this joint category, and the probability of selecting one is thus 0.40.

When we look at the individual elementary events, however, the multiplicative relation used for earlier joint events does not work. It appears that $p(\text{Freshman and Science}) \neq p(\text{Freshman})p(\text{Science})$, since $0.40 \neq (0.50)(0.72)$. The relation does not hold here, because the two events are **nonindependent.** Unlike the coin/die experiment, the outcome of one experiment *does* affect the outcome of the other. The probability that a student is in the Liberal Arts category depends somewhat on his or her class standing. Seniors are divided equally between the colleges, but four times as many freshmen are in Science as in Liberal Arts. If we draw a freshman, it is more likely that he or she will be a science student than if we draw a senior. This is another way of saying that the two simple experiments are nonindependent; the outcome of one influences the probabilities associated with outcomes in the other experiment. When this condition, nonindependence, exists, the probabilities of the joint event are unequal to the product of the simple probabilities.

Table 8.7 *Enrollment Data for Comparison with Table 8.6 (See text for explanation.)*

	LIBERAL ARTS	SCIENCE	TOTALS
Freshmen	1,400	3,600	5,000
Sophomores	700	1,800	2,500
Juniors	420	1,080	1,500
Seniors	280	720	1,000
Totals	2,800	7,200	10,000

This relation, in fact, serves as the definition of **independent events.** Events A and B are said to be independent if they obey the relation:

$$p(A \text{ and } B) = p(A)p(B)$$

To understand more completely the reason that joint events in Table 8.6 are described as nonindependent, consider the enrollment figures given in Table 8.7. Notice here that the numbers of Liberal Arts and Science students are the same as in Table 8.6: 2,800 and 7,200, respectively. And, there are also 5,000 freshmen, 1,800 sophomores, 1,500 juniors, and 1,000 seniors. Thus, just as with Table 8.6 data, the probabilities for elementary events are:

$$p(\text{Freshman}) = \frac{1,400 + 3,600}{10,000} = 0.50$$

$$p(\text{Science}) = \frac{3,600 + 1,800 + 1,080 + 720}{10,000} = 0.72$$

Now look at the probability associated with the joint event (Freshman and Science). There are 3,600 students in this joint category, so we can calculate:

$$p(\text{Freshman and Science}) = \frac{3,600}{10,000} = 0.36$$

If we check for independence by using the multiplicative rule, we find that these elementary events are truly independent, because for Table 8.7 data:

$$p(\text{Freshman and Science}) = p(\text{Freshman}) \, p(\text{Science})$$
$$0.36 = (0.50)(0.72)$$

Still another way of describing independence here is to say that, for each class, the proportions of students in the two colleges are the same as the overall proportions of students in the two colleges. This condition does not exist when the two simple experiments are nonindependent.

So, in order to state a general rule for the multiplicative "and" operation, we will need a means of handling nonindependent events as well as independent events. **Conditional probability** provides such a means.

The symbol for a conditional event is $p(A \mid B)$: the probability that A occurs, given that B occurred. Look again at Table 8.6; notice there are many possible conditional probabilities that could be constructed such as:

p(Freshman | Science)

In words, this stands for "the probability that a freshman is selected, given that he or she is a science student." That is, we are supposing that we have selected a science student; *then* what is the probability that this science student is a freshman? The "given" thus means that we are only concerned with the 7,200 Science students. The expression p(Freshman | Science) is the relative frequency of freshman *among Science students*, not among the whole student body. Here:

$$p(\text{Freshman} \mid \text{Science}) = \frac{4,000}{7,200} = 0.556$$

For practice, it might be useful to consider the conditional probability p(Science | Freshman). Here the order is reversed, and the probability in question is the relative frequency of Science students among freshmen. In this case:

$$p(\text{Science} \mid \text{Freshman}) = \frac{4,000}{5,000} = 0.80$$

Conditional probabilities make it possible to specify the joint probability relation for nonindependent events. The general rule is:

$$p(A \text{ and } B) = p(A)p \ (B \mid A)$$
$$= p(B)p \ (A \mid B)$$

The two ways of calculating the joint probability are equivalent. From Table 8.6 we can test these formulas:

$$p(\text{Freshman and Science}) = p(\text{Freshman})p(\text{Science} \mid \text{Freshman})$$
$$= (0.50)(0.80)$$
$$= 0.40$$

The equivalent equation can be computed as a check:

$$p(\text{Freshman and Science}) = p(\text{Science})p(\text{Freshman} \mid \text{Science})$$
$$= (0.72)(0.555)$$
$$= 0.40$$

As a final check, we can go directly to the place in Table 8.6 representing the joint event in question; we find that 4,000/10,000 of the students are represented there. Calculation of probabilities for joint events is summarized in Table 8.5.

It might be a good idea at this point to summarize the "and" and "or" probability operations.

In the general case, for events A and B:

$$p(A \text{ or } B) = p(A) + p(B) - p(A \text{ and } B)$$
$$p(A \text{ and } B) = p(A)\, p(B\,|\,A) = p(B)\, p(A\,|\,B)$$

When A and B are mutually exclusive, the probability of (A and B) occurring is zero; in this case:

$$p(A \text{ or } B) = p(A) + p(B) - 0$$
$$= p(A) + p(B)$$

By definition, mutually exclusive events cannot be independent. If one occurs, the other cannot possibly occur. Hence, they are completely dependent.

This last point should become especially clear to you if you test the multiplicative relationship for independence, using two mutually exclusive events. Using the data in Table 8.6, the probability of selecting a freshman student is 0.50; the probability of selecting a sophomore is 0.25. If these two events were independent, then the probability of selecting a student who is both freshman and sophomore (the joint event) would be equal to $(0.25)(0.50)$. But the probability of such a joint event is, of course, zero. Mutually exclusive events are never independent.

When A and B *are* independent events, the probability that (A and B) occurs is reduced to $p(A)p(B)$. If the components of a joint event are independent, the conditional probability of ($A \mid B$) is exactly equal to the probability of A occurring.

Discrete Probability Distributions

Variables are entities that can change value or kind from observation to observation. Since statistical questions are framed in terms of variables, we must consider probabilities associated with different values of variables.

If the outcome of an experiment is uncertain, it, too, is a variable, since results of the experiment can change value or kind from trial to trial. When all the outcomes of a simple experiment are considered as different values of a variable, we are dealing with a **random variable.**

The term "random" does not mean that all outcomes are equally likely; it simply means that a probability can be associated with each. The results of each experiment listed in Table 8.3, for instance, can be considered random variables.

The examples considered so far in this chapter have dealt with **discrete random variables.** Remember from Chapter 1 that a discrete variable can have only specific values, while a continuous variable can have an infinite number of values between any two points on the measurement scale. Selecting a student at random from Ficticia University and determining his or her class standing produces a discrete random variable, because there are only four possible outcomes. If, instead, we select a student and measure his or her nose length, we will have a continuous random variable, since there are an infinite number of different lengths possible between the longest and shortest lengths observed.

Probabilities associated with both kinds of variables are analyzed by means of **probability distributions.** The discrete probability distribution is simpler, so we will consider it first.

Figure 8.3 is a relative-frequency histogram showing the class standings of Ficticia University students, specifically the proportion of students in each (discrete) class. As demonstrated in Chapter 5, we can find the proportion of students that falls in any interval of the class variable by finding the area under the curve that lies over that interval (if all the histogram bars are of the same width). Thus, the question, What proportion of the students are upperclassmen (juniors and seniors)? can be answered by finding the area under the curve over the junior-senior interval. The total area under the curve is assigned a value of 1.00, and the area in question covers 0.25 of that. Hence, the relative frequency with which these observations occur is 0.25. Since relative frequency has been equated with probability, we can also say that the probability of selecting one student at random who is an upperclassman is 0.25.

Figure 8.4 is another relative-frequency distribution, this time showing a distribution of (hypothetical) hat sizes of major league baseball players. Here the histogram bars are all only $\frac{1}{8}$ unit wide, but we can still ask the question, What is the probability that a baseball player, selected at ran-

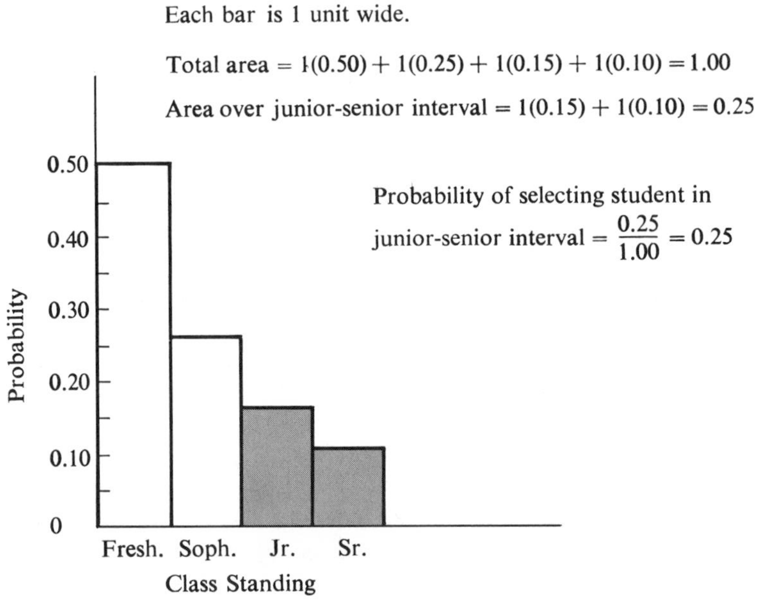

Each bar is 1 unit wide.

Total area = 1(0.50) + 1(0.25) + 1(0.15) + 1(0.10) = 1.00

Area over junior-senior interval = 1(0.15) + 1(0.10) = 0.25

Probability of selecting student in
junior-senior interval = $\frac{0.25}{1.00}$ = 0.25

Figure 8.3 Discrete probability distribution for the experiment of selecting a student and determining his or her class standing. The probability that a selected student falls in any interval on the horizontal axis equals the proportion of the total graph area over that interval.

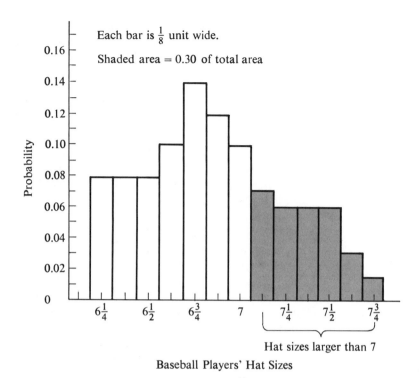

Figure 8.4   Discrete probability distribution for the experiment on baseball players' hat sizes. As in Figure 8.3, the probability of selecting one member of this distribution at random who falls into any interval is equal to the proportion of the total graph over that interval. When a baseball player is picked at random, the probability that his hat size is larger than 7 equals 0.30.

dom, will have a hat size larger than 7? The answer is found by determining the proportion of the area under the curve that lies over all hat sizes above 7. The calculations show that this area was determined by actually measuring the areas of the histogram bars involved. The desired probability is equal to that area, 0.30.

Pay special attention to the fact that with a *discrete* random variable we can find probabilities associated with each possible value of the variable. The probability that a randomly selected Ficticia student will be a freshman is 0.50. The probability that a randomly picked baseball player wears a hat of size $6\frac{5}{8}$ is 0.10. A different situation exists when we turn to probability distributions associated with continuous variables.

0̸ The Binomial Distribution

We are now ready to cover a very useful discrete probability distribution, the binomial distribution. The binomial distribution has many, many applications in both the natural and social sciences and we will be able to suggest only a few of these here. To make this distribution, we will

need only the concepts covered already in this chapter: the counting rule
for combinations and some of our basic probability operations.

Later on, we shall see how the binomial distribution can be used to help
answer such questions as, "Is this form of psychotherapy really effective?", "Do young drivers really account for a disproportionate number
of automobile accidents?", and "Does this drug really affect behavior?"
These weighty matters will have to wait until the later chapters on hypothesis testing, however; in this chapter, we will be more concerned with
learning the mathematical properties of the binomial distribution.

Even though the binomial distribution involves no new concepts, it will
be helpful here to introduce a new term. This is the **Bernoulli experiment,**
the building block for the binomial distribution. A Bernoulli experiment
is a simple experiment that has only two outcomes. Flipping a coin
and noting the side that lands up, choosing a person at random and
recording his or her sex, interviewing a psychotherapy patient and designating him as either "improved" or "not improved" all qualify as Bernoulli experiments. As you might guess, many kinds of observations in
the behavioral sciences that do not meet the requirements for ratio-,
interval-, or even ordinal-level measurement, *do* qualify as Bernoulli
experiments. Or, they can be made into Bernoulli experiments by dividing
all outcomes into two categories. This is one reason the binomial distribution (built on the Bernoulli experiment) will be so useful to us.

All we need to do now to make a Bernoulli experiment applicable to the
binomial situation is determine the probability associated with each of
the two Bernoulli outcomes. Typically, one outcome is called a "success" and the other a "failure," even if success and failure have little
meaning for the particular experiment at hand. In a study of library usage,
for instance, where the Bernoulli experiment consists of identifying library users as either "student" or "nonstudent," we might call these two
categories "success" and "failure," respectively, even though neither
outcome is "more successful" than the other; we might just as well
reverse the designations. At any rate, the probability that the "success"
outcome appears is symbolized p, and the probability that the "failure"
outcome appears is represented by q. Since the two outcomes in a
Bernoulli experiment are mutually exclusive and exhaustive, it is of
course true that:

$$p + q = 1.0$$

Figure 8.5 shows how some Bernoulli distributions could be graphed,
although the Bernoulli distribution is so simple that such graphing is rarely
done. Because p and q must sum to 1.0, the Bernoulli distribution is called
a one parameter distribution; knowing one value (either p or q) specifies
the whole distribution.

At this point, you may be wondering why we take so much trouble to
describe the Bernoulli distribution. Such concern is understandable, because the Bernoulli distribution is so simple that it is of little interest.
However, we are now ready to use it in building the binomial distribution.

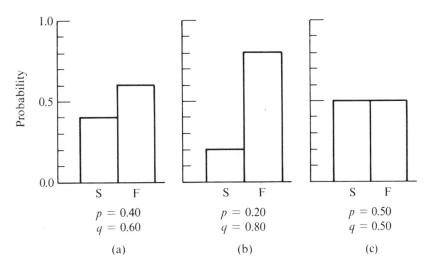

Figure 8.5   Three Bernoulli distributions. Notice that in each case, $p + q = 1.0$.

A **binomial experiment** consists of a number of *independent* Bernoulli experiments; the number of Bernoulli trials involved is designated N. Flipping 7 coins and noting the number of heads that turn up is a binomial experiment with $N = 7$; choosing 14 people at random and counting the number of males is a binomial experiment with $N = 14$; interviewing 39 psychotherapy patients and noting the number that can be rated as "improved" is a binomial experiment with $N = 39$. Notice two things about these examples. First, each Bernoulli trial is independent of the others (coin flips do not influence each other, and the people were selected in such a way that each selection had no influence on any other selection). And, secondly, the only item of information we paid attention to was the *number* of outcomes in each Bernoulli category: the number of heads, the number of males, the number of patients improved. (Again, if we know N, and the number of "success" outcomes, the number of "failure" outcomes is determined.)

The preceding paragraph contains the description of a binomial variable. If we conduct N independent Bernoulli experiments, and record the number of "success" outcomes, designated r, then r represents a binomial variable. What we are after now is the probability associated with each possible value of r, and, thus, the probability distribution for r. To show how such a distribution is derived, we will use the somewhat mundane (but easy to describe and easy to understand) example of the coin flip.

Suppose that you have a coin that, when flipped, has $p(\text{Heads}) = p(\text{Tails}) = 0.50$. Let us call the flip of the coin a Bernoulli experiment, and the appearance of heads will be called the "success" outcome. Thus, $p = 0.50$ and $q = 0.50$. Suppose also that you flip it four times in a row, and record the number of successes; call this number r. Our value of N is

4, and a brief consideration of the situation should show you that r can have $N + 1$ values (0, 1, 2, 3, or 4 in this case). The binomial distribution specifies the probability associated with each of these values.

Consider first the case where $r = 4$. With our coin-flipping example, there is only one way this can happen: you must obtain four successive heads, that is, four successes in a row. What is the probability of that happening? Since each Bernoulli event is independent of the others, we can multiply the probabilities involved:[5]

$$\text{prob}(r = 4) = \text{prob}(H \text{ and } H \text{ and } H \text{ and } H)$$
$$= (p)(p)(p)(p) = (p)^4 = (0.50)^4 = 0.0625$$

But what about situations where $r = 3$? Here, three heads can be obtained in four different ways:

H and H and H and T

H and H and T and H

H and T and H and H

T and H and H and H

Notice that the probabilities of each one of these ways are identical:

$$\text{prob}(H \text{ and } H \text{ and } H \text{ and } T) = (p)(p)(p)(q) = p^3q$$
$$= (0.50)^3(0.50) = 0.0625$$
$$\text{prob}(H \text{ and } H \text{ and } T \text{ and } H) = (p)(p)(q)(p) = p^3q$$
$$\text{prob}(H \text{ and } T \text{ and } H \text{ and } H) = (p)(q)(p)(p) = p^3q$$
$$\text{prob}(T \text{ and } H \text{ and } H \text{ and } H) = (q)(p)(p)(p) = p^3q$$

For any given sequence, it does not matter which order the heads and tail come in; only the numbers of them determine the probabilities. Since these four possibilities are mutually exclusive outcomes of the binomial experiment, the probability that any one of them occurs, that is, the probability that $r = 3$, can be found by adding:

$$\text{prob}(r = 3) = p^3q + p^3q + p^3q + p^3q = 4\,p^3q = 4\,(0.0625) = 0.2500$$

Having now done things the hard way, we are ready to simplify. Let us begin by figuring out how many ways there are of getting each value of r. In the above example, we found that for $r = 3$, there were four different sequences of coin flips possible. A shorter way than listing the outcomes would have been to apply the counting rule for combinations. We were

[5]This is another unfortunate situation where one symbol would normally have to represent two different things. In the first row here, the expression prob($r = 4$) would normally be written $p(r = 4)$, and mean, "The probability that the variable r equals 4." However, the symbol p must also represent the probability of a Bernoulli success; hence, to avoid confusion, this chapter will use "prob" in place of "p" when the latter would cause confusion.

asking, essentially, "given 4 trials (N trials), how many different ways can we distribute 3 (r) successes across the 4 trials?" Or, put another way, "assuming that 3 trials out of 4 will be designated as 'successes,' how many different groups of 3 trials can we select out of the 4?" The answer is given by the counting rule covered several sections earlier:

$$\binom{N}{r} = \frac{N!}{r!(N-r)!} = \frac{4!}{3!\,1!} = 4$$

This is the first part of our simplified process for finding probabilities associated with any value of the binomial variable r: For each value of r, find the number of sequences possible that produce that value of r. The counting rule is a simple way of doing this. Next, find the probability of *each* sequence. If you consider the example given you should see that given N, the probability of any *one* series containing r successes is $p^r q^{N-r}$. Thus, the probability of r Bernoulli successes, given N independent Bernoulli trials, is the product of (a) the number of possible sequences that have r successes, times (b) the probability of each sequence. In symbols: $\binom{N}{r} p^r q^{N-r}$. Now we can compute the probability associated with each value of r, starting at $r = 0$ and continuing through $r = N$.

$$\text{prob}(r = 0) = \binom{4}{0} (0.50)^0 (0.50)^4 = (1)(1)(0.0625) = 0.0625$$

(Remember that $0! = 1$, and any number raised to the zero power also equals 1.)

$$\text{prob}(r = 1) = \binom{4}{1} (0.50)^1 (0.50)^3 = (4)(0.50)(0.1250) = 0.2500$$

$$\text{prob}(r = 2) = \binom{4}{2} (0.50)^2 (0.50)^2 = (6)(0.25)(0.25) = 0.3750$$

$$\text{prob}(r = 3) = \binom{4}{3} (0.50)^3 (0.50)^1 = (4)(0.125)(0.50) = 0.2500$$

$$\text{prob}(r = 4) = \binom{4}{0} (0.50)^4 (0.50)^0 = (1)(0.0625)(1) = 0.0625$$

This is our entire distribution, and it is graphed in Figure 8.6a. Notice that all of the probabilities add up to one ($0.625 + 0.250 + 0.3750 + 0.250 + 0.625 = 1.000$). Since all given values of r constitute an exhaustive and mutually exclusive set of outcomes for the binomial experiment, this must be the case.

When we use the binomial distribution to answer research questions later on, we will ask questions such as the following: What is the probability of getting two or more heads in four coin flips? This question calls for us to add the probabilities for a number of different values of r:

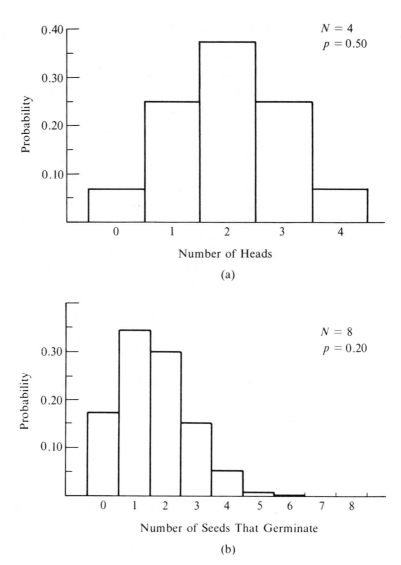

Figure 8.6   Two binomial distributions developed in the text.

$$\text{prob}(r \geq 2) = \text{prob}(r = 2 \text{ or } r = 3 \text{ or } r = 4)$$
$$= \text{prob}(r = 2) + \text{prob}(r = 3) + \text{prob}(r = 4)$$
$$= 0.3750 + 0.2500 + 0.0625$$
$$= 0.6875$$

The binomial distribution just presented is symmetrical, which you can easily verify if you examine Figure 8.6a. Symmetrical binomial distributions arise only when p and q equal 0.50. We will finish this section on the binomial distribution by considering one more example, a case where p and q differ.

Suppose that we have a very old packet of garden pea seeds. The seeds are four years old, and we learn from other sources that for seeds of this age, the probability of germination is only 0.20. If we plant eight seeds, what is the probability that we get at least four plants? To answer this, we need to construct a binomial distribution with $N = 8$, $p = 0.20$, and $q = 0.80$. (Notice that the binomial distribution is a *two*-parameter distribution; the entire distribution is determined after N and p are specified.) Here, r represents the number of success events, that is, plants that grow, and we note that r can have $N + 1$ values; it will be possible to have anywhere from 0 to 8 plants. Following the same procedure used previously, we can calculate the probability associated with each value of r:

$$\text{prob}(r = 0) = \binom{8}{0} (0.20)^0 \ (0.80)^8 = (1)(1)(0.167772) \quad = 0.167772$$

$$\text{prob}(r = 1) = \binom{8}{1} (0.20)^1 \ (0.80)^7 = (8)(0.20)(0.209715) \ = 0.335544$$

$$\text{prob}(r = 2) = \binom{8}{2} (0.20)^2 \ (0.80)^6 = (28)(0.04)(0.262144) = 0.293601$$

$$\text{prob}(r = 3) = \binom{8}{3} (0.20)^3 \ (0.80)^5 = (56)(0.008)(0.32768) = 0.146801$$

$$\text{prob}(r = 4) = \binom{8}{4} (0.20)^4 \ (0.80)^4 = (70)(0.0016)(0.4096) = 0.045875$$

$$\text{prob}(r = 5) = \binom{8}{5} (0.20)^5 \ (0.80)^3 = (56)(0.00032)(0.512) = 0.009175$$

$$\text{prob}(r = 6) = \binom{8}{6} (0.20)^6 \ (0.80)^2 = (28)(0.000064)(0.64) = 0.001147$$

$$\text{prob}(r = 7) = \binom{8}{7} (0.20)^7 \ (0.80)^1 = (8)(0.000013)(0.80) \ = 0.000082$$

$$\text{prob}(r = 8) = \binom{8}{8} (0.20)^8 \ (0.80)^0 = (1)(0.000003)(1) \quad = 0.000003$$

Sum of probabilities = 1.000000

This distribution is graphed in Figure 8.6b. Notice from the figure, or by adding the probabilities given above, that the probability of getting at least four plants (that is, $r \geq 4$) is pretty small:

$$\text{prob}(r \geq 4) = 0.045875 + 0.009175 + 0.001147 + 0.000082$$
$$+ 0.000003$$

$$= 0.056 \text{ approximately}$$

This concludes our introduction to discrete probability distributions — probabilities associated with variables that can only take on certain values. Now we will see how probability concepts are applied to continuous variables.

Probability Distributions for Continuous Variables

Figure 8.7 represents a relative-frequency distribution for a **continuous random variable** X. X might represent lengths, weights, times, voltages — any variable that can have an infinite number of different values between two points on the horizontal axis. Notice two things about this graph: (1) There are no numbers on the vertical axis, and (2) the curve is smooth.

The reason that there are no numbers shown on the vertical axis is that each individual value of a continuous random variable will have a very, very low relative frequency. Since there are an infinite number of different possible values, it is likely that no value will turn up more than once, even in a very large population. Although actual measuring devices are not precise enough to tell us so, it is possible that one human being might have a height of 5 ft 8.92837493082734958 . . . in., for instance. Thus, with continuous random variables we cannot ask for the probabilities associated with individual values occurring, since the probability of each value occurring is very close to zero. (Each single value is a *point* on the horizontal axis; the area over a point is represented by a line of zero width extending upward.) Instead of graphing the relative frequencies with which single values appear, we use a **probability density function.** This is a smooth curve (like Figure 8.7) that will yield probabilities in just the same way as histograms do, except that we can ask only for probabilities associated with specific *intervals* on the horizontal axis.

To be sure, numbers are associated with the vertical axis in probability density functions; the height of the curve is specified by a mathematical rule, such as that used to generate the normal curve (p. 124). These

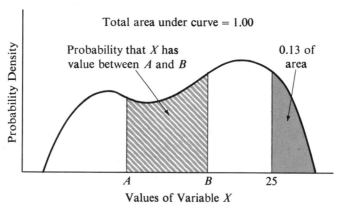

Figure 8.7   Probability density function for continuous random variable X. The area over any specific value of X is essentially zero, but the area over any interval on the X scale is still equal to the probability that X takes a value in that interval.

numbers, or vertical-axis heights, are not easily interpreted, and they would add nothing to our discussion here.

Using the probability density function in Figure 8.7, we can ask, What is the probability that a randomly selected member of this population has an X value greater than 25? The total area under the density curve is arbitrarily set at 1.00; the area under the curve over the interval "25 or greater" is equal to 0.13. Hence, the probability in question is 0.13. If we repeatedly select individuals from this population (being sure to replace them in the group after each observation so as not to alter the relative frequencies), in the long run 0.13 of them will have values greater than 25.

Just as we did for the relative-frequency curves presented in Chapter 5, we can ask three kinds of questions about intervals: (1) What is the probability that a randomly selected individual has a value greater than A? (2) What is the probability that a randomly chosen member of the population will have a value less than A? (3) What is the probability that a randomly selected individual will have a value between A and B? (A and B can be any points on the measurement scale.)

As you might expect at this point, this technique for finding probabilities will be used often with tabled values of areas under the normal curve. If a distribution can be assumed to have a nearly normal shape (see Chapter 5), we can use the unit normal table to find areas under the curve that lie over different intervals. Whereas these areas were interpreted as relative frequencies in Chapter 5, they are treated as probabilities here.

Figure 8.8 represents a frequency distribution of machine-part sizes produced by a hypothetical machinist (the same distribution as shown in Figure 5.2). Because differences in part sizes are due strictly to human and machine errors, we assume that the distribution is normal in shape. The mean, μ, is represented as 1.000 in., and the standard deviation, σ, as

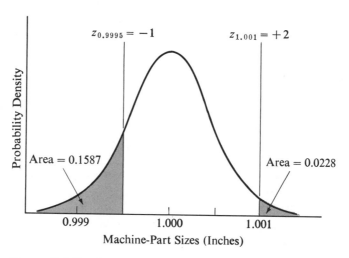

Figure 8.8   The probability that a selected machine part will be either greater than 1.001 in. or less than 0.9995 in. is equal to the sum of the areas over these intervals.

0.0005 in. What is the probability that one part, selected at random from all those produced, will be larger than 1.001 in.?

This probability is represented by the area shaded at the right side of the curve and can be found by computing the z-score of 1.001 in. and then using the unit normal table:

$$z = \frac{1.001 - 1.000}{0.0005} = 2.0$$

From Table A, Appendix 1, the area beyond z for $z = 2.0$ is 0.0228, and this is the probability that the part will be larger than 1.001 in.

Now, suppose that the part in question can be used only if it falls within the range 0.9995 to 1.001 in. Anything larger or smaller than these is considered a defective part. What is the probability that a randomly picked part is defective?

The question is actually an "or" question requiring the addition of two probabilities. If X is the part size, what is $p(X > 1.001 \text{ or } X < 0.9995)$? Since these two probabilities represent mutually exclusive events, the answer will be simply $p(X > 1.001) + p(X < 0.9995)$. We know that $p(X > 1.001)$ is 0.0228. We must now find $p(X < 0.9995)$:

$$z = \frac{0.9995 - 1.0000}{0.0005} = \frac{-0.0005}{0.0005} = -1.0$$

The unit normal table lists the area beyond a z of $+1.0$ as 0.1587. The normal curve is symmetrical, so the area beyond a z of -1.0 is also 0.1587. Thus:

$$p(X > 1.001 \text{ or } X < 0.9995) = 0.0228 + 0.1587 = 0.1815$$

SUMMARY

An element of uncertainty is involved whenever inferential statistics are applied. In the inferential process, estimates or conclusions about the population are based on observations of samples. The possibility exists that these are in error, because not every population member is included in the sample. Inferential statistics include not only the methods for making inferences about populations but also methods for evaluating the likelihood that they are in error. These methods require a means of describing and measuring uncertainty; this is the function of probability statements.

Events may be described as certain or uncertain. Certain events either absolutely will occur or absolutely will not occur. A probability of 1 means that an event is certain to occur, a probability of zero that it is certain not to occur. Probability values between 0 and 1 indicate different degrees of uncertainty.

One way of defining probability is to consider the probability of an event to be its relative frequency of occurrence. That is, on an infinite number of trials, the proportion of trials on which the event occurs is taken as its probability.

Many activities can be described as simple experiments, that is, processes or actions that lead to a single, well-defined outcome. In order to analyze probabilities in the simple experiment we must prepare a list of possible outcomes that is exhaustive (everything that might happen on a trial is included in an outcome category) and composed of mutually exclusive events (the results of each trial fit into one, and only one, outcome category). If these conditions are met, the probabilities associated with each outcome category can be determined, and the sum of these probabilities will equal 1.

Outcome probabilities may be determined by running many trials of the experiment and computing the relative frequency of occurrence of each outcome. Sometimes probabilities associated with outcomes can be determined a priori (by reason instead of experience) by dividing the number of elementary events in an outcome category by the total number of elementary events possible.

In computing a priori probabilities, it is sometimes helpful to use counting rules to count possible events. Two such rules discussed here are permutations (the number of orders in which N things may be arranged) and combinations (the number of different groups of r things that may be selected from a group of N things).

When events A and B are mutually exclusive, we can calculate the probability that either A or B occurs by adding their probabilities. When they are not mutually exclusive, the general rule for the "or" situation is:

$$p(A \text{ or } B) = p(A) + p(B) - p(A \text{ and } B)$$

Events A and B are said to be independent if the following relation applies:

$$p(A \text{ and } B) = p(A)p(B)$$

Thus, independence is defined as the situation where the probability that two events occur is equal to the product of their individual probabilities. When A and B are not independent, the "and" probability for the joint event must be calculated using conditional probability:

$$p(A \text{ and } B) = p(A) \, p(B \,|\, A) \qquad \text{or equivalently,}$$
$$p(A \text{ and } B) = p(B) \, p(A \,|\, B)$$

If probability values can be attached to the outcomes of a simple experiment, these outcomes constitute different values of a random variable. When a finite number of outcomes are possible, the probabilities associated with individual outcomes can be described in a discrete probability distribution. One such distribution discussed in this chapter is the Bernoulli distribution, which shows probabilities of outcomes in a simple experiment that has only two possible outcomes. The Bernoulli distribution is involved in constructing another discrete probability distribution, the binomial distribution. The binomial distribution shows the probabilities associated with different numbers of Bernoulli successes or failures occurring in a sequence of N independent Bernoulli trials. Both the binomial and Bernoulli distributions describe many kinds of experiments in the behavioral and natural sciences.

If an experiment has an infinite number of possible outcomes, it is not possible to associate a probability with each, so a probability density function is used to specify the probability that outcomes fall into different *ranges* of values. The normal curve is often used as a probability density function. If a relative-frequency distribution of events is normal in shape, we can use the unit normal table to find the probabilities associated with observations falling into different intervals.

KEY CONCEPTS

probability statement	mutually exclusive events	random variable
certain event	a priori probability	discrete random variable
uncertain event	elementary events	probability distribution
methods of expressing probability	permutations	Bernoulli experiment
probability as relative frequency	factorial of a number	binomial variable
	combinations	binomial experiment
simple experiment	joint events	binomial distribution
outcome	nonindependent events	continuous random variable
trial	independent events	probability density function
exhaustive events	conditional events	normal curve as a probability density function

SPOTLIGHT 8–a The Gambler's Fallacy

"When you're hot you're hot" is more than just a song title. It is an expression that summarizes some irrational feelings all of us — including nongamblers — occasionally fall prey to, known as the "gambler's fallacy."

Suppose you flip a coin six times and it turns up heads every time. What would you bet is the outcome of the next flip? Most people would probably choose heads; it is hard to fight the belief that there is a "run" or "streak" in progress. A few might choose tails for just the opposite reason, thinking that the laws of probability demand an evening up of the score. Both lines of reasoning are faulty, however. The probability that heads will appear on the next flip is exactly 0.50, just as it has been on every preceding flip. "But," someone might ask, "isn't it extremely unlikely that seven heads in a row will turn up?" It is. The probability of obtaining seven consecutive heads is only 0.0078. However, *once six consecutive heads have already occurred* (the probability of that happening is 0.0156), the probability that the next flip will be heads is 0.50. Each flip is statistically independent of all preceding flips. When events are independent, the probability that one occurs is in no way altered by the occurrence or nonoccurrence of the other.

If you still have trouble discounting the fallacy, consider the story of a man who was caught bringing a bomb on board an airplane. When questioned, he replied there was nothing to worry about — he was just a professor of statistics deathly afraid of being bombed on an airplane. His calculations showed that while the probability was very low that someone carrying a bomb would board any given airplane, it is much lower that *two* such people would board the same airplane. So he attempted to lower the probability of getting on board with a madman by bringing his own bomb.

SPOTLIGHT 8–b On Improbable Events

Leicester, England, June 22 — The congregation of more than 300 was singing "inflame, we pray, our inmost hearts, with fire from heaven above," when lightning struck the Church of St. James the Greater.

"The whole place was suddenly bathed in light," said the vicar, the Rev. Lawrence Jackson. Some dust fell on him, but no one was hurt, damage was negligible, and the service went on.

The event described above actually occurred in 1960. What do you suppose the probability is that lightning would strike that particular church *while* the congregation was singing that particular line? It is no doubt practically 0.0.

It is quite common after disasters or other momentous events for news commentators or other journalists to dwell on the unlikely string of events that preceded and led to the event. Many such events, such as the sinking of the *Titanic* or the explosion of the *Hindenburg*, were indeed very unlikely events. However, upon reflection you should be able to see that *any specific* event is really almost "impossible" in a probabilistic sense. With this in mind, imagine how you would have responded had you been on the jury in the 1968 trial described below:[6]

> **Trial by Mathematics.**   After an elderly woman was mugged in an alley in San Pedro, Calif., a witness saw a blonde girl with a ponytail run from the alley and jump into a yellow car driven by a bearded Negro. Eventually tried for the crime, Janet and Malcolm Collins were faced with the circumstantial evidence that she was white, blonde and wore a ponytail while her Negro husband owned a yellow car and wore a beard. The prosecution, impressed by the unusual nature and number of matching details, sought to persuade the jury by invoking a law rarely used in a courtroom — the mathematical law of statistical probability.

[6]"Trial by Mathematics," *Time* (April 26, 1968). Reprinted by permission from *Time*, The Weekly Newsmagazine. Copyright Time, Inc., 1968.

The jury was indeed persuaded, and ultimately convicted the Collinses (TIME, Jan. 8, 1965). Small wonder. With the help of an expert witness from the mathematics department of a nearby college, the prosecutor explained that the probability of a set of events actually occurring is determined by multiplying together the probabilities of each of the events. Using what he considered "conservative" estimates (for example, that the chances of a car's being yellow were 1 in 10, the chances of a couple in a car being interracial 1 in 1,000), the prosecutor multiplied all the factors together and concluded that the odds were 1 in 12 million that any other couple shared the characteristics of the defendants.

Only One Couple. The logic of it all seemed overwhelming, and few disciplines pay as much homage to logic as do the law and math. But neither works right with the wrong premises. Hearing an appeal of Malcolm Collins' conviction, the California Supreme Court recently turned up some serious defects, including the fact that not even the odds were all they seemed.

To begin with, the prosecution failed to supply evidence that "any of the individual probability factors listed were even roughly accurate." Moreover, the factors were not shown to be fully independent of one another as they must be to satisfy the mathematical law; the factor of a Negro with a beard, for instance, overlaps the possibility that the bearded Negro may be part of an interracial couple. The 12 million to 1 figure, therefore, was just "wild conjecture." In addition, there was not complete agreement among the witnesses about the characteristics in question. "No mathematical equation," added the court, "can prove beyond a reasonable doubt (1) that the guilty couple *in fact* possessed the characteristics described by the witnesses, or even (2) that only *one* couple possessing those distinctive characteristics could be found in the entire Los Angeles area."

Improbable Probability. To explain why, Judge Raymond Sullivan attached a four-page appendix to his opinion that carried the necessary math far beyond the relatively simple formula of probability. Judge Sullivan was willing to assume it was unlikely that such a couple as the one described existed. But since such a couple did exist — and the Collinses demonstrably did exist — there was a perfectly acceptable mathematical formula for determining the probability that another such couple existed. Using the formula and the prosecution's figure of 12 million, the judge demonstrated to his own satisfaction and that of five concurring justices that there was a 41% chance that at least one other couple in the area might satisfy the requirements.[7]

"Undoubtedly," said Sullivan, "the jurors were unduly impressed by the mystique of the mathematical demonstration but were unable to assess its relevancy or value." Neither could the defense attorney have been expected to know of the sophisticated rebuttal available to them. Janet Collins is already out of jail, has broken parole and lit out for parts unknown. But Judge Sullivan concluded that Malcolm Collins, who is still in prison at the California Conservation Center, had been subjected to "trial by mathematics" and was entitled to a reversal of his conviction. He could be tried again, but the odds are against it.

---

[7]The proof involved is essentially the same as that behind the common parlor trick of betting that in a group of 30 people, at least two will have the same birthday; in that case, the probability is 70%.

PROBLEMS

1. Summarize each of the following observations in a probability statement. In each case, assume that one element is to be selected from the larger group, according to a system that gives every element an equal chance of being selected; you should thus name an event (one outcome of the selection) and its associated probability.

   a. About two out of every three students who use the library are upperclassmen.

   b. I have bought 4 tickets for the raffle, but there are 496 other tickets besides mine in the drum.

   c. About one out of every twenty students in the introductory psychology class eventually majors in psychology.

   d. Of all the registered voters in Central City, a total of 78 percent is registered with the Democratic Party.

2. For each simple experiment given below, produce a list of outcomes (of course, there may be many possible different lists for each experiment, but make sure that yours has only mutually exclusive events, and that the events listed are exhaustive).

   a. Playing the state lottery.

   b. Recording the distance it takes a car to stop on wet pavement.

   c. Taking a final examination in chemistry.

   d. Trying to catch a mouse with a mousetrap.

   e. Buying $10,000 worth of stock in hope that it will increase in value during the next year.

3. An experimental psychologist is testing memory in a situation where people are given a list of 12 different letters and then asked to recall them in order. How many different orders are possible? How many orders are possible if 15 letters are used?

4. A box contains Christmas tree bulbs, one of each of the following colors: red, green, blue, yellow, orange, and white. If four bulbs are selected at random

   a. How many different color combinations are possible?

   b. What is the probability that the red, green, blue, and yellow bulb are selected?

   c. What is the probability that the white bulb is not selected? (Hint: determine the number of groups of 4 that can be selected from 5 available bulbs *and* the number of groups of 4 that can be selected from 6 available bulbs.)

5. How many different 5-man basketball squads can be selected from a group of 10 candidates? From a group of 8 candidates? If the coach chooses 5 men randomly from a group of 10, what is the probability that he selects the 5 tallest men?

The population of a certain city can be classified with respect to religion and political affiliation as shown in the table. Use this table to answer questions in problems 6, 7, 8, and 9.

	RELIGIOUS PREFERENCE			OTHER, OR NONE
	PROTESTANT	CATHOLIC	JEWISH	
Republican	600	300	150	200
Democrat	400	800	150	300
Other, or None	200	100	50	50

6. A pollster selects one citizen at random for an interview. What is the probability that the selected person will be
   a. A Protestant?   d. A non-Catholic?
   b. A Republican?   e. A Catholic Republican?
   c. A Catholic or a Protestant?

7. A citizen is selected at random for an interview; he or she is then replaced in the city's population and another random selection is made. (The replacement is done so that each selection will be truly independent of the other.) What is the probability that
   a. Both citizens selected are Republican?
   b. Neither citizen is a Catholic?
   c. One is Jewish and one is Protestant?
   d. Neither citizen is Protestant nor a Republican?

8. This city happens to have 1,980 women. Of these women, 900 are registered Democrats and 800 are registered Republicans. What is the probability that
   a. A randomly selected citizen is a female Democrat?
   b. A randomly selected citizen is a male who has no political affiliation?
   c. A person selected randomly from among the Democrats is a male?
   d. A person selected randomly from among the women is a Democrat?
   e. A randomly selected citizen is neither male nor a Republican?

9. Construct a probability distribution graph for the experiment "Selecting a citizen at random and determining his or her political affiliation"; make a similar graph for the experiment "Selecting a citizen at random and determining his or her religious affiliation." Are these discrete or continuous distributions?

10. Four coins are flipped, a process that may be viewed as four independent Bernoulli experiments with p(Heads) $= p$(Tails) $= 0.50$. What is the probability that

    a. All four coins land heads up?

    b. There are two heads and two tails facing up?

    c. There is at least one heads?

11. A certain pain-relieving drug does not ease headaches for everyone who takes it; for headaches, the probability of relief is about 0.40 for a randomly selected person. Suppose that ten people with headaches take the drug. Construct the probability distribution associated with this binomial experiment and then determine the probability that

    a. All people find relief.

    b. Five or more people are relieved.

    c. At least two people are relieved.

12. Scores on a nationally administered test are normally distributed, with a mean of 500 and a standard deviation of 100. What is the probability that a student selected at random from those taking the test

    a. Scores over 500?

    b. Scores between 700 and 800?

    c. Scores either in the top 10 percent of those taking the test or in the bottom 10 percent?

    d. Receives a score within fifty points of the mean?

    e. Receives a scholarship to Ficticia University, if a score of 780 or better is required to obtain one?

    f. Scores below 300?

13. Explain in your own words the use of a probability density function.

14. In a normal distribution with a mean of 417 and a standard deviation of 36, what is the probability of selecting one observation at random of the values indicated? (Round z values to 2 decimal places.)

    a. Greater than 450?   d. Greater than 500?

    b. Between 400 and 450?   e. Between 400 and 500?

    c. Less than 390?

15. For the distribution of test scores described in problem 12 above, one student is selected at random and his or her score is recorded. The process is then repeated. What is the probability that

    a. Both students have scores above 700?

    b. Neither student scored above 700?

9. Sampling: Choosing Some to Represent All

After reading this chapter you should be able to do the following:

1. *Give some instances requiring inferential rather than descriptive analysis.*

2. *Describe procedures used to ensure that the sample adequately represents the population.*

3. *Distinguish among population, sample, and sampling distributions. Discuss the essential characteristics of each.*

4. *Present the characteristics of the sampling distribution of the mean that are specified by the Central Limit Theorem.*

5. *Make probability statements about the values of sample means.*

6. *Use the binomial distribution as a sampling distribution and make probability statements about the value of the binomial variable r.*

Why We Sample

Who is going to win the next presidential election? Does violence on television actually influence children's behavior? Are women still being paid less than men for doing the same work? All of these questions have an important characteristic in common: they refer to populations that are so large that we cannot possibly hope to make an observation on every member. The descriptive techniques covered in Chapters 1 through 7 will not be of much help in approaching these questions because descriptive techniques do not allow us to generalize beyond the group we actually make observations on.[1] Descriptive techniques are adequate when interest focuses only on a group that is entirely accessible (such as when summarizing one class's test-taking records), but they are of limited use in

---

[1] Descriptive statistics may *suggest* that certain states of affairs exist in groups outside of those actually studied. However, we cannot properly draw conclusions about those groups until the appropriate inferential techniques have been applied.

evaluating research questions. Research questions almost always concern populations, only some of whose members can be brought under observation.

We might, for example, be interested in learning something about the verbal ability of fourth-grade children in the United States. Since we cannot reasonably expect to test the ability of every such child in the country, our study will have to use children we can actually test. A new category of statistical methods is required if we want to draw conclusions about a large group based on observations of only some of its members.

In statistical terms, the entire group we are ultimately interested in is called the **population** (all fourth-grade children in the United States), and the smaller group selected for actual testing is called the **sample.** Procedures that allow us to draw conclusions about populations from observations on samples are called **inferential statistics.**

In practice, the need for inferential statistics occurs frequently. Public opinion polling organizations, for example, study samples because they cannot interview the entire public. Inferential methods are almost always involved when laboratory experiments are evaluated statistically, since virtually all experiments must be performed on samples rather than on populations. Inferential techniques themselves are discussed in Chapters 10–15. This chapter covers the theory that makes inferential statistics possible.

Sampling Procedures

When using or interpreting inferential statistics, it is very important to know which **sampling procedure** was used to select population members for the sample. Inferences about the population may not be possible if the proper sampling procedure was not used. Proper sampling procedure ensures that the sample adequately represents the population. A sample that is not representative of the population is called a **biased sample.**

To illustrate with a simple example, suppose that we wanted to study the attitudes of American citizens toward the further use of nuclear power in generating electricity. We are going to sample (select) 100 people for interview, and generalize a summary of their opinions to the nation as a whole. Would it be permissible to select all 100 from among college students on a particular campus? From unemployed construction workers in a particular district? From among nuclear engineers? No one would think of generalizing results from such unrepresentative samples to the population as a whole. Such samples would undoubtedly be biased.[2]

How then should one go about getting a representative sample of a very large population? The answer to that question is still undergoing refinement by professional pollsters. Even though polls of attitudes and voter preferences are getting to be more accurate than they used to be, they are

---

[2]Biased in a statistical sense, that is. Saying that a sample is biased does not imply a value judgment of any kind about sample members.

still not perfect. In 1948 the Gallup Poll (Conducted by the American Institute of Public Opinion) predicted that Presidential candidate Dewey would receive 49.5 percent of the popular vote, and that candidate Truman would receive 44.5 percent. The poll was taken on November 2, and as everyone knows, on November 7, Truman defeated Dewey (Truman obtained 49.5 percent of the popular vote, Dewey 45.1 percent). There are two possible reasons for such a prediction error: (1) Either the populace changed preferences somewhat between November 2 and November 7, or, more likely, (2) the sample was not representative of the population. In eleven national elections since 1948, the Gallup Poll has registered an average error of only 1.6 percentage points.[3] The increase in accuracy is no doubt due to better sampling procedures that insure a more representative sample.

Discussed here are two commonly used sampling procedures designed to ensure that the sample is an accurate, unbiased representation of the population from which it was drawn: **random sampling** and **stratified random sampling.**

Random Sampling

Random sampling is probably the most important sampling procedure used in statistical applications. Random sampling means that population members were chosen for the sample under two conditions. First, every member of the population had exactly the same chance of being selected for the sample as every other population member, and second, selection of one individual for the sample did not influence the chances of any other individual of being chosen for the sample.

If, for example, we wanted a random sample of American citizens for our study of attitudes on nuclear energy, we would have to choose sample members in such a way that every person in the population would have exactly the same probability of being selected as every other person.

As you might guess, random sampling is simple to explain but difficult to achieve in practice. Sometimes many individuals in the population are not accessible for observation; at other times, the population cannot be specified or known as completely as we would like. Since true random sampling is so difficult to obtain for many experiments and studies, a conscientious statistician will specify the nature of his sample and how it was selected. That way, the reader can decide for himself if conclusions drawn about the population are valid.

Stratified Random Sampling

Stratified random sampling is another important sampling procedure. This method is used to pick sample members when the researcher has reason to believe that the population is composed of distinct subgroups, or strata.

---

[3]G. Gallup, "Opinion Polling in a Democracy," in *Statistics: A Guide to the Unknown*, ed. J. M. Tanur et al. (San Francisco: Holden-Day, 1972).

Consider, for instance, a study to assess the attitudes of American citizens toward reinstituting the military draft. In this case, it would be reasonable to expect that different groups in the population of American citizens would differ with regard to the variable under study (attitudes toward the draft). Before selecting a sample for study it would be useful to consider what these strata might be and what proportion of the population falls into each. If the population stratification is not reflected in a corresponding stratification of the sample, the sample will probably be biased. It might, therefore, be reasonable to consider this population as being composed of two strata: veterans and nonveterans; or perhaps, of five strata: people aged 1–15 years, 16–30 years, 31–45 years, 46–60 years, and 61 or more years. You can probably think of many other ways to divide up the population into strata that might have been meaningful for this study. Deciding which stratification scheme to use is the task of the researcher; that decision is based on previous studies of the same population or other a priori knowledge. The important point is that the researcher must decide which stratification scheme to use before selecting his sample.

When stratified random sampling is used, members from each stratum in the population are included in the sample. The proportion of the sample from each stratum is made equal to the proportion of the population in each stratum.

Suppose, in another study, the population of interest is composed of a certain city's residents. For this study, the population can be divided into four meaningful subgroups based on religion: Catholic (composing 30 percent, or 0.30, of the population), Protestant (0.40 of the population), Jewish (0.10 of the population), and those with other preferences (including no preference) (0.20 of the population). Within each stratum the required number of people are selected randomly for the corresponding sample stratum. Thus, if the total sample is of size 50, the sample should be composed of fifteen Catholics, twenty Protestants, five Jews, and ten people with other preferences. The fifteen Catholics are selected randomly from the city's Catholic stratum, the twenty Protestants are selected randomly from the city's Protestant stratum, and so on.

When should you use random sampling, and when should you use stratified random sampling? This, again, is one of those questions that *can* be answered definitively, but only after much more material is discussed, and that material is better left to larger statistics texts. In general, we can say that a very large and truly random sample will automatically reflect population stratification, even with respect to strata that the experimenter does not know about before selecting a sample. When the sample is small, the chances of incorporating bias are greatly increased, and the use of a stratification plan offers a means of counteracting some potential bias. Statisticians use sophisticated methods not covered here to place a numerical value on "very large" and "small" in specific situations.

If you are called upon to evaluate somebody else's inferential statistics, it is a good idea to look for possible bias in the sample as a routine precaution. A biased sample can invalidate a researcher's conclusions, and a biased sample is likely to result if one of the above procedures is not used. You may notice that even advertisements are beginning to give you information about how samples were selected for demonstrations used in the ad. Statements such as "Eight out of ten doctors surveyed recommend the ingredients of Brand A pain reliever," with no other information about how the doctors were chosen, are increasingly rare in the media.

In the behavioral sciences it is particulary difficult to avoid biased samples when experiments are taken out of the laboratory into the field. For example, a perfectly random sample of people may be selected to receive a questionnaire in the mail. But responses to the questionnaire do not necessarily represent a random sample if some people choose not to return them. In these cases, the experimenter must anticipate that some sample members will not respond and prepare to evaluate the data accordingly.

Distributions Involved in Sampling

We will put off for the moment any discussion of what kind of inferences are made about populations from sample data. That material follows in the following chapters. In the rest of this chapter we will be concerned with the factors that make those inferences possible — in particular, some very important characteristics of samples and populations and of the relation between them.

Whenever we draw a sample from a population and compute statistics from sample observations, three different distributions are involved: the population distribution, the sample distribution, and the sampling distribution of the statistic. These distributions are related to each other and have some common characteristics that cause some beginning statistics students to confuse them. To understand the important differences among these distributions, it is not necessary to learn difficult new concepts — all three may be thought of as different types of frequency distributions. The difficult aspect of this area consists of learning new terminology to apply to special cases of old concepts. Keep this in mind as we consider the three distributions individually.

The Population Distribution

The first distribution we consider in the sampling situation is the **population distribution.** In experiments or studies we are usually concerned with observations on a variable in the population. A frequency distribution for this variable in the population is the population distribution.

We have already discussed some characteristics of frequency distributions, including the mean and the standard deviation. *If* the mean and standard deviation could be computed for the entire population, that is, *if* every member of the population contributed to the calculated values of mean and standard deviation, these characteristics would be the known

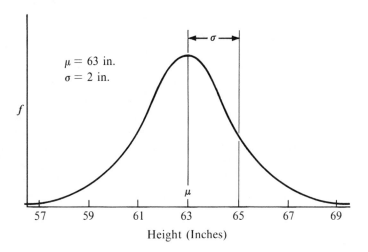

Figure 9.1 Possible frequency distribution of heights for the population adult female residents of England. The population is represented as having a mean of 63 in. and a standard deviation of 2 in.

population parameters. However, as we have just indicated, we are using inferential statistics because it is not possible to actually carry out these computations. Thus, in inferential statistics, we never compute parameters — if we did, we would be doing descriptive statistics. The best we can do in inferential statistics is make inferences about parameter values that must remain unknown to us.

To use a possibly mundane but probably clear example, suppose that the variable of interest is height and that the population is adult female citizens of England. The mean height of this population is a (population) parameter symbolized by μ, the Greek lowercase mu.[4] In order to compute the value of μ, we would have to measure the height of every adult female citizen of England and then compute the mean of these measurements. Suppose that Figure 9.1 represents a frequency distribution for this population. (Data presented here are simplified and in some cases manufactured for illustrative purposes.) From the graph we can get a rough idea of the typical height and the spread of the distribution. The population mean, μ, is given as 63 in. (Remember, of course, that it would be realistically impossible to actually compute this value. In this example, we are assuming sort of an omnipotence in order to show how different values are related.)

The population standard deviation is another parameter of this distribution in which we may be interested. It is symbolized by σ, the Greek lowercase sigma. In order to calculate σ, we would have to measure every

---

[4]This is, of course, the same μ used for the mean of the normal curve (Chapter 5). Parameters of theoretical mathematical distributions and parameters of populations are expressed by some of the same symbols.

member of the population and compute the standard deviation of these measurements. In the example, $\sigma = 2$ in.

Population parameters are thus characteristics of the entire population. As research workers or research consumers, we are ultimately interested in population parameters. In fact, the first step in designing an experiment is usually to formalize basic questions into statements about parameters. What is the *mean age* at which (the population of human) children begin walking? Does the population of smokers have a lower *mean age* at death than the population of nonsmokers? Since we can rarely make an observation on every member of the population, however, we can seldom know parameter values from direct calculation. Usually, characteristics of populations must be inferred from observations made on samples.

The Sample Distribution

After choosing a population and a variable, and after deciding which parameters of the population will tell us important items of information about the way the variable is distributed in the population, the next step in the inferential process is to apply sampling procedures and choose a sample. The **sample distribution** is simply the frequency distribution of the sample.

Figure 9.2 shows one possible sample distribution that might occur with a sample of 100 individuals from the population of adult English women. Notice that the horizontal axis of the graph represents the variable under study (just as in the graph of the population distribution) and the vertical axis represents the frequency in each measurement class. This graph, however, represents a total group of 100 individuals, while the population graph represents a group of millions.

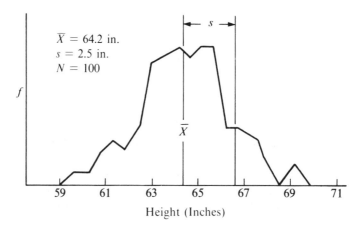

Figure 9.2   Distribution of heights in one possible sample of size $N = 100$ taken from the population shown in Figure 9.1. This particular sample happened to have a mean of 64.2 in. and a standard deviation of 2.5 in.

We can also compute a mean and standard deviation for the sample. These characteristics of the sample would not be parameters (parameters are characteristics of populations) but rather **sample statistics.** Symbols used for sample statistics are usually letters of the Roman alphabet. As indicated in Chapter 3, the most commonly used symbol for the sample mean is the familiar $\bar{X}$. As indicated in Chapter 4, the sample standard deviation is represented by the symbol s. Notice that every member of the sample contributes to the value of the sample statistics and that every member of the population contributes to the value of the population parameters.

You may be asking why it is necessary to distinguish between sample statistics and population parameters. Shouldn't the sample standard deviation be the same as the population standard deviation?

Compare Figures 9.1 and 9.2. Notice that the population mean μ is given as 63 in. and the sample mean $\bar{X}$ is given as 64.2 in. The two values differ, and it is likely that if we took a different sample of another 100 women, the value of $\bar{X}$ would be something else for the second sample. The same situation occurs with the standard deviation. The population standard deviation σ is shown as 2 in. and the sample standard deviation is shown as 2.5 in. Here again, s and σ differ, and it is likely that an s computed on a different sample would differ from the first s and from σ.

Population parameters are treated as if they do not change during the sampling process. They are assumed to be constant and unchanging. But the values of sample statistics vary from sample to sample. Some samples will include both short and tall English women, one sample will include the 100 shortest English women, and so on. The only way to ensure that $\bar{X} = \mu$ and that $s = \sigma$ is to include every member of the population in the sample.

As indicated, the purpose of many research studies can be described as an attempt to learn something about population parameters. But this kind of question must be answered from sample statistics, and we have just seen that sample statistics vary from sample to sample, even though parameters are fixed. It is because of this characteristic of sample statistics — variability from sample to sample — that we must consider a third type of frequency distribution, the sampling distribution.

The Sampling Distribution

Of the three types of distributions discussed in this chapter, the sampling distribution is the most complex. It is related to both the sample and the population and thus makes possible the inferential connection between them. A **sampling distribution** is a frequency distribution of a statistic's value, that *could* be obtained by taking a very, very large number of samples (all of the same size) from a given population, then computing the statistic's value in each sample, and arranging these values in a frequency distribution. In practice, this is never done. However, the crucial assumption is that our sampling distributions *would* arise, if this were done.

Consider, for instance, the process of selecting 100 English women from the population in Figure 9.1. The sample's height could be determined, and we could call this $\bar{X}_1$. This value might or might not equal the parameter μ (remember that values of statistics can vary from sample to sample). Now select another 100 women from the same population and compute a mean for that sample, $\bar{X}_2$. Continue this process over and over until an infinite number of sample means have been computed. The frequency distribution of sample means obtained is a **sampling distribution of the mean.** Even though in practice no one will collect an infinite number of samples, knowing what would happen *if* that were done is extremely important.

The sampling distribution of the mean is thus the *theoretical* frequency distribution of sample means, for an infinite number of samples of a certain size from a given population. Notice that all samples involved are the same size — there will be a different sampling distribution of the mean for samples of size $N = 16$, another for samples of size $N = 137$, and so on.

We could just as well construct sampling distributions for other statistics besides the sample means. The sampling distribution of the standard deviation, for example, would be the frequency distribution of sample standard deviations from many samples of a given size from the population.

Sampling distributions are highly useful in inferential statistics because they tell us what values a statistic may have (when computed from sample observations from a specified population) and the relative frequency with which those different values occur.

Fortunately, we can obtain the sampling distribution of the mean without going through the laborious procedure just described. If the population is very large at all, this would be a lifetime project for any human statistician. In fact, the time and effort required to compute an entire sampling distribution in this manner would even require a substantial amount of computer time in most cases. Mathematicians have long since discovered alternative ways of determining characteristics of sampling distributions.

Incidentally, it is an understatement to say that you will understand very little of inferential statistics unless you first understand what sampling distributions represent and how their characteristics are determined. It is also true that other concepts to be introduced shortly will be relatively easy to grasp if you have a clear notion of the sampling-distribution concept. In other words, it will be downhill from here on if you know what a sampling distribution is.

Characteristics of Sampling Distributions

Two sampling distributions of the mean are illustrated in Figure 9.3. There are several important characteristics of the graph to point out. First, the horizontal axis of the graph represents the variable of interest, height. *In the sampling distribution of the mean, however, the values located on this*

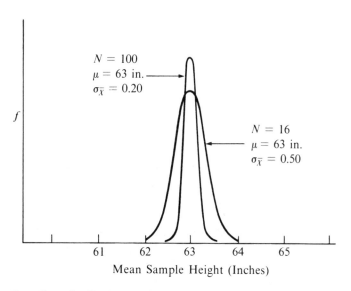

$N = 100$
$\mu = 63$ in.
$\sigma_{\bar{X}} = 0.20$

$N = 16$
$\mu = 63$ in.
$\sigma_{\bar{X}} = 0.50$

Mean Sample Height (Inches)

Figure 9.3 Sampling distribution of the mean for two different sample sizes ($N = 16$ and $N = 100$) from the population shown in Figure 9.1. Notice the effect that sample size has on the shape of the distribution.

axis represent the mean of the heights of people in the sample, not the heights of individuals. The vertical axis represents frequencies of sample means. One sampling distribution of the mean is for samples of size $N = 16$ and one is for samples of size $N = 100$. Two distributions are shown to emphasize the point that there is a different sampling distribution of the mean for each possible sample size.

In the preceding example we drew a sample of 100 from the population, measured each sample member's height, and computed the mean of the sample measurements. The mean for one sample turned out to be 64.2 in., but we have just seen that this number is only one possible value from a whole distribution of possible values for the sample mean. Other samples (if we had actually drawn them) would have had other means. In order to proceed with the business of making inferences about μ we need to know some characteristics of the sampling distribution to which our sample mean belongs. Specifically, we need to know the mean, the standard deviation, and the shape of this distribution of means. These characteristics are not necessarily the same as the shape, mean, and standard deviation of the population distribution or of the sample distribution. Comparison of Figures 9.1, 9.2, and 9.3 suggests that each of these distributions can have its own shape, mean, and standard deviation.

By way of review, remember that the population mean μ and the population standard deviation σ are not necessarily the same as the sample mean $\bar{X}$ and sample standard deviation s. Now we are seeking still more values, the mean and standard deviation of the sampling distribution of the mean.

These two values are related to the population parameters μ and σ. It can be shown mathematically that the mean of the sampling distribution of the mean is exactly equal to the population mean μ. Thus, the population mean is 63 in. and the mean sample mean is also 63 in. This holds for all sampling distributions of the mean, regardless of sample size.

Another name for the mean of the sampling distribution of the mean is the **expected value of the mean,** symbolized $E(\bar{X})$. Without straying too far from the thread of our discussion, we should emphasize that other statistics besides the mean also have sampling distributions; the mean of any statistic's sampling distribution is called the **expected value** of that statistic. The expected value of the mean is the mean of the sampling distribution of the mean (if necessary, take a minute to ponder the *meaning* of that sentence). The expected value of the range would be the mean of the sampling distribution of the range, and so on.

The sampling distribution of the mean also has a standard deviation that we need to know. The standard deviation of the sampling distribution of the mean is known as the **standard error of the mean,** symbolized $\sigma_{\bar{x}}$. The symbol includes σ, to indicate that it is a standard deviation, and a subscript $\bar{X}$, to indicate that it is a characteristic of a distribution of sample means.

The term **standard error** also applies to other sampling distributions. The standard error of any statistic is the standard deviation of its sampling distribution. Thus, we could have a standard error of the standard deviation, a standard error of the median, and many other standard errors.

For the sampling distribution with which we are concerned here, the standard error may be determined quite simply if we know the population standard deviation σ and the number of individuals (N) in each sample. The standard error of the mean is related to the population standard deviation and sample size in the following way:

$$\sigma_{\bar{x}} = \frac{\sigma}{\sqrt{N}}$$

In our example, $\sigma = 2$ and $N = 100$. So, if we took all possible samples of size 100 (again, a very high number of samples), computed the mean of each sample, and plotted a frequency distribution of these sample means, the standard deviation of this distribution would be:

$$\sigma_{\bar{x}} = \frac{2}{\sqrt{100}} = \frac{2}{10} = 0.20$$

Suppose instead that we had taken all possible samples of size $N = 16$. Then our standard error of the mean would be:

$$\sigma_{\bar{x}} = \frac{2}{\sqrt{16}} = \frac{2}{4} = 0.50$$

Notice two things about these relations: (1) For any population, the standard error of the mean is always smaller than the population standard

Table 9.1 *Characteristics of the Population Distribution, the Sample Distribution, and the Sampling Distribution of the Mean*

	MEAN	STANDARD DEVIATION
Population distribution	μ	σ
Sample distribution	$\bar{X}$	s
Sampling distribution of the mean	μ	$\sigma_{\bar{X}}$

deviation (unless samples of size $N = 1$ are used), and (2) the larger the sample size, the smaller the standard error. The effect of sample size on spread of the sampling distribution is clearly illustrated by comparing the two sampling distributions in Figure 9.3. This should suggest something to you about estimating parameter values from sample statistics: if you want the statistic's value to come close to the parameter value, use the largest sample size possible. Values of statistics vary more from sample to sample when small samples are used.

We now know the mean and standard deviation of the three distributions involved in sampling: the population, sample, and sampling distributions. Table 9.1 summarizes the symbols used for these characteristics. We are almost ready to use this information in inferential statistics. But first we must consider one more characteristic of the sampling distribution of the mean: shape.

Shape of the Sampling Distribution of the Mean

Remember from Chapters 5 and 8 that knowing the shape of frequency distributions allows us to perform important calculations; if the distribution's shape is known, we can find the proportion of the distribution that falls in certain intervals by finding the area under the curve that lies over those intervals. If the shape is normal, those areas are listed in the unit normal table.

In the case of the sampling distribution of the mean, we are fortunate. For large samples (about $N = 30$, or larger), the sampling distribution of the mean is usually considered normal. This happy state of affairs is specified by the **central limit theorem,** which states that the sampling distribution of the mean will be normal in shape for large samples, even if the original population distribution is not.

To see the central limit theorem at work, take, for example, the frequency distribution of annual incomes of American citizens. The population is "American citizens," and the variable of interest is "annual income." A frequency distribution for this variable is not normal in shape, but rather highly skewed in a positive direction. There are many people grouped together at the low-income end and at the middle of the distribution, and very few at the ultra-high income end. Now, suppose we construct a sampling distribution of the mean by taking every possible sample of size 47 from this population and compute the mean salary of each sample. A few samples will include forty-seven millionaires; other samples will consist entirely of unemployed people. If we construct a frequency distribution of these sample means, we will have the sampling distribution of the mean for $N = 47$, which will be normal in shape even though the original population distribution is heavily skewed. The normality of this distribution is assured by the central limit theorem.

What about sampling distributions of the mean for sample sizes smaller than $N = 30$? In these cases, the sampling distribution of the mean may no longer be normal. With small samples (usually taken as $N = 29$ or smaller), the shape of the sampling distribution of the mean is known only in special cases, such as when the underlying population distribution *is* normal. If the population distribution is normal, and small samples are drawn from it, the shape of the sampling distribution of the mean is described by the *t*-distribution. Tables of areas under the *t*-distribution are also available, and its use similar to the normal distribution. We will discuss the *t*-distribution in more detail in the following chapters.

At this point, however, you should be able to answer and distinguish between two types of questions involving probability and normal distributions. The first type was introduced in Chapter 8: Given a normally distributed population of heights, with $\mu = 63$ in., and $\sigma = 2$ in. what is the probability that a randomly selected individual from this population is taller than 64 in.? For this question, we find a z-score for one individual in the *population*, specifically, one who is 64 in. tall.

$$z_X = \frac{X_i - \mu}{\sigma} = \frac{64 - 63}{2} = \frac{1}{2} = +0.50$$

From Table A we see that the "Area Beyond z" for a z value of $+0.50$ is 0.3085. Thus, the probability of drawing one individual randomly from that population, and coming up with someone 64 in. or taller is 0.3085.

The second type of question uses the sampling distribution. What is the probability of selecting 16 individuals from that population and obtaining a sample with a sample mean of 64 in. or greater? Now we are asking about a statistic ($\bar{X}$) from a distribution of statistics. Its z-score is obtained as follows:

$$z_{\bar{X}} = \frac{\bar{X} - \mu}{\dfrac{\sigma}{\sqrt{N}}} = \frac{64 - 63}{\dfrac{2}{\sqrt{16}}} = \frac{1}{\dfrac{2}{4}} = +2.00$$

From Table A, we find that the "Area Beyond z" for a z value of $+2.00$ is 0.0228. This, then, is the probability of obtaining a sample mean 64 in. or greater, using a sample size of 16. In the first case above, we took an element from a distribution of elements (the population distribution); in the second case, we took a statistic from a distribution of statistics. In each case, the z was found by taking the difference between the value of interest and the distribution mean, and dividing it by the standard deviation *of that distribution*.

The sampling distribution is the foundation of inferential statistics. As we will see, it is the inferential bridge between sample statistics and population parameters that gives us a means of determining the likelihood that different values of statistics occur.

Ⓞ The Binomial Distribution as a Sampling Distribution

The preceding examples all dealt with a continuous random variable (height). We will close this chapter by considering a situation involving a discrete random variable, the binomial variable introduced in the last chapter.

Just as height measurements represent a random variable, so do values of the binomial variable r. Suppose, for instance, that we are working with a large population of psychotherapy patients; suppose also that for this group, the probability that any one patient will show improvement after a month of therapy is 0.50. We can thus consider each patient's therapy a Bernoulli trial, with $p = 0.50$ and $q = 0.50$. When we advance to hypothesis testing in the next chapters, we will be interested in the answers to questions like: What is the probability that in a group of 10 patients, randomly selected from the population, 8 or more will show improvement? We have already answered such questions in the preceding chapter, but let us approach them once more, this time in the context of a sampling problem.

Our variable of interest is the number of patients, r, who show improvement out of the group of N. Because of sampling variability, we might find a group of 10 in which everyone is improved, another group where only 2 people improve, and so on. In short, r will no doubt vary from sample to sample, and the binomial probability distribution for r can also be taken as its sampling distribution. This particular distribution is graphed in Figure 9.4. The sampling distribution has a long-run average called the expected value of r, and it means the same thing as did the sampling distribution of $\bar{X}$, in that if we take an infinite number of samples of size N, the mean value of r will equal its expected value. For the binomial distribution, the expected value of r is very easy to compute:

Expected value of $r = E(r) = Np$

For the example at hand:

$$E(r) = (10)(0.50) = 5.0$$

Thus, if we take many, many samples of 10 patients from this population, (selecting people randomly, and in such a way that all samples are independent of one another), we will average 5 patients improved per sample.

You may have noticed, in comparing the sampling distribution of the binomial with the sampling distribution of the mean, that when we used

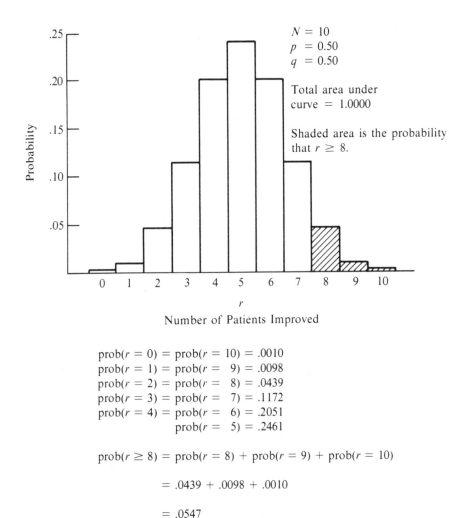

$$prob(r = 0) = prob(r = 10) = .0010$$
$$prob(r = 1) = prob(r = 9) = .0098$$
$$prob(r = 2) = prob(r = 8) = .0439$$
$$prob(r = 3) = prob(r = 7) = .1172$$
$$prob(r = 4) = prob(r = 6) = .2051$$
$$prob(r = 5) = .2461$$

$$prob(r \geq 8) = prob(r = 8) + prob(r = 9) + prob(r = 10)$$

$$= .0439 + .0098 + .0010$$

$$= .0547$$

Figure 9.4   Binomial distribution for $N = 10$, $p = 0.50$, showing the probability that 8 or more patients improve, out of a group of 10.

the binomial, we did not bother to determine the standard error of the sampling distribution. With the statistic $\bar{X}$ we had to do so, because we had to find a z-score for $\bar{X}$ in a distribution of sample means. With r, however, we did not use a table, but rather computed the probabilities directly for each value of r. (The values shown in Figure 9.4 were obtained using the same methods given in Chapter 8.) There are situations, however, where we will want to know the standard error of the binomial distribution.

Consider, for instance, the situation where N is very large. Such a situation is shown in Figure 9.5, where the graphed distribution is a binomial distribution with $N = 40$ and $p = 0.50$. Since the total area under the curve (i.e., the sum of all of the individual r-value probabilities)

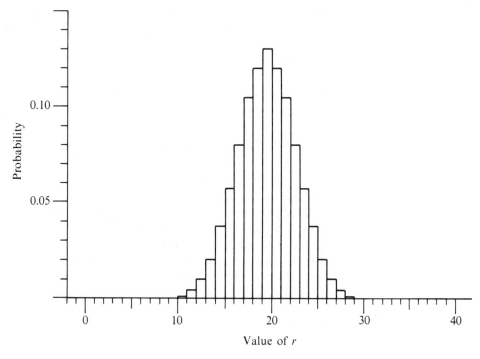

Figure 9.5   Binomial distribution with $N = 40$ and $p = 0.50$. When this distribution is taken as a sampling distribution for the binomial variable r, it approximates a normal distribution with a mean of Np and a standard error of $\sqrt{Npq}$.

must equal 1.0, each of the many individual r values has a rather small probability. This should suggest to you that, as N increases in value, the discrete binomial distribution becomes more and more like a continuous distribution.

Notice also that the binomial distribution also begins to look like a normal distribution. Indeed, when N is infinity, the binomial and normal distributions are identical. However, to save much laborious calculation, the normal distribution may be taken as a reasonably close approximation of the binomial when N is equal to 40 or larger. We may then treat the binomial variable as a normally distributed variable with the following parameter values:

Expected value of r: $E(r) = Np$

Standard error of r: $\sigma_r = \sqrt{Npq}$

This allows us to compute a z-score for any value of r:

$$z_r = \frac{r - E(r)}{\sigma_r} = \frac{r - Np}{\sqrt{Npq}}$$

Let us finish this chapter by showing how such a z value can be used to answer probability questions about r. Suppose again that we return to the population of psychotherapy patients visited earlier and select 40 at random. What is the probability that 25 or more will show improvement? We *could* calculate the probability that r is equal to or greater than 25 by individually calculating prob($r = 25$), prob($r = 26$), . . . , prob($r = 40$) and then adding all of these values. However, since $N = 40$, the binomial distribution can be treated as a normal distribution, and we can instead determine a z-score for the value $r = 25$:

$$z = \frac{r - Np}{\sqrt{Npq}} = \frac{25 - 20}{\sqrt{(40)(0.50)(0.50)}} = \frac{5}{3.1623} = 1.58$$

From Table A we find that the "Area Beyond z" for $z = 1.58$ is 0.0571. Thus, our calculations tell us that the probability of finding 25 or more improved patients from a group of 40 is 0.0571. Had we actually calculated the probabilities for each value of r, and used the discrete binomial rather than its normal approximation, we would have calculated a value of 0.0769 as an answer. You can see that this differs somewhat from the normal approximation; however, such a difference becomes even smaller as N increases beyond 40. It is the statistician's decision, of course, whether to use the discrete probabilities or the normal approximation in any case.

SUMMARY

There are many times when we want to study population characteristics but cannot actually make an observation on every member of the population. In these cases, some of the population members are selected for study; that is, a sample is drawn from the population.

When sampling is necessary, it is very important to ensure that the chosen sample accurately represents the population. Otherwise, the sample is biased, which means that conclusions drawn about the population from sample statistics may not be valid. Two procedures designed to ensure that samples will be representative are random sampling and stratified random sampling. Random sampling means, essentially, that every member of the population has an equal chance of being selected for the sample. Stratified random sampling may be necessary when there are identifiable subgroups (strata) in the population. With this procedure, the proportion of the sample from each subgroup (stratum) is made equal to the proportion of the population in each subgroup.

When sampling procedures are used, and statistics are computed on sample observations, three different distributions are involved: the population distribution, the sample distribution, and the sampling distribution of the statistic. The sample distribution and population distribution are, respectively, frequency distributions of the sample and population. Characteristics of the population are parameters; characteristics of the sample are statistics. Usually our interest lies in finding information about parameters; and usually we have to infer this information from sample statistics. The third type of distribution, the sampling distribution, is the inferential bridge between sample and population. The sampling distribution is the frequency distribution of all values a statistic will have for all possible samples of a given size from the population. Characteristics of sampling distributions are related to both the population characteristics and sample size. The virtue of the sampling distribution is that it allows us to calculate the probability that different values of sample statistics occur.

One highly useful sampling distribution is the sampling distribution of the mean. Its expected value (mean) and standard error (standard deviation) are easily determined if the population mean and standard deviation are known and the sample size is specified. Furthermore, the shape of the sampling distribution of the mean may be considered normal for large samples ($N = 30$ or more) and described by a t-distribution for smaller samples. For large samples, the sampling distribution of the mean will be normal even when the underlying population distribution is not.

The binomial distribution is also used as a sampling distribution when the statistic involved is r, the number of Bernoulli successes in N trials. The binomial distribution is a discrete sampling distribution, but it approximates a continuous normal distribution when N is large. It may be used to answer questions about the probability of r assuming different values.

KEY CONCEPTS

population	stratified random sampling	sample statistics
sample	population distribution	expected value
inferential statistics	population parameters	standard error
sampling procedure	sample distribution	central limit theorem
biased sample	sampling distribution	t-distribution
random sampling	sampling distribution of the mean	sampling distribution of the binomial variable

SPOTLIGHT 9   **Whom Does the Sample Represent?**

"Everyone knows that a representative sample of lemon pie, for instance, must include the meringue on the top, the lemon in the middle, and the crust on the bottom."

— George Gallup and Saul Rae[5]

As you read reports of experiments in the behavioral sciences you may notice that in some instances researchers selecting people for samples take the greatest pains to ensure that the sample is representative of the population. The Gallup Poll, for instance, is based on a highly complex stratified random sample, which takes into account "considerations of geography, occupation, age, sex, political affiliation, race, religion, and general cultural background."[6] On the other hand, a psychologist studying the ability of human beings to understand speech in the presence of distracting noise may use only female college students between the ages of 18 and 20 as subjects; other psychologists interested in general principles of learning may base their conclusions entirely on data obtained from laboratory-raised albino rats. When are representative sampling procedures needed, and when aren't they? To answer that question, one needs to think very carefully about (1) what is being studied, and (2) to what population will sample results be generalized. W. L. Hays, in his book *Statistics for Psychologists*, offers some guidelines on the matter in a section entitled "To What Populations Do Our Inferences Refer?"

Most psychologists who use inferential statistics in research rely on the model of simple random sampling. Yet how does one go about getting such a "truly" random sample? It is not easy to do, unless, as in all probability sampling, each and every potential member of the population may somehow be listed. Then, by means of a device such as random number tables, individuals may be assigned to the sample with approximately equal probabilities.

However, in behavioral sciences such as psychology, interest often lies in experimental effects that, presumably, should apply to a very large population of men or other living organisms. Such a listing procedure is simply not possible. Still other experiments may refer to all possible measurements that *might* be made of some phenomenon under various experimental conditions, where estimated true values may be sought from the experimental observation of a few instances. Here, the population is not only infinite, it is hypothetical, since it includes all *future* or *potential* observations of that phenomenon under the different conditions. In sampling from such experimental populations, where there is no possibility of listing the elements for random assignment to the sample, the only recourse of the experimenter is to draw his basic experimental units in some more or less random, "haphazard," way, and then *make sure that in his experiment only random factors determine which unit gets which experimental treatment*. In other words, there are two ways in which randomness is important in an experiment: the first is in the selection of the sample as a whole, and the second is in the allotment of individuals to experi-

---

[5]G. Gallup and S. Rae, *The Pulse of Democracy* (New York: Glenwood Press, 1968), p. 64.
[6]Ibid., p. 60.

mental treatments. Each kind of randomness is important for the "generalizability" of the experimental results, so that when one does an experiment he usually takes pains to see that both kinds of randomness are present. However, even given that individual cases are assigned to experimental manipulations at random, the possible inferences are still limited by the fundamental population from which the total sample is drawn.

How does one know the population to which the statistical inferences drawn from a sample apply? If random sampling is to be assumed, *the population is defined by the sample and the manner in which it is drawn. The only population to which the inferences strictly apply is that in which individuals have equal likelihood of appearing in the sample.* It should be obvious that simple random samples from one population may not be random samples of another population. For example, suppose that some one wishes to sample American college students. He obtains a directory of college students from a midwestern university and, using a random number table, takes a sample of these students. He is not, however, justified in calling this a random sample of the population of American college students, although he may be justified in calling this a random sample of students at *that* university. *The population is defined not by what he said, but rather by what he did to get the sample.* For any sample, one should always ask the question, "What is the set of potential cases that could have appeared in my sample with equal probability?" If there is some well-defined set of cases which fits this qualification, then inferences may be made to that population. However, if there is some population whose members could not have been represented in the sample with equal probability, then inferences do not *necessarily* apply to that population when methods based on simple random sampling are used. Any generalization beyond the population actually sampled at random must rest on extrastatistical, scientific, considerations.[7]

PROBLEMS

1. Describe a procedure that might be used to select a random sample of
   a. Three students from a class of twenty;
   b. 100 residents from a city's population;
   c. 1,000 people from among the nation's television viewers;
   d. 1,000 veterans;
   e. Thirty students on a college campus.

2. Out of a city's population, 100 people are to be selected for an interview designed to assess the citizens' attitudes toward various issues. What strata (if any) might it be important to identify before sampling if the study were concerned with finding out attitudes toward:
   a. A labor-union strike currently under way?
   b. Busing of students to achieve integration?
   c. A sales tax?
   d. Candidates for mayor?

3. Many advertisements present the results of surveys among consum-

---

[7] From *Statistics* by William L. Hays. Copyright © 1963 by Holt, Rinehart and Winston, Inc. Reprinted by permission of Holt, Rinehart and Winston and the author.

ers, or "tests" in which consumers compare one brand with another. Find an example in print of such an advertisement where the sampling procedure is clearly described and where an unbiased sample has apparently been used. Find another example in print that used such a group but where no information is given regarding sample selection.

4. In most inferential applications it is advisable to use as large a sample as is practical. From your knowledge of the relation between sample size and the sampling distribution of the mean, explain why.

5. A number of laboratory rats are being used in studies of nutritional value of several food products. The animals are selected from a population whose mean weight is 400 g and a standard deviation weight of 40 g. The population is normally distributed.

   a. One rat is selected at random from the population. What is the probability that it weighs more than 420 grams?

   b. A sample of 40 rats is selected randomly. What is the probability that the sample mean is greater than 420 grams?

   c. A sample of 36 rats is selected randomly. What is the probability that the sample mean is greater than 420 grams?

   d. For a sample size of 64, what is the probability that the sample mean is either greater than 410 g or less than 390 g?

6. Scores on a law school entrance examination are distributed normally with a mean of 500 and a standard deviation of 70.

   a. An applicant taking the test is selected at random. What is the probability that he or she will have a score of 350 or less?

   b. 100 test-taking applicants are selected at random. What is the probability that this group has a mean score between 490 and 520?

   c. 1,600 test-taking applicants are selected at random. What is the probability that this group has a mean of 505 or more?

7. A very large population of measurements is normally distributed with a mean of 20 meters and a standard deviation of 4 meters.

   a. Sixteen measurements are selected at random from the population for closer scrutiny. What is the probability that this sample has a mean greater than 21?

   b. One hundred measurements are then selected, randomly, from the population. What is the probability that this sample of 100 has a mean greater than 21?

   c. What is the probability that a sample of size 4, randomly selected from this population, has a mean greater than 21?

8. Find a newspaper summary of a nationwide opinion poll. How much information are you given in the newspaper item about the sampling procedures used? Did the pollsters attempt to obtain a truly random sample? A stratified random sample? Indicate whether the polling

organization expressed any limits that should be applied to the generalizations of the conclusions based on sample data. Do you think the results of the poll truly reflect the opinions of the indicated population?

9. Suppose that you have a biased coin: it lands with heads up with a probability of 0.60. What is the probability that, in flipping the coin six times, you obtain

   a. No heads?

   b. 6 heads?

   c. 3 or more heads?

10. Suppose that you flip the coin described in problem 9 a total of 100 times. Using the normal approximation to the binomial, determine the probability of obtaining

    a. 75 or more heads.

    b. 50 or more tails.

    c. 55 or fewer heads.

10. Introduction to Hypothesis Testing: Tests with a Single Mean

After reading this chapter you should be able to do the following:

1. Explain the general purpose and logic involved in testing hypotheses.

2. Conduct a hypothesis test to decide whether $\bar{X}$ differs by some specified value because of chance factors or for some other reason.

3. Interpret the results of the hypothesis test in terms of actual variables.

4. Describe the two kinds of error possible in a hypothesis test and ways to reduce the likelihood of each; explain how statistical power operates in the test and how to increase it.

The Basic Question

Hypothesis testing is a term applied to a whole family of inferential methods that help one draw conclusions about populations based on observations of samples. This does not begin to suggest the real utility of these techniques, however. If statements about populations are framed in the right way, hypothesis testing can be used to help answer questions about cause and effect: Does smoking *cause* lung cancer? The tests are also used to determine whether or not groups are really different or just apparently different: Do children trained in the "new math" really do better in high school than their traditionally taught contemporaries? The various hypothesis tests are all intended to help the researcher reach a **statistical decision** about a state of reality in the population. They never prove anything absolutely, but when used in conjunction with a probability statement, they allow one to estimate the likelihood that those decisions are in error. Indeed, they come as close to being proof as statistical methods can.

The kind of hypothesis testing presented in this book, developed by Jerzy Neyman and E. S. Pearson in the early part of this century, is the kind of hypothesis testing currently used in virtually all behavioral science research. It is, however, not the only approach used in answering questions about cause and effect. In the next twenty years researchers may possibly incorporate more practices from the so-called Bayesian approach[1] to decision making. For the present, we shall focus only on the Neyman-Pearson approach, simply because it is almost the only kind of decision-making system you will encounter in current experimental reports.

We will begin with a simple form of hypothesis testing that addresses a specific kind of question. Consider the following examples, and see if you can determine the basic *type* of question involved in each.

Sludge Oil Company has developed a gasoline additive that, Sludge contends, will increase gasoline mileage in any car. To demonstrate the product, a representative sample of 100 American cars is chosen and run exclusively on the new Sludge fuel. It is known that the current average miles/gal obtained by American cars is, say, 18, but the 100 cars in the test average 19 miles/gal. Has Sludge proved anything? Is the difference in fuel efficiency due to the fuel additive, or simply to sampling slightly better cars for the test?

As part of an educational experiment, twenty-six school children, randomly chosen from among the city's fourth-graders, spend an hour a day working with a teaching machine (a device that presents programmed questions and informs children if their answers are right or wrong). At the end of the school year all fourth-graders in the city take a comprehensive math exam. Scores obtained vary substantially from child to child, but the twenty-six who received programmed instruction average 81, while other children average 75. Does the test prove the utility of teaching machines? Is the small group's average higher because of the teaching machine, or did they just happen to be a superior group?

The principal of Adams H.S. argues that Adams students are academically superior. On a national college entrance exam, Adams students average 575, while the national average on the same exam is 500. Is the principal's assertion justified?

Each of these questions asks, essentially, whether or not a small group is really different from some population. Or, more accurately, each asks whether the difference between the mean of the sample $(\bar{X})$ and the

---

[1]In Bayesian statistics, one attempts to make decisions about the state of reality by combining information from many sources and producing something called a "subjective" or "personal" probability distribution for parameter values. Bayesian methods differ in many ways from the Neyman-Pearson approach we cover. The Neyman-Pearson approach is more formal in that all information contributing to the decision is derived from the data, whereas Bayesian methods incorporate information from other sources as well. Those who wish to explore some introductions to Bayesian statistics are referred to the following readings:

Novick, M. R. "Introduction to Bayesian Inference." Chapter 16 of Blommers, P. J., and Forsyth, R. A. *Elementary Statistical Methods in Psychology and Education*, 2nd ed. (Boston: Houghton Mifflin, 1977).

Hays, W. L. "Some Elementary Bayesian Methods." Chapter 19 of *Statistics for the Social Sciences*, 2nd ed., (New York: Holt, Rinehart and Winston, 1973).

population mean μ is due to chance factors or to a real superiority. (Notice that we assume a value for μ, the population mean, in each case. For the moment, assume that this value was determined from other statistical procedures, and that we are quite certain that the population means are accurate.) The methods in this chapter are designed to suggest an answer to these questions.

The Hypotheses

Consider the Sludge Oil mileage test again. One hundred cars using Sludge gasoline averaged 1 mile per gallon above the national average. There are many factors that might have caused the difference: perhaps drivers in the Sludge test were extremely careful not to waste gas with fast starts or speeds too high or low; or, possibly, cars selected for the test were simply the makes that obtain better fuel efficiency. We assume, however, that such factors were not operating and that Sludge made every effort to conduct a fair test. This would mean that an unbiased sample of cars was selected from the population (see Chapter 9) and that the miles-per-gallon figures were obtained under exactly the same conditions used to obtain the 18 miles/gal national average figure. A fair test, then, is one in which there are only two possible sources for the difference between population and sample: (1) the effects of the fuel additive, and (2) random variability. Various uncontrollable error factors are still going to produce differences in miles per gallon among cars, and these could conceivably be responsible for the difference between μ and $\overline{X}$.

Hypothesis testing provides a means of making a **statistical decision,** that is, deciding which of the two alternatives is responsible for the difference. The test begins by posing the question in a formal manner, so that it can be answered in terms of what we know about probability and sampling distributions. Two mutually exclusive and exhaustive states of reality are proposed; these **hypotheses** are stated so that one or the other must be true. For the gasoline test, they might take the following form:

$$H_0: \mu_s = k$$
$$H_1: \mu_s \neq k$$

where H_0 = the null hypothesis;
H_1 = the alternative hypothesis;
μ_s = the mean of the population from which the sample comes (a parameter);
k = some constant value.

We will consider each element in these expressions carefully. Let us begin with μ_s. This is the mean of the population from which our sample comes after it has been given the fuel additive. This population may be large or small, or it may not even exist (if the additive has not yet been placed on the market). It is the population that would exist, if all cars were run with the Sludge additive. Even though this population is more potential than it is reality, it is its mean (μ_s) that we are really interested in. We want to know if μ_s differs from 18 mpg. Because our sample cars all had Sludge fuel, they may be considered elements from this population.

Now consider k. This is a constant value that we are comparing with μ_s. Specifically, we want to decide whether or not μ_s differs from k. Here, we have assumed that the national mpg average is 18, and this is the value of k that we use in the test; the important point is that k may be arbitrarily chosen by the experimenter to provide an interesting comparison with the sample mean, or it may actually be computed from observations on some population. The test we are performing does not "care" where k's value comes from.

H_0 is the symbol for the **null hypothesis,** which presents one possible state of reality: the mean of the population from which the sample comes (μ_s) is equal to the specified value of k. If H_0 is not true, the **alternative hypothesis,** H_1, must be true: the mean of the population from which the sample comes is not equal to that value. Notice that the test states hypotheses in terms of a parameter, μ_s. For this reason, the test is known as a **parametric test.**

Notice also that the alternative hypothesis does not propose a value for μ_s. H_1 says only that μ_s is different from k. The states of reality specified by the hypotheses are shown in Figure 10.1. The solid curve illustrates the situation indicated by the null hypothesis, and the other dashed curves indicate some of the possibilities under H_1. The null hypothesis says that the observed difference between Sludge-fueled cars and the assumed population mean, 18 mpg, was due to sampling error, that is, random variability in the selection of sample cars. The alternative says, in effect, that the difference between the sample mean and k is real. The test will give us a means of choosing between these two possibilities.

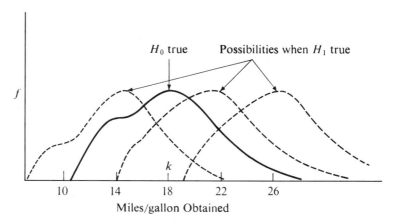

Figure 10.1 States of reality specified by H_0 and H_1. The null hypothesis says that the distribution of Sludge-fueled cars has the same mean as the constant k (solid curve). The alternative hypothesis says simply that the Sludge population has a different mean, and any one of the dashed curves might describe that population.

We must now survey the information known to us. If the population standard deviation σ is known, we can perform the hypothesis test using the z-test (pp. 276–280). If, however, we must perform the hypothesis test without knowing the exact value of σ, we must use the t-test (pp. 280–285). We will begin with the slightly simpler case, in which we know (from some source other than the sample data) the value of σ. Then, in the following section we will consider the more usual case, where the statistician has only his sample data to go on.

The Hypothesis Test (σ Known)

The null hypothesis specifies that sample cars come from a population with a mean of 18 miles/gal, while the alternative hypothesis allows an infinite number of possible values for the mean mpg. We must decide whether to assume that H_0 is true or false; we must choose between existence or nonexistence of a very specific state of affairs. If it is highly unlikely that the observed statistics appear when H_0 is true, we will *reject* H_0 and assume that the alternative H_1 is true. The results of the test will thus be a decision to accept or reject H_0.

To start, assume that we know the following items of information:

$$k = 18 \text{ miles/gal}$$
$$\sigma = \ 4 \text{ miles/gal}$$
$$\overline{X} = 19 \text{ miles/gal}$$
$$N = 100 \text{ cars}$$

Our decision will be based on an answer to the question, How likely is it that a sample of size $N = 100$ has a mean value of 19 or more when drawn from a population with a mean of 18 and standard deviation of 4? In conducting the test, we will not phrase the question in exactly this form, but it is presented in this way here to introduce you to the logic involved in the test. To study the probability associated with values of a sample mean, we will have to utilize the sampling distribution of the mean.

If the null hypothesis is true, then the means of all possible samples of size $N = 100$ from this population will take the form of Figure 10.2. From the Central Limit Theorem we know that the sample means produce an approximately normal distribution (since the sample size is larger than 30). As indicated in Chapter 9, this distribution will have a mean equal to the population mean 18 and a standard error of $\sigma_{\bar{x}}$ of $\sigma/\sqrt{N} = 4/\sqrt{100} = 0.40$.

From Figure 10.2 we can see that once in a while a sample will be chosen that has a mean of 19.5 miles/gal, and occasionally one will have a mean of 16.5 miles/gal, but these extreme deviations from the value of μ will be relatively rare. The procedure in the hypothesis test is to find values of the sample mean that are highly unlikely using this sampling distribution. We will use the value 18 miles/gal as our k value and decide whether μ_s equals k or some other value.

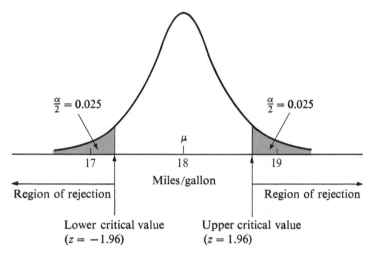

Figure 10.2   Sampling distribution of the mean specified by H_0. With $\mu_s = 18$ and $\sigma = 4$, the sampling distribution for $N = 100$ has a mean of 18 and a standard error of 0.40 (drawn slightly out of scale). Regions of rejection are shown for $\alpha = 0.05$.

The hypothesis test goes as follows: we decide that if we obtain a value of $\overline{X}$ so far from k that it would appear only with a low probability when H_0 is true, we will reject H_0; rejecting H_0 means that we assume H_1 to be true.

A crucial step in this test is deciding precisely what is meant by a "low probability." Here, we may select any probability between 0.0 and 1.0, but usually we will pick a low value such as 0.05 or 0.01. If we arbitrarily pick 0.05 as our value, then 0.05 is called the **level of significance**, or equivalently, the **alpha level** of the test. Having settled on a value of alpha, $\alpha = 0.05$, we can find the values of $\overline{X}$ that occur with a relative frequency of α or less when H_0 is true.

Remember that the probability of an observation falling within a given interval of the measurement scale can be specified by finding the area under the relative-frequency distribution that lies over that interval (or probability density function with continuous variables). Here, we will proceed in the opposite direction, starting with a known probability and looking for values. Suppose for the moment that H_0 is true; we will find the values of $\overline{X}$ that are so far above μ or so far below μ that they occur with a relative frequency of 0.05 or less. Since this sampling distribution is symmetrical, equal numbers of these will lie above the mean and below the mean. So we need the highest 0.025 values of $\overline{X}$ and the lowest 0.025 values, that is, the intervals on the horizontal axis that lie under the upper 0.025 area of the curve and the lower 0.025 area of the curve. The probability that an observation falls either in one end or the other is thus $0.025 + 0.025$, or 0.05. These regions are shaded on Figure 10.2; the intervals under them are called **regions of rejection**, because a value of $\overline{X}$ there leads to rejection of H_0. The process of conducting the test can be rephrased again as the business of determining whether or not our observed statistic falls within the regions of rejection.

The lowest border of the upper region of rejection is called the **upper critical value,** and the uppermost value of the lower region of rejection is the **lower critical value.** Since the sampling distribution is normal, we can use the unit normal table to find these values. In practice, we do not find the critical values themselves, but rather the z-scores associated with them, and each of these is called *z*-**critical.** We then find a z-score associated with our sample mean, called *z*-**observed** and compare it to *z*-critical in a **decision rule:** If z-observed is equal to or greater than the upper z-critical, reject H_0; or, if z-observed is equal to or less than the lower z-critical, reject H_0.

In hypothesis testing, the z-critical is obtained from the table and the z-observed is computed from our sample statistic and knowledge of the sampling distribution. When these values are inserted in the decision rule, the test is completed.

First we find the values of the two z-criticals. For $\alpha = 0.05$, we need the point on the horizontal axis of the unit normal distribution that has 0.025 of the area *beyond* it on the upper end. This is found in Table A, Appendix 1, by looking in column 3 for 0.025; Table A indicates that an area beyond z of 0.0250 is associated with a z value of 1.96. This is the upper z-critical. Similarly, to find the lower z-critical, we merely insert a minus sign in front of the z value and find that a z of -1.96 has an area of 0.025 beyond it on the lower end of the distribution. Since upper and lower z-criticals have the same magnitude, we can use an absolute value sign to shorten the decision rule:

If $|\text{ z-observed }| \geq 1.96$, reject H_0

In the sampling distribution of the mean specified by the null hypothesis, what z-score is associated with a sample mean of 19? Remember the general formula for z-scores:

$$z = \frac{\text{Value of interest} - \text{Mean of the distribution}}{\text{Standard deviation of the distribution}}$$

Here the value of interest is the sample mean, 19, and the mean of the distribution of means is k, or 18. The standard deviation of this distribution is $\sigma/\sqrt{N}$, not σ. Hence:

$$z\text{-observed} = \frac{\bar{X} - k}{\sigma_{\bar{x}}} = \frac{\bar{X} - k}{\sigma/\sqrt{N}} = \frac{19 - 18}{4/\sqrt{100}} = \frac{1}{0.4} = 2.50$$

Since $|\text{ z-observed }|$ is greater than 1.96, we reject H_0 and accept H_1. We have decided that the difference between $\bar{X}$ and k is due to something other than random variability because our obtained sample mean (or one more deviant) would appear with a probability of 0.05 or less if H_0 were true. Further interpretation of the decision will be taken up on pages 285–286.

Suppose instead that we wanted to be *very sure* that the difference between $\bar{X}$ and k was not due to random variability. The alpha level of 0.05 used in the preceding example may leave some room for doubt, because if the null hypothesis is true, one out of twenty (0.05) samples taken from the population would have values in the regions of rejection. We might test the hypothesis at a different level of significance, say, 0.01. We would reject H_0 only if it were even more unlikely that the obtained sample mean came from the sampling distribution specified by H_0. Changing the level of significance affects only the value of z-critical; z-observed remains 2.50 regardless of the chosen alpha. Figure 10.3 illustrates the situation with the more stringent conditions. z-criticals have become the points that cut off 0.005 (that is, $\alpha/2$) on each end of the distribution. From Table A we find that the area beyond z is 0.005 for a z of 2.576. So, with $\alpha = 0.01$, the decision rule becomes:

If $\mid z\text{-observed} \mid \geqslant 2.576$, reject H_0

With a z-observed of 2.50, we do not reject H_0. Had we rejected the null hypothesis at this alpha level, it would have been even more convincing evidence that Sludge's additive produced the difference between $\bar{X}$ and k. As it is, Sludge will have to be content with saying that the difference in miles per gallon was significant at the 0.05 level.

Applying the same principles in another example, consider the third situation described on page 273. It is known from experience that a certain college entrance exam produces a mean score of 500 and a standard deviation of 100 when administered to large groups of American high-school seniors. Thirty-six Adams H.S. students take the test and obtain a mean score of 575. Is the principal correct in assuming that the higher-than-average score of Adams students is due to a genuine superiority? Or, are the higher-than-average scores just due to selecting unusually bright students from the Adams H.S. student body? That is, is the mean of this sample *significantly different* from the larger mean (500)?

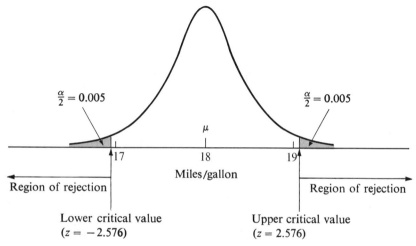

Figure 10.3 The same hypothesis test as illustrated by Figure 10.2, but with $\alpha = 0.01$. Changing the alpha level changes only the regions of rejection and values of z-critical.

Our k value in this case will be 500. Again, the same pair of hypotheses are used:

H_0: $\mu_s = k$

H_1: $\mu_s \neq k$

Either (1) the population from which the sample comes has a mean equal to 500, or (2) the sample belongs to a population with a different mean.

Consider very carefully the population whose mean is μ_s. This does not represent the population of all Adams H.S. students. It includes only students who attended Adams and who chose to take the test.

z-criticals are obtained from Table A in the same manner as before. For $\alpha = 0.05$ the decision rule is again:

If $|z\text{-observed}| \geq 1.96$, reject H_0

Here,

$$z\text{-observed} = \frac{\bar{X} - k}{\sigma_{\bar{x}}} = \frac{\bar{X} - k}{\sigma/N} = \frac{575 - 500}{100/\sqrt{36}}$$

$$= \frac{75}{100/6} = \frac{75}{16.7} = 4.50$$

The z-observed clearly exceeds z-critical, so the difference is significant at the 0.05 level. Most researchers would then check to see if it is significant at the 0.01 level.[2] As shown above, z-critical at the 0.01 level is 2.57, so the difference is significant at the 0.01 level. We might even check to see if it is significant for $\alpha = 0.001$. From Table A we find that a z of 3.30 cuts off $\alpha/2$ on each end of the distribution; this too is less than z-observed, so the difference is reported as being significant at the 0.001 level.

The conclusion to be drawn, then, is that the difference between $\bar{X}$ and k is not due simply to random variability.

The Hypothesis Test (σ Unknown)

In the preceding examples we were given values for k and σ. Usually, when hypothesis tests are applied to questions about actual means, we are not so fortunate. σ is usually unknown. In such cases, we must modify slightly the procedures shown in the preceding section and perform a t-test.

There are many reasons why one might know k but not σ. Information published by the Census Bureau, private polling organizations, and news sources is often furnished with means and medians, but not so often with

---

[2]The strict rules of Neyman-Pearson hypothesis testing require one to set an alpha level *before* conducting the experiment and to stick to it after the results are in. However, most researchers today feel that their results are not biased by checking in the manner described, to determine the lowest alpha value obtained.

population standard deviations. Often, too, a researcher wants to know if his or her sample mean differs from some known value, even though that value does not necessarily represent a population mean that has actually been computed. In these cases, we must use the sample standard deviation, s, to *estimate* $\sigma_{\bar{x}}$. The following *estimated* standard error of the mean can be used:

$$\text{estimated } \sigma_{\bar{x}} = \frac{s}{\sqrt{N-1}} = s_{\bar{x}}$$

Remember that sample statistics vary from sample to sample. In testing hypotheses about the mean when σ is unknown, the statistician has two sample statistics to work with ($\bar{X}$ and s) rather than just one. This produces a sampling distribution of the mean that differs in shape from the normal distribution. As if this weren't bad enough, a different-shaped sampling distribution of the mean is involved for every different sample size. Collectively, sampling distributions of the mean using the estimated $\sigma_{\bar{x}}$ are known as **student's t-distribution,** or simply as the **t-distribution.**[3] Fortunately, the t-distribution is used in the same way as the normal distribution in making the test.

Taking another example, from page 273, suppose that the city's fourth-grade children all take a standardized math exam and produce a mean score of 75. At the beginning of the school year, however, twenty-six students were chosen at random to take part in an experiment using a teaching machine. At the end of the year this sample produced a mean score of 81. The population standard deviation σ is unknown, but the sample standard deviation is 16 points. Did the group score significantly higher than the population? Or, is the difference between $\bar{X}$ and 75 due only to chance factors in selecting sample members? This test calls for the same pair of hypotheses:

H_0: $\mu_s = k$

H_1: $\mu_s \neq k$

The decision rule now uses values of the t-distribution instead of z values:

If $\mid t\text{-observed} \mid \geq t\text{-critical}$, reject H_0

The obtained mean is now seen as a value in the t-distribution rather than in the z-distribution. So, in order to complete this test we must set the example aside temporarily and look at some characteristics of the t-distribution.

Figure 10.4 shows some t-distributions; each is the shape of the sam-

[3]"Student" was the pseudonym used by W. S. Gossett, who published the first description of the t-distribution in 1908. It is interesting that before working out the mathematical properties of the distribution, Gossett obtained the distribution empirically using height and left middle finger measurements of 3,000 criminals. You may want to read the original paper on the t-distribution: Student, "The Probable Error of a Mean," *Biometrika*, Vol. 6, pp. 1–25, 1908.

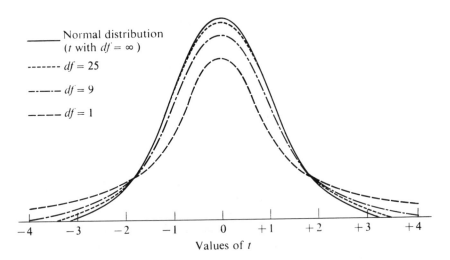

Values of t

Figure 10.4   Four different t-distributions. As the number of degrees of freedom increases, the curve approaches a normal distribution.

pling distribution of the mean for samples of a particular size. Notice that the values on the horizontal axis are t's rather than z's. In order to choose the correct t-distribution, we must find the **degrees of freedom** involved. This value, symbolized df, cannot be discussed in depth here.[4] It suffices to say that in the kind of sampling distribution we are using here, $df = N - 1$ and that it determines which t-distribution is appropriate. Here, $N = 26$, so $df = 25$. Thus, we will use the t-distribution with 25 degrees of freedom in setting the critical values for the test.

Notice that the "tails" are fatter when df is a small number. This means that extreme values of t are more likely to be obtained when samples are small.

Our sample mean (81) is a value in a t-distribution with 25 degrees of freedom, and this distribution is shown in Figure 10.5, together with a t-distribution for $df = 4$ for comparison purposes. We need the critical values along the horizontal axis that cut off $\alpha/2$ in the former distribution. With $\alpha = 0.05$, t-criticals cut off 0.025 of the area under the t-distribution at each end. Notice that this t-critical is different for each t-distribution.

Critical t-values can be found in Table B, Appendix 1. An abbreviated t-table is presented here as Table 10.1. Since there are many t-distributions, it would take up substantial page space to publish complete tables showing areas under all parts of the curve, as is done for the unit normal table. Thus, t-tables typically contain only values of t that cut off a few specific areas, corresponding to the most frequently used alpha

---

[4]The following example might help you understand why df for the one sample case are $N - 1$. Our statistic is $\bar{X}$ and we have determined its value for our sample to be 81. If we want to hold that value of 81 constant, in a sample of size 26, the first 25 values may be any numbers at all, but the 26th value is then determined; that is, to arrive at a given value for $\bar{X}$, the first $N - 1$ sample values are free to vary, but the last is then determined.

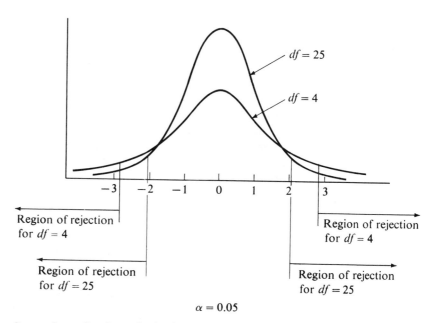

$$\alpha = 0.05$$

Figure 10.5 Comparison of regions of rejection for $df = 4$ and $df = 25$ under the t-distribution.

levels. For the moment pay attention only to the columns under the heading "Level of Significance for Two-Tailed Test." To find t-critical, proceed down the df column until you reach 25. The critical t, which cuts off 0.025 of the area, is listed under the alpha level 0.05. Here, t-critical = 2.060. Had we used a sample of only five students ($df = 4$), t-critical would have been 2.776.

Returning to the example, the decision rule is now:

If $|\, t\text{-observed}\,| \geq 2.060$, reject H_0.

Table 10.1 *Abbreviated Table of Critical Values of* t *for Two-Tailed Tests*

| | LEVEL OF SIGNIFICANCE FOR TWO-TAILED TEST | | | | | |
df	0.20	0.10	0.05	0.02	0.01	0.001
$\vdots$	$\vdots$	$\vdots$	$\vdots$	$\vdots$	$\vdots$	$\vdots$
4	1.533	2.132	2.776	3.747	4.604	8.610
5	1.476	2.015	2.571	3.365	4.032	6.859
6	1.440	1.943	2.447	3.143	3.707	5.959
$\vdots$	$\vdots$	$\vdots$	$\vdots$	$\vdots$	$\vdots$	$\vdots$
24	1.318	1.711	2.064	2.492	2.797	3.745
25	1.316	1.708	2.060	2.485	2.787	3.725
$\vdots$	$\vdots$	$\vdots$	$\vdots$	$\vdots$	$\vdots$	$\vdots$

The value for t-observed is found in the same way as was z-observed, except that the estimated $\sigma_{\bar{x}}$ is used in place of $\sigma/\sqrt{N}$. Thus:

$$t\text{-observed} = \frac{\bar{X} - k}{\text{est } \sigma_{\bar{x}}} = \frac{\bar{X} - k}{s/\sqrt{N-1}} = \frac{81 - 75}{16/\sqrt{25}} = \frac{6}{3.2} = 1.88$$

Therefore, we do not reject the null hypothesis.

You may be wondering why the estimated $\sigma_{\bar{x}}$ is computed as $s/\sqrt{N-1}$ and not as $s/\sqrt{N}$. The reason is that the sample standard deviation s we learned to compute in descriptive statistics tends on the average to underestimate σ. Thus, the estimated $\sigma_{\bar{x}}$ would tend to be too small unless we divide by $N - 1$ instead of N.

Most people who use t-tests prefer to compute an **unbiased sample standard deviation,** symbolized $\hat{s}$, instead of s; we, too, will have to compute $\hat{s}$ in the next chapter. The only computational difference between s and $\hat{s}$ occurs when the variation is divided to produce the mean square. Specifically:

$$s = \sqrt{\frac{\sum (X_i - \bar{X})^2}{N}} \qquad \hat{s} = \sqrt{\frac{\sum (X_i - \bar{X})^2}{N-1}} = \sqrt{\frac{\sum X_i^2 - \frac{(\sum X_i)^2}{N}}{N-1}}$$

Using $\hat{s}$ to estimate the standard error of the mean, we divide by the square root of N. Thus, the following relations hold:

$$\text{estimated } \sigma_{\bar{x}} = \frac{s}{\sqrt{N-1}} = \frac{\hat{s}}{\sqrt{N}}$$

One standard deviation may be computed from the other using the following relationships among variances:

$$\hat{s}^2 = \left(\frac{N}{N-1}\right)s^2 \qquad \text{and} \qquad s^2 = \hat{s}^2\left(\frac{N-1}{N}\right)$$

In the t-test illustrated above, $s = 16.000$. For the same sample, $\hat{s} = 16.317$. Thus, t-observed would be:

$$t\text{-observed} = \frac{\bar{X} - k}{\hat{s}/\sqrt{N}} = \frac{81 - 75}{16.371/\sqrt{26}} = \frac{6}{3.2} = 1.88$$

When computing a t-test from raw data, you will probably want to compute $\hat{s}$. However, standard deviations for groups are sometimes obtained from electronic calculators or prewritten computer programs, which provide you with s instead of $\hat{s}$. So be sure you know which sample standard deviation you have.

At this point it might be useful to review the steps in the single mean hypothesis test. The hypothesis tests on pages 278 and 281 are known, respectively, as the z-test and the t-test. The procedures in each are nearly the same; what differs is the shape of the test statistic sampling distribution, depending on whether the population standard deviation is known or estimated.

If you have looked at Table B, Appendix 1, you may have noticed that it lists every possible *df* value up to 30, and then only critical *t* values for *df* = 40, 60, 120, and infinity. The larger the value of *df*, the more the *t*-distribution approximates a normal distribution. In fact, when *df* = 30 or more, the *t* and normal distributions are so similar that many people obtain *t*-critical from the unit normal table as if they were looking for *z*-critical. That does not change the fact that when σ is unknown, the test is still called a *t*-test. A better procedure, however, is always to use the *t*-table when σ is unknown, even when *df* are high. If the exact *df* are not listed, you may interpolate between given values or simply use *t*-critical for the next lower *df* listed. For instance, if *df* = 65, it would be appropriate to use *t*-critical for *df* = 60.

One caution must be mentioned. Remember that the Central Limit Theorem was invoked on page 261 to assume that the sampling distribution of the mean was normal in shape. The theorem specifies that the sampling distribution of the mean is approximately normal for samples of size 30 or more regardless of the shape of the original population. Nonnormal population distributions were purposely drawn in Figure 10.1 to emphasize this fact. However, the central limit theorem does not apply to small samples (usually defined as $N = 30$ or less). The *t*-distribution is appropriate for small-sample use only when the population distribution is normal.

Interpreting the Decision

What does it mean when a null hypothesis is rejected? Or not rejected? What do the different alpha levels mean with regard to the original question? It is well to keep all these considerations in mind when presented with statistical "proof" derived from hypothesis tests.

We found, on pages 279–280 that the difference between the mean test score for Adams H.S. students and the national mean was significant at the 0.001 level. This finding could be reported as follows: "The sample mean differed significantly from the national mean ($z = 4.50$, $p \leq 0.001$)." This report includes the *observed* *z* value and the significance level so that readers can judge for themselves the meaning of "significant." Readers should know that the expression $p \leq 0.001$ means that a *z*-critical was exceeded, which would be exceeded by *z*-observed in only one out of a thousand such situations where H_0 is true. That is, if the results observed here were due to chance error, the observed difference between $\overline{X}$ and k would appear only once out of every one thousand times a sample is taken. When a *t*-test is reported, the degrees of freedom are also included in the parenthetical information: $t = 2.41$, $df = 16$, $p \leq 0.05$.

Finding a significant difference means only that it is likely that something else besides chance sampling factors produced the difference. It does not necessarily support the conclusions that Sludge gas is actually better, or that Adams H.S. students are truly superior. These conclusions are warranted only if there is no other way to account for the results.

When a hypothesis is rejected, the lower the numerical value of alpha, the more convincing the evidence that the difference is real. If the difference is significant at the 0.05 level, the probability that it was produced by chance sampling factors is 0.05 or less; if at the 0.001 level, the probability that it was produced by chance sampling factors is 0.001 or less. By tradition, alpha values above 0.05 are rarely used.

Interpreting the results when H_0 is not rejected is more difficult. In the teaching-machine example, for instance, failure to reject H_0 does not prove the machines had no effect. It simply means that the difference between $\bar{X}$ and k could have been caused either by chance sampling or by a real effect of the machine, but if the machine effect was there, the test did not find evidence strong enough to reveal it.

A statistical decision is similar in some ways to a jury's verdict. A decision of "guilty" means that jurors found the evidence compelling enough to conclude beyond a reasonable doubt that the defendant was *not* innocent. A guilty verdict is thus comparable to a null hypothesis rejected at a very low alpha level. Something must be established by the evidence in either case; guilt or significance must be proved, at least to the satisfaction of reasonable minds. A verdict of "not guilty," on the other hand, does not depend on proof of innocence. As with the nonrejected null hypothesis, the not guilty decision usually means, simply, that guilt was not demonstrated in the trial.

Another important point to be made here is that a significant difference is not necessarily an important difference. Suppose that we had another situation where a class of grade-school students is given extra practice with a teaching machine designed to help them learn arithmetic skills. Suppose also that the class produces a mean score of 82 on an arithmetic achievement exam, and that the national mean score is 80 on this test. Finally, suppose that the difference between national mean and class mean is significant at the 0.001 level. It is safe to conclude that the difference is *real*, and not due to sampling error. But should the school system buy teaching machines? Are they worth the expense and trouble involved in obtaining them? Perhaps not, if the real difference in scores is tiny. Thus, when reading results of experiments, it is important to remember that a highly significant result may still be relatively unimportant.

One-Tailed Versus Two-Tailed Tests

Perhaps you noticed an apparent incongruity between the original questions in the preceding examples and the hypotheses used to answer them. The alternative hypothesis in each case simply said that μ_s and k were different. However, in each case, the researchers were interested in differences in a specific *direction*. That is, Sludge wanted to demonstrate that miles-per-gallon averages were *higher* with the fuel additive; the teaching-machine test was really intended to demonstrate *improvement* in math ability due to the machines. But the tests used would have rejected the null hypothesis had there been a substantial *decrease* in fuel efficiency or math scores as well.

The preceding hypothesis tests are known as **nondirectional tests,** or more descriptively, as **two-tailed tests.** Occasionally it is appropriate to use a different kind of alternative hypothesis that puts the region of rejection all at one end of the sampling distribution. This is known as a **directional,** or **one-tailed test.**

As an example, suppose we develop the teaching-machine illustration a little further. It was not indicated on pages 281–284 whether the twenty-six children in the sample used the machine in addition to their regular math instruction or in place of it. Let's assume here that it was used in addition to regular instruction and so would be unlikely to lower any child's math exam score. Furthermore, there is only one condition under which we would consider buying the machines for the school system: if the machines demonstrated a significant *increase* in arithmetic ability (not simply a *difference* in ability). Thus, the possibility of a decrease in children's scores because of machine use is unlikely and uninteresting and a one-tailed test would be appropriate. A two-tailed test is always appropriate; the one-tailed test is appropriate only under special conditions, such as those just described. For this example, the hypotheses would be:

H_0: $\mu_s \leq k$

H_1: $\mu_s > k$

Here we don't consider the possibility that μ_s is less than k. If $\bar{X}$ turned out to be less than k, we would still have to accept the null hypothesis, no matter how much lower the value of $\bar{X}$.

The advantage of using a one-tailed test can be seen from Figure 10.6. In the one-tailed test, alpha represents an area under the sampling distribution all at one end of the curve. For $\alpha = 0.05$, t-critical is exceeded by 0.05 of the sample means *above* k (when H_0 is true). Thus, t-critical is closer to k, and it will be easier to reject H_0 under the directional test. The appropriate t-critical is found in Table B, Appendix 1, using the columns under the heading Level of Significance for One-Tailed Test. With $df = 25$, t-critical is 1.708. The one-tailed decision rule is simply:

If t-observed > 1.708, reject H_0

On page 284, the t-observed for this sample mean was found to be 1.88 — a value that leads to rejection of H_0 in the one-tailed test but not in the two-tailed test.

One-tailed tests in the opposite direction are also possible. If one is willing to reject H_0 only when $\bar{X}$ is *less* than μ_0, the appropriate pair of hypotheses is:

H_0: $\mu_s \geq k$

H_1: $\mu_s < k$

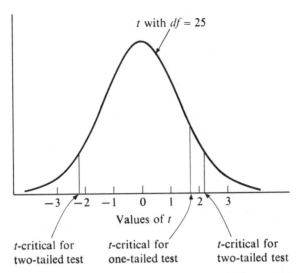

Figure 10.6   Comparison of *t*-critical in one- and two-tailed tests. In the one-tailed test alpha represents an area all at one end of the distribution. This "pushes" *t*-critical closer to zero. In the example in the test, *t*-observed fell between *t*-critical for the one-tailed test (1.708) and *t*-critical for the two-tailed test (2.060). (The figure is drawn out of scale for clarity.)

The hypotheses, regions of rejection, and decision rules for one-tailed and two-tailed tests are summarized in Table 10.2.

Making Errors

Regardless of the decision reached, there is always the possibility that an error has been made because sampling could produce, by chance, samples that argue either for or against the null hypothesis regardless of the state of reality. Table 10.3 (p. 290) summarizes the four possibilities we must consider with respect to error in every hypothesis test. First consider the situation in which the null hypothesis describes the true state of affairs. Suppose that the Sludge fuel additive did absolutely nothing to change fuel efficiency. If we fail to reject the null hypothesis in this case, no error is made and the decision is correct. However, suppose that for unknown reasons, we had chosen a sample with high $\bar{X}$, even though H_0 is true — so much higher, in fact, that we reject H_0. This is an erroneous decision known as a **Type I error:** rejecting a true null hypothesis. The probability of making such an error is directly under the control of the statistician, who determines which alpha level he or she will use to set the critical values. Since alpha equals the relative frequency of sample means that turn up in the region of rejection under a true H_0, it is also the probability of a Type I error. Thus, the lower the value of alpha, the less chance there is of rejecting a true null hypothesis.

Why then don't we simply refuse to accept a statistical decision unless

Table 10.2 *Hypotheses, Regions of Rejection, and Decision Rules for One-Tailed and Two-Tailed Tests*

	ONE-TAILED TESTS		TWO-TAILED TEST

Regions of rejection

	ONE-TAILED TESTS		TWO-TAILED TEST
Hypotheses			
	H_0: $\mu_s \geq k$	H_0: $\mu_s \leq k$	H_0: $\mu_s = k$
	H_1: $\mu_s < k$	H_1: $\mu_s > k$	H_1: $\mu_s \neq k$
Decision rules (*t*-test)			
	If *t*-observed is less than *t*-critical, reject H_0.	If *t*-observed is greater than *t*-critical, reject H_0.	If $\mid$ *t*-observed $\mid$ is greater than *t*-critical, reject H_0.

significance is obtained at a very low alpha value, say, 0.001? The reason is that decreasing the probability of a Type I error *increases* the probability of another kind of error.

Consider now the case when the alternative hypothesis is true, illustrated in Figure 10.7. Here it appears that Sludge's additive really does increase fuel efficiency; samples of cars from the Sludge population produce a sampling distribution of the mean shown with a solid line. The dashed line indicates the sampling distribution of the mean specified by the null hypothesis. A vertical line extends upward from the critical *z* value; values of *z*-observed to the right of *z*-critical lead to rejection of H_0, and those to the left lead to acceptance of H_0. Now, the lower the alpha value, the further to the right this line moves. This reduces the probability of a Type I error when H_0 is true, but that is not the case here. Moving the line to the right also *increases* the area under the H_1 curve that lies to the left of *z*-critical, labeled β, or beta. Beta represents the probability of not rejecting H_0 when H_0 is false, that is, the probability of making a **Type II error.** Unlike the value of alpha, the value of beta is rarely known. Rejecting H_0 does not tell us what μ_s is, so we do not know how much of the H_1 sampling distribution lies to the left of *z*-critical. Nonetheless, we *do* know that whatever the value of beta, it is increased by lowering alpha.

So it looks like we are caught in a double bind. Decreasing the probability of a Type I error increases the probability of a Type II error, and vice versa. It is at this point that the statistical producer or consumer must evaluate the relative cost or consequences of each type of error. In the

Table 10.3 *Level of Significance and Power (from a chart prepared by Dr. Leona S. Aiken)*

| | | YOUR STATISTICAL DECISION | |
		REJECT H_0	DON'T REJECT H_0
True state of affairs	H_0 true	Error $p(\text{Error}) = \alpha$ TYPE I ERROR	No error $p(\text{No error}) = 1 - \alpha$
	H_0 false	No error $p(\text{No error}) = 1 - \beta$	Error $p(\text{Error}) = \beta$ TYPE II ERROR

α = Probability of Type I error, or level of significance; the probability of rejecting the null hypothesis when in fact it is true.

$1 - \alpha$ = Probability of not rejecting a true null hypothesis.

β = Probability of a Type II error; the probability of not rejecting the null hypothesis when in fact it is false. β can be known only for a *specific H_1*, i.e., an alternative hypothesis is specified with a specific mean; since this is rarely done, the value of β is usually unknown.

$1 - \beta$ = Power. Probability of rejecting the null hypothesis when in fact it is false.

This is the only distribution that is known

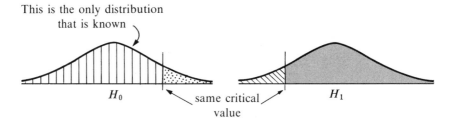

H_0 same critical value H_1

Only one of these distributions represents the true state of reality

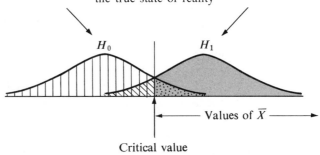

H_0 H_1

Values of $\overline{X}$

Critical value

α / $1.0 - \alpha$ } on H_0 distribution β / $1.0 - \beta$ } on H_1 distribution

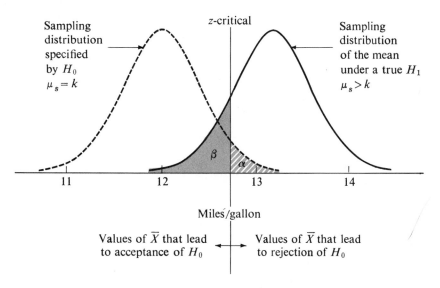

Figure 10.7   Situation that exists when H_1 is true. Some sample means from the H_1 distribution will fall in the region of acceptance for H_0. When this occurs, a Type II error has been made. The probability of a Type II error (β) is increased when alpha is lowered.

gasoline test a Type II error would mean simply that the real effect of the additive would be overlooked. This might not be as damaging to the company's interests as a Type I error — proclaiming an improvement in fuel efficiency when in fact there is none. On the other hand, consider a statistical decision of this type that might be made by a doctor diagnosing a patient. According to the results of a blood test, say, the doctor might decide that a patient differs enough from normal to be considered ill; if in fact the patient is not ill, the physician has committed a Type I error. This might result in some unnecessary treatment or further tests, which might be unimportant compared to the consequences of a Type II error — declaring a patient to be healthy when in fact he is not. So, Sludge might use a low value of alpha in order to minimize the chances of a Type I error; the doctor might use a higher alpha value to minimize the chances of a Type II error.[5] Happily, there is a way to reduce the probabilities of both types of error.

Figure 10.8(a) shows two sampling distributions, one specified by H_0 and one specified by H_1. It also shows alpha and beta regions for a hypothesis test at the 0.05 level. Notice that the remaining area under the H_1 curve is labeled $1 - \beta$. This is the probability of rejecting H_0 when H_1 is true, that is, the probability that the sample mean falls in the region of rejection when H_1 indicates the true state of affairs. $1 - \beta$ is known as the

---

[5]Applying the language of statistical decision theory to the jury verdict situation described earlier, we can say that an incorrect guilty verdict represents a Type I error and an incorrect not guilty verdict represents a Type II error.

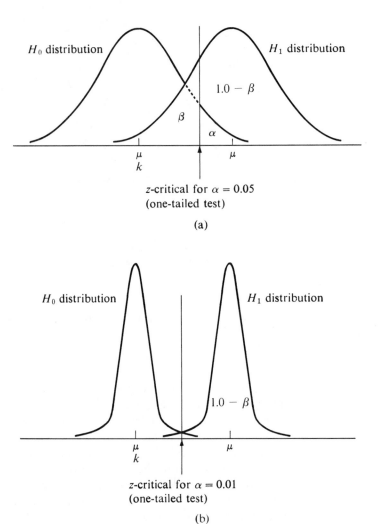

Figure 10.8   The effect of sample size on error. In part (a) samples of size $N = 36$ were used. Both sampling distributions overlap substantially, producing a relatively large probability of a Type II error (β) and low power ($1 - \beta$). In part (b) samples of size $N = 200$ produce less overlap between distributions. Thus, alpha can be lowered from 0.05 to 0.01, and the probability of a Type II error is still reduced.

power of the test. Compare these distributions with the pair shown in Figure 10.8(b). The measurement scale is the same, but the sampling distributions are not so spread out. Even though the alpha level in part (b) is 0.01, the value of beta is much lower because there is less overlap between distributions. Proportionally more of the H_1 distribution falls in the $1 - \beta$ region, so part (b) shows the more powerful test.

How can we change the situation found in part (a) into the part (b) case? The answer is surprisingly simple: use more subjects or more observations. Remember that the spread of the sampling distribution is specified by $\sigma_{\bar{x}}$, which is equal to $\sigma/\sqrt{N}$ and is estimated as $s/\sqrt{N-1}$ or $\hat{s}/\sqrt{N}$. The larger the value of N in any case, the smaller the value of $\sigma_{\bar{x}}$ (or $s_{\bar{x}}$, the estimated $\sigma_{\bar{x}}$) and the more statistical power in the test.

Statistical power is particularly important where μ_s and k are truly different but very close together. In this case, there will be substantial overlap between the two sampling distributions, and it will be very difficult to reject a false H_0, since sampling variability will overshadow the small but real difference between means. This is why researchers who attempt to investigate extrasensory perception phenomena (ESP) use thousands and sometimes hundreds of thousands of trials in their experiments. The presumed effects of psychokinesis or telepathy, for instance, are very weak. The mean number of cards identified by blindfolded subjects with ESP abilities might not be very different from the mean number of correct guesses for people without ESP abilities. Therefore, in trying to demonstrate that the difference between k and μ_s observed in a card-guessing experiment is not due to luck, these researchers run many, many thousands of trials before obtaining enough statistical power to reject H_0. And, even though the results may be highly significant, say at an alpha of 0.00001, the presumed effects are so tiny as to be uninteresting to most other scientists. At any rate, many scientists feel that the obtained high significance in such experiments is probably a better demonstration of statistical power than of ESP. Of course, now that you know what statistical power and significance are, you can form your own opinion.

It should be emphasized that the values of β and $1 - \beta$ are almost never known. The null hypothesis distribution is the only distribution whose location on the horizontal axis is known (that is specified in the null hypothesis). The alternative hypothesis says only that the H_1 distribution is located somewhere else, i.e., that its mean is different from the H_0 mean. In other words, the alternative hypothesis is *inexact*. In order to know values of β and $1 - \beta$, we would have to propose a mean for the H_1 distribution, thereby testing the H_0 against a *specific exact alternative*. This is rarely done. However, even though β's value is usually unknown, we still know that it can be increased or decreased by changing either alpha level or sample size.

SUMMARY

Hypothesis tests form a highly useful family of inferential methods. When questions are posed in the right way, hypothesis tests can be used to examine cause and effect relations; they can be used to determine the likelihood that groups are really different or just appear to be different through accidents of sampling.

Essentially, hypothesis tests are methods for choosing between two possible states of reality in the population. The methods in this chapter are designed to determine whether or not a sample comes from a known population. The test begins with a proposal of two hypotheses: (1) a null hypothesis, which says that the sample belongs to the population with a known mean, and (2) an alternative hypothesis, which says that the sample comes from some other population. The statistical decision in favor of one or the other is based on a consideration of the sampling distribution of the mean specified by the null hypothesis. If the probability is very low that the obtained sample mean (or one more extreme) would appear when the null hypothesis is true, the null is rejected and the alternative accepted.

Rejecting the null hypothesis does not prove absolutely that the sample comes from some population other than the known population; it only means that it is highly unlikely. Regardless of the statistical decision reached, there is a chance that it is in error. We have discussed methods for reducing the likelihood of each type of error as well as other factors to be considered in interpreting results of hypothesis tests.

KEY CONCEPTS

hypothesis testing	region of rejection	nondirectional test (two-tailed test)
statistical decision	critical value	
null hypothesis	decision rule	directional test (one-tailed test)
alternative hypothesis	z-test	Type I error
parametric test	t-test	Type II error
alpha level (level of significance)	student's t-distribution	statistical power
	degrees of freedom	unbiased sample standard deviation

SPOTLIGHT 10   Where Are All of Those Tests?

Most of the text in this chapter implies that hypothesis testing is a widely used approach to actual research problems. And yet, this chapter may represent the first contact with formal statistical decision making for many readers. If statistical decisions in government, industry, and science have indeed had such a profound impact on our lives, why haven't we heard of hypothesis testing and statistical decision making before undertaking a systematic study of statistics?

Dr. Irwin D. J. Bross answered this question in the introductory chapter of his delightful book, *Design for Decision*. He briefly discussed the revolution in many fields of science that occurred with the application of sophisticated statistical methods in the period from 1920 to 1940. Then, turning to the matter of public awareness of these methods, he commented:

I cannot blame you if, at this point, you scratch your head and murmur, "All this looks suspiciously like the old ballyhoo. If Statistical Decision is such a world-shaking affair why haven't I felt some of the tremors?'' You may not have heard of the statistical ''revolution'' that I mentioned earlier, and, to digress a bit, let me explain *why* you may not have heard of these matters. The main reason is that publications on the subject are written only for fellow specialists (and even these worthies have trouble understanding them). It may take twenty years before these ideas reach other *scientists* in a comprehensible form and even longer before they are taught to students. Specific techniques (in cookbook form) may be transmitted more rapidly, but the *ideas* diffuse very slowly.

A few scientists, it is true, have tried to write for the public. But while the public has eagerly accepted the television sets, wonder drugs, and bigger strawberries that scientific research has produced, they have been profoundly uninterested in the fundamental ideas, the *Scientific Method*, that have made this research fruitful. People must have the very *latest* electronic gadget, but they cling tenaciously to ideas and methods of thinking that were obsolete three hundred years ago.

This delay in the transmission of *ideas* is, I believe, one of the factors which has led our civilization to its present crisis. Moreover, the already dangerous situation is steadily getting worse because it is increasingly difficult to translate the language of science — a symbolic one — into everyday English.[6]

Take another look at the way scientific results are reported in the magazines and newspapers. Occasionally you may find reference to probability, but rarely, if ever, will you see a reference to a hypothesis test or to a *statistical* decision. Rejected null hypotheses in the scientific journals have a way of turning into ''proof'' in the papers.

PROBLEMS

1. Compute both s and $\hat{s}$ for the following groups of data.
   a. 4, 2, 6, 3
   b. 85, 87, 48, 57, 71
   c. 575, 535, 565, 602, 550, 490
   d. 19, 20, 21, 18, 17, 17, 16, 18, 25, 21

2. Interpret verbally the meaning of the hypotheses

   H_0: $\mu_s = k$

   H_1: $\mu_s \neq k$

   in each of the following studies.

   a. A medical researcher conducts a study to determine if returning astronauts have abnormal bloodcounts.
   b. Do children who grow up with certain vitamin deficiencies score differently on IQ tests from children who weren't deprived?
   c. Do teenage drivers have higher accident rates than drivers in general?

---

[6]From I. D. J. Bross, *Design for Decision* (New York: Macmillan, 1953), pp. 3–4. Reprinted by permission from Macmillan Publishing Co. Copyright 1953.

3. Compute the standard error of the mean for the following situations:
   a. $s = 24.3, N = 10$
   b. $\hat{s} = 14.4, N = 12$
   c. $s = 251.75, N = 8$
   d. $SS = 125, N = 12$
   e. $SS = 12.36, N = 12$
   f. $\hat{s} = 247, N = 38$

4. Does just one cocktail alter human reaction times? To answer this question, a number of college sophomores were tested in a driving simulator machine. From past experience, it was known that the sophomore population had a mean reaction time of 310 ms (ms = milliseconds; a millisecond is one one-thousandth of a second) and that these times were distributed with a standard deviation of 30 ms. A sample of 100 students were tested in the machine, each after having a cocktail containing 50 ml of 80-proof liquor. This group of 100 produced a mean reaction time of 320 ms. Perform the appropriate two-tailed z-test to determine if alcohol significantly affected reaction time; interpret the results of your test.

5. For the situation described in problem 4 above, what would a Type I error be? A Type II error? Discuss as many ways as you can to reduce the standard error of the mean in this study to produce more statistical power.

6. For the situation described in problem 4, suppose instead that the population standard deviation had been unknown, and that a smaller sample was used. A group of 25 students produced a mean reaction time of 320 ms. Their standard deviation ($\hat{s}$) was 35 ms. Again, conduct the appropriate two-tailed hypothesis test. Interpret the results of the test.

7. Suppose that the population of adult males has a mean height of 1.70 m, but that the standard deviation of that population is unknown. An archaeologist unearths 100 male skeletons from a primitive culture and determines the mean height to be 1.67 m, with $\hat{s} = 0.12$. Can he conclude that this group's mean height differed from that of the known population? (Use a two-tailed test.)

8. A sample of size $N = 16$ has a mean of 108 and a variation (SS) of 1,257. Use the two-tailed t-test to decide whether or not this sample comes from a population with a mean of 110.

9. Referring to the situation described in problem 8, suppose instead that a sample of size $N = 950$ produced a mean of 108 and a variation of 1,257. Again using the two-tailed t-test, decide whether or not this sample comes from a population with a mean of 110. If your results for this problem differ from those in problem 8, discuss the reason for this difference in terms of statistical power.

10. Eight hundred students in a large chemistry class took an exam, but one section of 25 students was accidentally given an extra ten minutes to finish the exam. Did the smaller group score significantly higher? It is reasonable to suggest that the extra time could only raise a student's score, and thus a one-tailed test would be appropriate. The large group produced a normal distribution of scores with a mean of 74. The small group received the following scores:

76, 72, 78, 80, 73, 70, 81, 75, 79, 76, 77, 79, 81,

74, 62, 95, 81, 69, 84, 76, 75, 77, 74, 72, 75

Conduct the appropriate hypothesis test.

11. In each decision-making situation below, describe the nature of a Type I error and a Type II error. Which type of error in each case would be more costly?

a. Most of the blips on a national defense radar screen represent migrating birds. A slightly unusual blip appears, and the radar operator must decide whether it represents another bird or an enemy aircraft.

b. A dollar-bill changer must decide whether a piece of paper inserted in its tray is real or counterfeit.

c. A medical researcher must decide whether or not a new experimental drug is superior to existing forms of treatment.

d. A coal mine safety device detects an abnormally high level of poisonous gas in the mine's atmosphere.

e. A study of violence in television shows suggests that children imitate aggressive models.

12. Did the noon-time temperatures for a given week have a mean value significantly above 40°F? The temperatures were:

45° 49° 52° 36° 42° 53° 39°

Perform the two-tailed t-test to make your decision.

13. After a year of carefully monitoring my car's fuel efficiency, I have determined that when I drive alone on the highway, it averages 24.5 miles per gallon. Is this efficiency reduced by carrying a 120-lb passenger? To find out, I determine the fuel consumption rate again on each of 14 days while carrying the passenger. These readings were:

23.2 25.8 21.4 22.6 24.0 24.9 21.5

20.6 19.9 26.3 23.5 24.6 22.2 21.7

a. Which is appropriate, the one-tailed test or the two-tailed test? Explain why.

b. Perform the test and decide whether or not mean mileage is reduced below 24.5 when carrying a passenger.

14. A nationally administered reading test for children was given across the country last year, and the mean score obtained was 730. This

year, a class at Willow Bend Elementary school took the exam and produced a mean score of 700, with $\hat{s} = 60$.

a. Suppose the class at Willow Bend consisted of 9 students. What decision can you make on the basis of the two-tailed t-test regarding the difference between the class's mean score and last year's national average?

b. Suppose instead that the same statistics were obtained by a class of 100 students. What can you now conclude about the difference between the class's mean and the national average?

15. Is the mean of the following group significantly different from a value of 50? Or is the difference between this mean and 50 due simply to sampling variability? Perform the two-tailed t-test for the following scores to decide:

34 67 43 57 42 67 71 36 52 48 49 36 67 45 38
    72 45 42 57 43 36 51 42 68 44 49 38 59

11. Hypothesis Testing: The Difference Between Two Means

After reading this chapter you should be able to do the following:

1. *Explain the steps in the following kinds of two-sample experiments: independent samples, matched pairs, repeated measures.*

2. *Read an account of an experiment and describe its research hypothesis, the independent variable, and the dependent variable.*

3. *Conduct a test for the significance of the difference between two means in the independent-samples t-test.*

4. *Conduct a test for the significance of the difference between two means in the related-samples t-test, using the direct-differences method.*

5. *Explain the advantages of using related-samples designs (when possible) over independent-samples designs.*

6. *Explain the conditions under which a significant difference between means constitutes proof that variable A has an effect on variable B.*

The Basic Question

If you understood the process of statistical decision making presented in Chapter 10 — the roles of the sampling distribution, probability, the decision rule — then you are almost ready to interpret the results of any hypothesis test you come across. If you can reconstruct the hypotheses tested and relate them to the issue under study, you can understand the implications of a rejected or accepted null hypothesis. A rejected H_0 always means that the probability of getting what you got, that is, of obtaining the obtained statistic (or one more deviant), is less than some specified probability if the null hypothesis is true. Of course, the many hypothesis tests used in statistics are sometimes valid only if some prerequisite conditions are met, and you may not always be capable of determining whether or not these have been satisfied. So, if the test is unfamiliar to you, you may have to have faith in the statistician's integrity and competence. Nonetheless, you can still evaluate his arguments based on the results of those tests.

This chapter discusses a test that is very similar to the one presented in Chapter 10; again we will use the z- and t-distributions and the same decision rules. While this test is computationally a little more complex, it is more generally applicable, particularly in experiments and other research applications. Whereas in Chapter 10 we were concerned with a single sample and a known population, we will be concerned here with two-sample experiments.

Given two sample means, $\bar{X}_1$ and $\bar{X}_2$, we will want to decide whether or not the samples come from populations with equal means. It turns out that the two-sample test can help answer important questions about cause and effect relations between variables. In the one-sample test we could in some cases decide that a sample differed from a population as a result of factors not due to sampling error (that is, reject the null hypothesis), but the implications we could draw from that decision were severely limited. In order to suggest that one single factor caused the difference, we had to assume that the sample was essentially identical to the population in every way except one. Under most actual conditions, this assumption is difficult to substantiate. In the two-sample test, we can place the cause and effect argument on much stronger ground — if the test is applied to data from a properly designed and conducted two-sample experiment.

The Two-Sample Experiment

The chapter on correlation emphasized the point that just because two variables are correlated, you cannot assume that changes in one cause changes in the other. Often, however, that is exactly what a researcher wants to know. Does changing the dosage rate of an antibiotic drug (variable 1) change the amount of time required for an infection to be cured (variable 2)? Does the number of hours an experimental rat has been deprived (variable 1) have an effect on the rate at which he learns to find his way through a maze (variable 2)? In a courtroom trial does a defendant's apparent socioeconomic status (variable 1) have an influence on his chances of obtaining an acquittal from the jury (variable 2)?

When a researcher wants to know if one variable has an effect on another, he or she can conduct a multisample experiment. The design and analysis of experiments with more than two samples are discussed under "Analysis of Variance" in Chapters 14 and 15. This chapter covers only two-sample experiments.

All experiments begin with a **research hypothesis,** a suspicion on the part of the researcher that changes in one variable will produce changes in another. For example, if he or she suspects that the answer to any of the questions in the preceding examples is yes, he or she is entertaining a research hypothesis. Before the experiment can begin, the research hypothesis must be framed in terms of two observable, measurable variables. One of these is then identified as the **independent variable** and the other as the **dependent variable.** The purpose of the experiment will be to determine whether the independent variable has an effect on the dependent variable. In the above examples, dosage rate, hours of food deprivation, and socioeconomic status would be independent variables; dependent variables would be amount of time required for cure, rate of maze learning, and outcome of the jury trial.

From a statistical point of view, there are basically two types of two-sample experiments: the **independent-samples design** and the **related-samples design.** The steps in an independent-samples experiment are summarized in Table 11.1. First a group of N subjects is selected. Strictly speaking, the results of this experiment may be generalized only to the population defined by the sampling procedure. (It might be useful to reread Dr. Hays's comments in the *Spotlight* for Chapter 9.) Whether the subjects are rats, machine parts, sophomore students in a psychology class, or voters, the population under consideration consists only of those subjects who had an equal chance of being selected for the sample. Whether or not the results are also applicable to other populations is a decision the researcher and statistical consumer will have to make on the basis of any other information they may have.

After the N subjects are selected, they are divided *randomly* into two groups. In the independent-samples experiment each subject has an equal probability of being assigned to either group.

Suppose, for instance, we conduct an experiment to investigate the effects of exercise (the independent variable) on weight loss in middle-aged men. Twenty such men volunteer to serve as subjects, and these are randomly assigned to two groups. There are many ways random assignment might be accomplished; one would be to number each person's name and then to read numbers from a random-number table, such as Table E in Appendix 1.[1] We proceed through the table until a number between 1 and 20 is encountered. The corresponding person is assigned to group I. The next number (between 1 and 20) determines who will be assigned to group II, and so on. It is not absolutely necessary to assign an equal number of subjects to each group, but computations in the hypothesis test are simplified if the number in group I (n_1) is equal to the number in group II (n_2).

A simpler method of deciding who goes into each group would be to flip a coin for each man: "Heads," the subject goes into group I, "Tails" he goes to group II. Another simple method would be to line up the men and assign every other man to group II; this would also produce a random distribution of subjects to groups *if* they were lined up in a random order.

The independent variable in this study, "exercise," is called *independent* because the experimenter is free to administer any amount, or **level,** of that variable to each group. Each group receives a different level

---

[1] To use a table of random numbers, like Table E, begin at any row or column and proceed up, down, left, right, or diagonally. The numbers are grouped in the table only to help you keep your place. They may be regrouped as single-digit, five-digit, or any-digit clumps as necessary. In this example, one could begin at, say, row 6. Moving to the right, the numbers listed are 77, 92, 10, 69, 07, 11, 00, 84, ... The first number encountered between 1 and 20 is 10, so person 10 is assigned to group I; then person 7 is assigned to group II, person 11 to group I, and so on until all twenty have been assigned.

Table 11.1 *Independent-Samples Experiment to Examine the Effects of Variable* A *on Variable* B

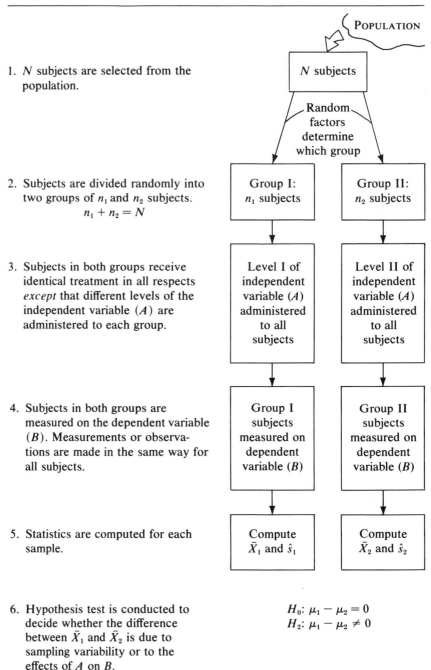

1. N subjects are selected from the population.

2. Subjects are divided randomly into two groups of n_1 and n_2 subjects.
   $$n_1 + n_2 = N$$

3. Subjects in both groups receive identical treatment in all respects *except* that different levels of the independent variable (A) are administered to each group.

4. Subjects in both groups are measured on the dependent variable (B). Measurements or observations are made in the same way for all subjects.

5. Statistics are computed for each sample.

6. Hypothesis test is conducted to decide whether the difference between $\bar{X}_1$ and $\bar{X}_2$ is due to sampling variability or to the effects of A on B.

of the exercise variable. Subjects in one group might even receive a zero level of the independent variable, in which case this group is called a **control group.** Here, let group II be a control group. Group I might receive a level specified as "jog one mile and do twenty minutes of calisthenics each day." The important points are (1) members of both groups are treated exactly the same in all ways throughout the experiment *except* with regard to the independent variable; and (2) *within* each group, all subjects receive equal treatment on the independent variable. A carefully conducted study of this type would ensure that other factors that might influence the dependent variable (weight loss), such as daily caloric intake, are equalized for all subjects.

The importance of treating all subjects alike in all respects except the independent variable cannot be overemphasized. In fact, the strength of the cause-and-effect argument depends on the extent to which this condition has been satisfied. Whenever possible, experiments are conducted so that the subjects do not even know which group they are in and are thus **blind** to the level of the independent variable received and the expected results. In drug studies subjects are kept blind by administering **placebo** pills or shots to members of a control group, for instance. If the experiment can be conducted so that neither the subjects nor the experimenter know (during the experiment) which group a subject is in, the experiment is a **double blind** study. The reason for keeping subjects and experimenters blind is that simply knowing how an experiment is "supposed" to turn out sometimes causes even well-meaning experimenters or subjects to bias experimental conditions so as to produce expected results.

The final step involving the subjects themselves consists of measuring every individual on the dependent variable, "weight loss." Here, this would be recorded as the number of pounds lost since the experiment began, and the figures might read 15, 9, −2 (a loss of −2 would represent a gain of 2), 3, and so on. It is important that everyone be measured on the dependent variable in exactly the same way, according to the same measurement scale.

When the data are in, we can compute $\bar{X}_1$ and $\bar{X}_2$. Most likely they will differ, but is that difference due to sampling variability or to the effects of exercise? To decide that question, we must conduct the hypothesis test concerned with the difference between two means.

This procedure could no doubt be used in a real experiment; however, we would expect a great deal of random variability from individual differences among subjects. This would be reflected in relatively large standard deviations for each group. Random variability is the hobgoblin of hypothesis testing; it produces wide, flat sampling distributions that, in turn, produce high error probabilities. Thus, when possible, experimenters turn from the independent-samples experiment to a design that attempts to separate and remove individual differences from random variability: the **related-samples experiment.** As we will see, related-samples experiments require a slightly different *t*-test than independent-samples experiments. Although there are two kinds of related-samples designs, the same t-test is used in both (related-samples) situations.

One type of related-samples experiment is the **matched-pairs design.** This is an attempt to equate the two groups as much as possible before starting the experiment, so that any obtained difference between $\bar{X}_1$ and $\bar{X}_2$ would be more likely due to the effects of the independent variable than to individual differences among subjects. The procedure is identical to that in the independent-samples experiment (Table 11.1) except that random factors *do not* determine which subject goes in each group. In the matched pairs experiment, the N subjects are divided into $N/2$ *pairs* so that the members of each pair are similar in characteristics that might influence the dependent variable. In the weight loss experiment, for instance, it would be reasonable to assume that a person's initial weight might be an important consideration. A 250 lb man could be expected to lose more pounds than a 140 lb man in a given time period. We could create a matched-pairs experiment by lining up people in order of weight before the experiment begins. The two heaviest would be the first pair, the third and fourth heaviest the second pair, and so on. Then one member of each pair would be assigned to each group. (*Which* member of each pair goes to group I and which goes to group II is determined randomly.)

Or suppose that an experiment was designed to test the impact of teaching machines on fourth-grade children's math ability. A better procedure than that used in Chapter 10 would be to select N children from the population and give them all a test of general math ability before starting the experiment. The two highest scorers would constitute the first pair, and one would go to group I, the other to group II. The procedure could be repeated for all N children, producing a matched-pairs, two-sample experiment. One group would receive no experience with the machine (level I of the independent variable), and the other group would receive daily practice on the device (level II).

After subjects have been assigned to their respective groups, the matched-pairs experiment is conducted exactly like the independent-samples experiment. The results, however, are evaluated with the related-samples t-test.

In some experiments it is possible to use the most extreme form of the related samples design by using the *same* subjects in both groups. This is known as a **repeated-measures design.** This design is shown schematically in Tables 11.2 and 11.3. In the simplest form of this experiment (Table 11.2) all N subjects are administered level I of the independent variable, measured on the dependent variable, administered level II of the independent variable, and then measured again on the dependent variable. The mean from the first set of dependent-variable measurements is $\bar{X}_1$ and the mean from the second set of measurements is $\bar{X}_2$. Using repeated measures reduces variability due to differences among subjects, because the same subjects are used in each group.

An experiment that would fit this plan would be one to study the effects of alcohol consumption on reaction time. Level I might consist of "no

Table 11.2 *Simple Repeated-Measures Experiment to Examine the Effects of Variable* A *on Variable* B

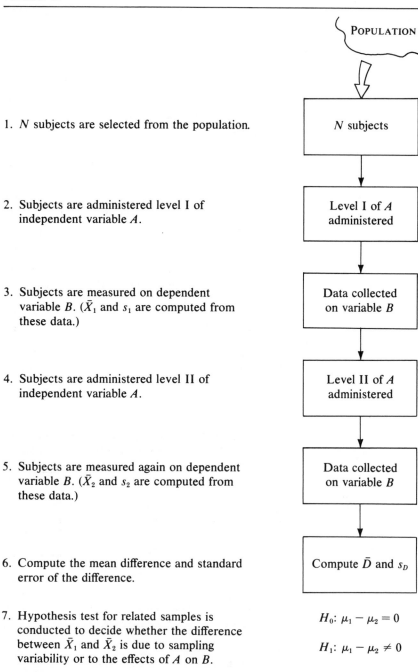

POPULATION

1. N subjects are selected from the population.

   N subjects

2. Subjects are administered level I of independent variable A.

   Level I of A administered

3. Subjects are measured on dependent variable B. ($\bar{X}_1$ and s_1 are computed from these data.)

   Data collected on variable B

4. Subjects are administered level II of independent variable A.

   Level II of A administered

5. Subjects are measured again on dependent variable B. ($\bar{X}_2$ and s_2 are computed from these data.)

   Data collected on variable B

6. Compute the mean difference and standard error of the difference.

   Compute $\bar{D}$ and s_D

7. Hypothesis test for related samples is conducted to decide whether the difference between $\bar{X}_1$ and $\bar{X}_2$ is due to sampling variability or to the effects of A on B.

   $H_0: \mu_1 - \mu_2 = 0$

   $H_1: \mu_1 - \mu_2 \neq 0$

Table 11.3 *Counterbalanced Repeated-Measures Experiment to Examine the Effects of Variable* A *on Variable* B

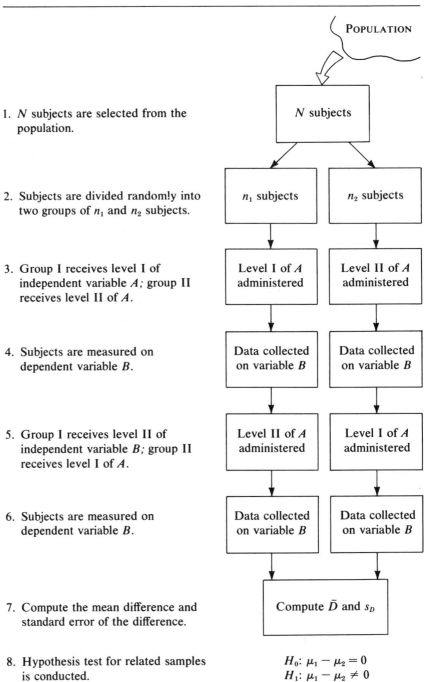

1. *N* subjects are selected from the population.

2. Subjects are divided randomly into two groups of n_1 and n_2 subjects.

3. Group I receives level I of independent variable *A;* group II receives level II of *A*.

4. Subjects are measured on dependent variable *B*.

5. Group I receives level II of independent variable *B;* group II receives level I of *A*.

6. Subjects are measured on dependent variable *B*.

7. Compute the mean difference and standard error of the difference.

8. Hypothesis test for related samples is conducted.

$H_0: \mu_1 - \mu_2 = 0$
$H_1: \mu_1 - \mu_2 \neq 0$

alcohol'' and level II might be 80 ml of 100-proof whiskey. All N subjects would take a reaction-time test after level I of the independent variable had been administered (here, nothing) and then, after a suitable delay, take the same test after level II had been administered. We could expect $\bar{X}_1$ to differ from $\bar{X}_2$. The related-samples hypothesis test would help us decide whether that difference was due to sampling variability or to the effects of alcohol.

A slightly better procedure would be to use the **counterbalanced design** shown in Table 11.3. Practice on the reaction-time task could change a subject's performance on the dependent variable. Hence, half the subjects could receive level I of the independent variable first and half level II first. This way, any practice effects that would tend to make the second test session faster, regardless of treatment level, would favor each treatment level equally. Of course, $\bar{X}_1$ would still represent the mean of all reaction times after level I, $\bar{X}_2$ the mean of all reaction times after level II.

In summary, two samples receive different levels of the independent variable. The procedure used to determine which subjects are in sample I and which are in sample II determines the basic design of the experiment. If random factors only make the determination, the independent-samples t-test is required. If the groups are equalized before the experiment begins through use of a matched-pairs design, or if the same subjects are used in both samples (repeated-measures design), the related-samples t-test is appropriate.

In Chapter 10 we were concerned with the values of individual means. In this chapter, however, the statistic of interest will be the *difference* between two means. In order to test the significance of that difference, we must first consider the pertinent test statistic and its sampling distribution.

The Sampling Distribution of the Difference Between Two Means

The two-sample experiment has been conducted, and $\bar{X}_1$ differs from $\bar{X}_2$. Suppose that the weight-loss/exercise study produces a mean weight loss of 15 lb in group I and a mean weight loss of 3 lb in group II. The statistic to be tested is the difference between $\bar{X}_1$ and $\bar{X}_2$, that is,

$$\bar{X}_1 - \bar{X}_2 = 15 - 3 = 12 \text{ lb}$$

As in any hypothesis test, we assume that this statistic is just one value in a sampling distribution of that statistic. Here, this is the **sampling distribution of the difference between two means.** Theoretically, this distribution could be found as follows. Start with *two* populations. From population I, draw a sample of n_1 observations and from population II draw a sample of n_2 observations. Make an observation on every individual in each sample and compute the difference between the two sample means. Repeat this process over and over until an infinite number of pairs of samples of size n_1 and n_2 have been observed. Then make a frequency distribution of the obtained differences between sample means. As with the sampling distribution of the mean (Chapter 9), we are fortunate that mathematicians have already determined the characteristics of this distribution. As with the sampling distribution of the mean, you will never have to create a sampling distribution of the difference between two means by such a long process; however, it is important to know that such a process would yield the distribution if it could be done.

Figure 11.1 represents the sampling distribution. As you might expect, its characteristics depend upon the population parameters, μ_1, μ_2, σ_1^2, and σ_2^2. The spread of the distribution will also depend on the sample sizes, n_1 and n_2. Specifically, the distribution will have the following parameters:

$$\mu_{\bar{X}_1 - \bar{X}_2} = \mu_1 - \mu_2$$

$$\sigma_{\bar{X}_1 - \bar{X}_2} = \sqrt{\frac{\sigma_1^2}{n_1} + \frac{\sigma_2^2}{n_2}}$$

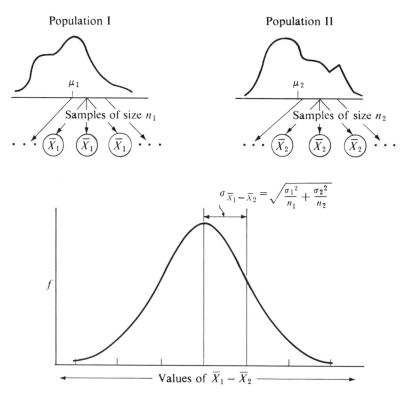

Figure 11.1   Sampling distribution of the difference between two means. All possible samples of size n_1 are drawn from population I, and all possible samples of size n_2 are drawn from population II. The large distribution represents the frequency distribution of all *differences* between every possible $\bar{X}_1$ and $\bar{X}_2$. The mean of this sampling distribution will equal $\mu_1 - \mu_2$. When $n_1 + n_2$ equals 30 or more, the distribution can be considered approximately normal, even when the original populations are not.

where $\mu_{\bar{X}_1 - \bar{X}_2}$ = mean of the sampling distribution of the difference between two means (or, the expected value of the statistic $\bar{X}_1 - \bar{X}_2$);

$\sigma_{\bar{X}_1 - \bar{X}_2}$ = standard error of the sampling distribution.

Moreover, the distribution will be normal in shape regardless of the sample sizes if the population distributions are normal. It will be normal in shape regardless of the population shapes if large enough samples are used (the Central Limit Theorem at work again).

To use a numerical example, suppose that we have two populations, and population I has $\mu_1 = 120$ and $\sigma_1 = 12$, and population II has $\mu_2 = 110$ and $\sigma_2 = 15$. Samples of size $n_1 = 36$ are drawn from population I, and samples of size $n_2 = 25$ are drawn from population II. When an infinite number of pairs of such samples have been drawn, and when the differences between $\bar{X}_1$ and $\bar{X}_2$ have been determined, the frequency distribution of these differences will have the following parameters:

$$\mu_{\bar{X}_1 - \bar{X}_2} = \mu_1 - \mu_2 = 120 - 110 = 10$$

$$\sigma_{\bar{X}_1 - \bar{X}_2} = \sqrt{\frac{\sigma_1^2}{n_1} + \frac{\sigma_2^2}{n_2}} = \sqrt{\frac{(12)^2}{36} + \frac{(15)^2}{25}} = \sqrt{4 + 9} = 3.61$$

Knowing that this distribution is normal, we could ask, What is the probability that the difference between means for two such samples drawn randomly from these populations is equal to or greater than 11? The answer, of course, is obtained by finding the z-score of this difference:

$$z = \frac{\text{Value of interest} - \text{Mean of the distribution}}{\text{Standard deviation of the distribution}}$$

For this difference,

$$z = \frac{11 - 10}{3.61} = +0.28$$

The unit normal table shows an area beyond z of 0.3897 for $z = 0.28$. This is the probability that the difference between $\bar{X}_1$ and $\bar{X}_2$ would exceed 11 when μ_1 and μ_2 differ by 10.

In the hypothesis test we ask similar questions. Although the preceding sampling distribution is the basis of the tests that follow, calculating the obtained statistics becomes a little more complicated, because in the real world one rarely knows the values of σ_1^2 and σ_2^2.

The Test-Independent Samples

Returning to the weight-loss experiment, suppose that the following statistics were obtained.[2]

---

[2] Here we are using $\bar{X}_1$ and $\bar{X}_2$ to symbolize the means of samples 1 and 2, respectively. At this point, we *could* have introduced a double subscript notation to distinguish between

GROUP I	GROUP II
$\bar{X}_1 = 15$ lb lost	$\bar{X}_2 = 5$ lb lost
$\hat{s}_1 = 6$ lb	$\hat{s}_2 = 3$ lb
$\hat{s}_1^2 = 36$	$\hat{s}_2^2 = 9.00$

Is the difference between means due to exercise or random variability? The pair of hypotheses that address this question can be stated in two equivalent forms:

$$H_0: \mu_1 = \mu_2 \qquad \text{or} \qquad H_0: \mu_1 - \mu_2 = 0$$
$$H_1: \mu_1 \neq \mu_2 \qquad\qquad H_1: \mu_1 - \mu_2 \neq 0$$

The null hypothesis states that after the experiment the two samples come from populations with equal means. The alternative hypothesis says that they come from populations with different means. Although either pair of hypotheses may be used, the pair on the right is the preferred form, because it emphasizes that we are working with a statistic that is a difference between means. The equations in this pair also permit more general applications of this hypothesis test by allowing substitution of other values than zero on the right side of the equations.

Since the standard deviations of these populations (σ_1 and σ_2) are unknown, we cannot compute the test statistic as easily as we computed z in the previous section. We will have to estimate σ_1 and σ_2 so that, in turn, $\sigma_{\bar{X}-\bar{X}_2}$ can be estimated.[3]

Although the sampling distribution of the difference between two means is normal, we cannot use the z-test or compute z-obtained values if σ_1^2 and σ_2^2 are unknown. If $\hat{s}_1^2$ and $\hat{s}_2^2$ are used to estimate $\sigma_{\bar{X}_1-\bar{X}_2}$, the sampling distribution is shaped as a t-distribution with $df = (n_1 - 1) + (n_2 - 1)$. This expression for df can be written in simpler form: $df = n_1 + n_2 - 2$. Here, $df = 10 + 10 - 2$, so the observed difference of 12 lb between sample means must be interpreted as one value in a t-distribution with 18 degrees of freedom. The same general logic underlying the one-sample test applies here. We select an alpha level, determine the values of t that are exceeded by the alpha of the observed t-values, and then compute t-observed for our statistic. The decision rule is the same as that used for the single mean test: If $|t\text{-observed}| \geq t\text{-critical}$, reject H_0.

We will first test the hypotheses about means at $\alpha = 0.05$. From Table

---

individuals within the group and group number. However, in the interest of clarity, the double subscript is probably better left to analysis of variance as covered in Chapter 14.

[3] Many people familiar with the independent samples t-test will recognize this as the point at which the test for homogeneity of variances is called for. However, since the t-test is very robust with respect to violations of assumptions of homogeneity of variances and assumptions of normality (see, for example, C. A. Boneau, "The Effects of Violations of Assumptions Underlying the t-Test", *Psychological Bulletin*, Vol. 57, pp. 49–64, 1960), discussion of that test will be postponed until Chapter 14 where it is discussed in context with the analysis of variance. In most experimental situations, violations of the homogeneity assumption will lead to a type I error rate no more than one or two percent different from the specified alpha.

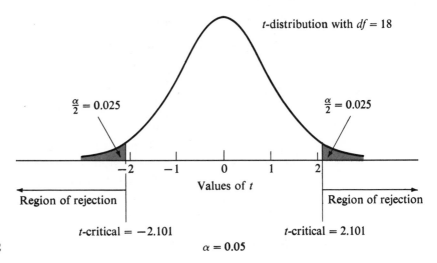

Figure 11.2

B, Appendix 1, we determine that t-critical for a two-tailed test, with $\alpha = 0.05$ and $df = 18$ is equal to 2.101. The sampling distribution and critical t-values are shown in Figure 11.2.

All that remains to be found is t-observed. This is very similar to finding z-obtained:

$$t\text{-observed} = \frac{(\bar{X}_1 - \bar{X}_2) - (\mu_1 - \mu_2)}{\text{estimated } \sigma_{\bar{x}_1 - \bar{x}_2}}$$

where $X_1 - X_2$ is the observed statistic and $\mu_1 - \mu_2$ is the mean of the sampling distribution specified by H_0.

Let's take care of the numerator first. $\bar{X}_1 - \bar{X}_2$ is the difference between sample means, 12. $\mu_1 - \mu_2$, under the null hypothesis, is zero, so this term disappears from the equation. This leaves us with:

$$t\text{-observed} = \frac{12 - 0}{\text{estimated } \sigma_{\bar{x}_1 - \bar{x}_2}}$$

To find the estimated standard error in the denominator, we compute what amounts to a weighted mean of the two sample variances[4] called the **pooled variance** and symbolized $\hat{s}_p^2$. When sample sizes are equal ($n_1 = n_2$), then each variance receives equal weight and $\hat{s}_p^2$ is simply the arithmetic mean of the two variances. The pooled variance will be used in the following way:

$$\text{estimated } \sigma_{\bar{x}_1 - \bar{x}_2} = \sqrt{\frac{\hat{s}_p^2}{n_1} + \frac{\hat{s}_p^2}{n_2}}$$

The quantity $\hat{s}_p^2$ thus occupies a position similar to that of σ^2 in the true standard error of this distribution presented on page 309.

---

[4]Variances can be averaged, but standard deviations cannot. This is why we have used variances so often in this chapter.

We obtain $\hat{s}_p^2$ as follows:

$$\hat{s}_p^2 = \frac{(n_1 - 1)(\hat{s}_1^2) + (n_2 - 1)(\hat{s}_2^2)}{n_1 + n_2 - 2}$$

For the example here:

$$\hat{s}_p^2 = \frac{(10 - 1)(36) + (10 - 1)(9)}{10 + 10 - 2} = \frac{405}{18} = 22.5$$

The pooled variance will always have a value somewhere between the values of the two individual variances, here 36 and 9.

The pooled variance is then used to compute the estimated $\sigma_{\bar{X}_1 - \bar{X}_2}$:

$$\text{estimated } \sigma_{\bar{X}_1 - \bar{X}_2} = \sqrt{\frac{22.5}{10} + \frac{22.5}{10}} = \sqrt{2.25 + 2.25} = \sqrt{4.50} = 2.12$$

This value is inserted in the formula for t-observed:

$$t\text{-observed} = \frac{12}{2.12} = 5.66$$

Since this is greater than the t-critical of 2.101, we reject the null hypothesis. Checking the t-critical values for other alpha levels, we also find that this t-observed is significant at the 0.001 level. So the written report on this experiment could assert that the difference between mean weight losses for the two samples was significant ($t = 5.66$, $df = 18$, $p \leq 0.001$).

The Test-Related Samples

As indicated earlier, using related-samples designs eliminates some random variability from the data by removing the effects of individual subject differences (not due to experimental treatment) from the variability in the data. Since the groups are equated to some extent, the difference between $\bar{X}_1$ and $\bar{X}_2$ is less likely to be due to irrelevant differences among subjects. This produces a smaller estimated $\sigma_{\bar{X}_1 - \bar{X}_2}$, which, in turn, produces a larger t-observed. This can be seen in the *general* formula for estimating the standard error of the difference between two means (the formula used on page 311 applies only to independent samples):

$$\text{estimated } \sigma_{\bar{X}_1 - \bar{X}_2} = \sqrt{\frac{\hat{s}_1^2}{n_1} + \frac{\hat{s}_2^2}{n_2} - (2)(r)\left(\frac{\hat{s}_1}{\sqrt{n_1}}\right)\left(\frac{\hat{s}_2}{\sqrt{n_2}}\right)}$$

Although we will not use this formula to compute the estimated $\sigma_{\bar{X}_1 - \bar{X}_2}$, it illustrates the consequences of using related instead of independent samples. The r in the expression is the Pearson r, describing the correlation between dependent variable scores for groups I and II. In the independent-samples experiment this correlation is *assumed* to be zero, since subjects are randomly assigned to groups; hence, the entire subtracted term is zero. In the related-samples experiment, however, members of one group are paired with members of the other (matched-pairs design) or each person has a pair of scores (repeated-measures design), and in this case a nonzero r has meaning. So, subtracting the term

$(2)(r)\left(\dfrac{\hat{s}_1}{\sqrt{n_1}}\right)\left(\dfrac{\hat{s}_2}{\sqrt{n_2}}\right)$ reduces the size of the estimated standard error

and increases t-observed. The net result, then, of using related samples is to increase the statistical power of the test, since we get larger t-observed values and are more likely to reject H_0 when H_1 is true.

There is one drawback to the related-samples design, however. Whereas in the independent-samples test $df = n_1 + n_2 - 2$, the degrees of freedom in the related-samples test is simply $n - 1$, where n is the number of paired observations. As you can see from Table B, Appendix 1, a lower degrees of freedom produces higher values of t-critical. But when random variability is relatively high, when individual differences among subjects are substantial, the advantage of the reduced error term outweighs the disadvantage of the lower degrees of freedom.

Fortunately, the related-samples t-observed can be computed via a shortcut known as the **direct-differences method.** To illustrate this technique, we will use the teaching-machine example described on page 304 as a matched-pairs experiment.

Prior to the experiment, eighteen children were given a comprehensive test of math skills. On the basis of those test scores, nine pairs of students were formed, and one child of each pair was assigned to group I and the other to group II. Group I received a year's training on the machine in addition to regular classroom lessons (level I of the independent variable), and group II received only regular classroom training (level II). At the end of the school year another comprehensive arithmetic exam was given; the resulting scores are listed in Table 11.4 as data for groups I and II.

From these data we find that $\bar{X}_1 = 114.55$ and $\bar{X}_2 = 109.00$. Children trained with the machines scored higher, *on the average*, than their counterparts in group II. Is the difference between means significant?

We could conduct the hypothesis test in a manner very similar to that used for the independent-samples test discussed in the preceding section, using the estimated $\sigma_{\bar{X}_1-\bar{X}_2}$ formula presented on page 312. However, the direct-difference method offers a simpler computational route.

First, the hypothesis test concerns essentially the same hypotheses used in the independent-samples test:

$H_0: \mu_1 - \mu_2 = 0$
$H_1: \mu_1 - \mu_2 \neq 0$

However, in the direct-differences method, these hypotheses are restated in a slightly different form:

$H_0: \mu_{\bar{D}} = 0$
$H_1: \mu_{\bar{D}} \neq 0$

where $\mu_D =$ the mean difference between members of each pair.

In terms of the basic question, the pairs of hypotheses are identical. However, the lower pair suggests that we use a different test statistic, the mean difference between paired observations. That is, after the experiment, dependent variable scores for the members of each pair will be different, but under the null hypothesis, these pairs come from a population where the mean of such difference scores is zero. The alternative hypothesis says, of course, that the mean difference is some nonzero value.

The fact that these pairs of hypotheses are equivalent can be seen by returning to Table 11.4. In the column labeled D, the difference between the members of each pair is computed. Group I's member of pair 1 scored 130, group II's member of pair 1 scored 123. The difference (D) between them is $130 - 123 = 7$. Notice that the group II score is always subtracted from the group I score, even when the second is larger than the first. The mean of these difference scores ($\overline{D}$) for all nine pairs is 5.56. This is exactly equal to the difference between $\overline{X}_1$ and $\overline{X}_2$. Thus, $\overline{X}_1 - \overline{X}_2$ and $\overline{D}$ are equivalent statistics. It can be shown mathematically that $\mu_{\overline{D}}$ and $\mu_1 - \mu_2$ are equivalent parameters. The different pairs of hypotheses are used only to emphasize the fact that we are computing the statistics in slightly different ways.

The observed statistic ($\overline{D}$) is assumed to come from a sampling distribution shaped as a t-distribution, with $n - 1$ degrees of freedom, where n is the number of paired observations. Here, we have 8 df.

We will make this test at an alpha level of 0.05. From Appendix Table B t-critical for $\alpha = 0.05$ in the two-tailed test is 2.306. Thus, the decision rule is: If $|t\text{-observed}| \geq 2.306$, reject H_0.

The value of t-observed is found by dividing $\overline{D}$ by the estimated standard error of the mean difference ($s_{\overline{D}}$). That is:

$$t\text{-observed} = \frac{\overline{D}}{s_{\overline{D}}}$$

The value of $s_{\overline{D}}$ is much easier to compute than the estimated $\sigma_{\overline{X}_1 - \overline{X}_2}$ used for the independent-samples test, and the process for this example is also summarized in Table 11.4. First, we need the standard deviation of the D values, $\hat{s}_D$. This is computed in the same way as any unbiased standard deviation except that D scores are used instead of X scores. As the table suggests, one first finds the difference between each D and $\overline{D}$. This is a deviation score in the same way that $(X - \overline{X})$ is a deviation score representing the difference between X and $\overline{X}$. These deviation values are listed under the column headed $(D - \overline{D})$, and their squared values are listed under $(D - \overline{D})^2$. Then, the sum of the $(D - \overline{D})^2$ values is divided by $n - 1$; the square root of this quotient is taken as $\hat{s}_D$. That is:

$$\hat{s}_D = \sqrt{\frac{\sum (D - \overline{D})^2}{n - 1}}$$

Table 11.4 *Direct-Differences Method, Related-Samples t-Test*

PAIR NO.	GROUP I DATA	GROUP II DATA	D	$(D - \overline{D})$	$(D - \overline{D})^2$
1	130	123	7	1.44	2.0736
2	131	125	6	0.44	0.1936
3	120	112	8	2.44	5.9536
4	121	125	−4	−9.56	91.3936
5	111	113	−2	−7.56	57.1536
6	118	115	3	−2.56	6.5536
7	100	91	9	3.44	11.8336
8	102	90	12	6.44	41.4736
9	98	87	11	5.44	29.5936

$$\sum = 1031 \qquad \sum = 981 \qquad \sum = 50 \qquad \sum = 246.2224$$

$$\overline{X}_1 = \frac{1031}{9} = 114.56 \qquad \hat{s}_D = \sqrt{\frac{\sum (D - \overline{D})^2}{n - 1}} \; (n = \text{number of pairs})$$

$$\overline{X}_2 = \frac{981}{9} = 109.00 \qquad = \sqrt{\frac{246.224}{8}}$$

$$\overline{X}_1 - \overline{X}_2 = 114.55 - 109.00 = 5.56 \qquad = \sqrt{30.7778}$$

$$\overline{D} = \frac{50}{9} = 5.56 \qquad = 5.55$$

$$t\text{-observed} = \frac{\overline{D}}{s_{\overline{D}}} = \frac{5.56}{1.88} = 2.957 \qquad s_{\overline{D}} = \frac{\hat{s}_D}{\sqrt{n}} = \frac{5.55}{\sqrt{9}} = 1.85$$

The *standard error* of the $\overline{D}$ statistic is then computed from this as follows:

$$s_{\overline{D}} = \frac{\hat{s}_D}{\sqrt{n}} = \frac{\sqrt{\dfrac{\sum (D - \overline{D})^2}{(n - 1)}}}{\sqrt{n}}$$

In the example summarized, $s_{\overline{D}} = 1.85$. Now we can find t-observed:

$$t\text{-observed} = \frac{\overline{D}}{s_{\overline{D}}} = \frac{5.56}{1.85} = 3.01$$

The value of t-observed is thus greater than t-critical of 2.306, so we reject H_0. We then check to see if this is significant at the 0.01 alpha level. The two-tailed t-critical for $\alpha = 0.01$ with 8 degrees of freedom is 3.355 so the difference is not significant at the 0.01 level. Rejecting the null hypothesis, of course, means that we decide the observed difference was not due simply to sampling variability.

One-Tailed Versus Two-Tailed Tests

It is possible to conduct one-tailed tests for the significance of the difference between two means, just as it was for the single-sample test situation. Again, however, use of the one-tailed test is not justified unless the researcher is interested only in differences between means in one direction and unless differences between means in the other direction are unlikely and have little meaning in terms of the experimental treatment with the independent variable.

The one-tailed t-test would not be appropriate for the weight-loss/exercise experiment, for instance. There is a *possibility* that members of the "exercise" group would actually lose *less* weight than their nonexercising counterparts (through changed appetite, improved general health, and so on). So the alternative hypothesis ($\mu_1 - \mu_2 \neq 0$) allows the difference between μ_1 and μ_2 to be in either direction.

In the teaching-machine example, however, it is reasonable to suggest that children's math abilities are only going to improve through additional help from the machine. So, when one is interested only in the possibility that μ_1 may be greater than μ_2, the appropriate pair of hypotheses for the one-tailed test is:

$$H_0: \mu_1 - \mu_2 \leq 0$$
$$H_1: \mu_1 - \mu_2 > 0$$

The sampling distribution and region of rejection for this test are shown in Figure 11.3(a). The test is conducted just like the two-tailed test, except that a different t-critical and decision rule are used. In the teaching-machine example, $df = 8$, so the one-tailed t-critical for alpha = 0.05 is 1.860. The decision rule is: If t-observed ≥ 1.860, reject H_0.

Similarly, if μ_1 can only be expected to be less than μ_2 the hypotheses are:

$$H_0: \mu_1 - \mu_2 \geq 0$$
$$H_1: \mu_1 - \mu_2 < 0$$

The sampling distribution and region of rejection for this test are shown in Figure 11.3(b). (In the one-tailed tests all the necessary plus and minus signs are likely to be confusing, so it is always a good idea to sketch the sampling distributions, region of rejection, and location of the obtained statistic.) The decision rule in this case is: If t-observed $< -t$-critical, reject H_0.

Interpreting the Decision

After a two-sample experiment, $\bar{X}_1$ is found to differ significantly from $\bar{X}_2$. What does this "prove" about the effects of variable A on variable B? It is meaningless unless the requirements for the two-sample experiment (Tables 11.1, 11.2, and 11.3) were rigidly met. The two groups must have

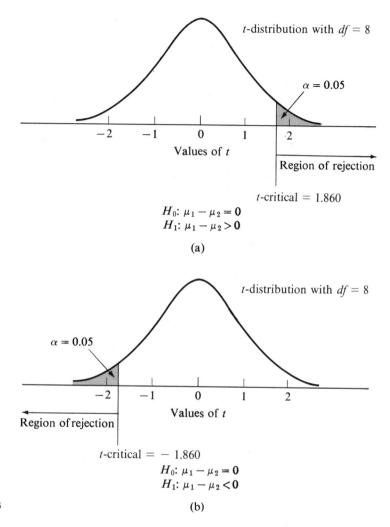

Figure 11.3

Figure (a) shows a t-distribution with $df = 8$, with $\alpha = 0.05$ and Region of rejection on the right, t-critical $= 1.860$, and hypotheses $H_0: \mu_1 - \mu_2 = 0$, $H_1: \mu_1 - \mu_2 > 0$.

Figure (b) shows a t-distribution with $df = 8$, with $\alpha = 0.05$ and Region of rejection on the left, t-critical $= -1.860$, and hypotheses $H_0: \mu_1 - \mu_2 = 0$, $H_1: \mu_1 - \mu_2 < 0$.

been treated identically throughout the experiment in every way except for the levels of the independent variable administered.

It is just as important that assignment of individuals to the different groups not be biased. In the examples discussed here, N subjects are selected from the population and *then* randomly assigned to groups. If, instead, group I subjects were selected from the population and then group II subjects were selected, the two groups would differ in ways that could be important. Suppose, for instance, that we want to find out if regular exercise leads to a longer life span. We examine the life histories of a number of people who are no longer living and divide them into two categories: "regular exercisers" and "nonexercisers." Suppose also that the active group has a mean life span significantly longer than the nonexercise group. Does this difference prove that exercise has an effect on life span? We cannot answer the question from this kind of evidence. These people were not randomly assigned to "exercise" and "nonexercise" groups. In fact, each person *chose* his group affiliation. It is possible that people who are healthier to begin with choose to lead more active lives than those who are less healthy. In other words, sample assignment could have been biased. In this case, it probably wouldn't be possible to do the properly controlled experiment, since a researcher would have to select a number of young people, assign them to groups (randomly, or through a matched-pairs design), and then require one group to exercise for a lifetime and the other not to exercise. Similarly, we must be careful when judging experiments that show heavy smokers having a higher rate of lung cancer than nonsmokers; presumably, people choose the group they wish to belong to, and other factors *could* produce the difference between groups. We cannot assume that because drivers of sports cars have more accidents than drivers of luxury sedans, sports cars are more dangerous than sedans. Again, different types of drivers are likely to choose different types of cars. Such observations are interesting, but they do not constitute "proof" unless a properly controlled experiment has been conducted.

When, as far as one knows, all conditions for the two-sample experiment have been satisfied, and the obtained difference is significant, then it is proper to *suggest* that variable *A* has an effect on variable *B*. If other researchers replicate the experiment and obtain similar results, then the cause-and-effect argument is strengthened. Remember, however, that such statistical decisions always include the possibility of error.

Before basing any action on the result of a significant difference, however, it is wise to consider the size of the effect and other nonstatistical factors. Remember from the discussion on statistical power (pp. 291–293) that if enough trials or enough subjects are used, even a small effect can be shown to be significant. To return to an example used in this and the previous chapter: If the teaching-machine experiment were repeated with, say, 10,000 children in each group, and if the difference between μ_1 and μ_2 were real but small, it is a good bet that we would obtain a difference between means that would be significant even at the 0.0000001 level. But suppose that this highly significant difference were obtained between $\bar{X}_1$ and $\bar{X}_2$ of only two points on the math exam. The difference is no doubt real, but whether or not such a small improvement in learning would justify the cost of buying the machines is another matter.

The utility of hypothesis testing, like all other statistical methods, depends ultimately on the ability of the statistical producer and consumer to *interpret* the statistics.

SUMMARY

The principles developed in Chapter 10 were extended here to test the significance of a difference between *two* sample means. This test, when applied to results of a two-sample experiment, can be used to answer questions about cause and effect, such as, Do changes in variable *A* have an effect on the value of variable *B*?

In the two-sample experiment *N* subjects are selected from the population and divided into two groups. If random factors only determine into which group each subject goes, the experiment is of the independent-samples variety. If the groups are matched before the experiment, or if the same subjects are used in both groups, the experiment is a related-samples design. During the experiment both groups are treated identically in all respects except that one group receives one level of the independent variable and the other group a different level of the independent variable. Both groups are then measured on the dependent variable. When this is done, the group I mean on the dependent variable usually differs from the group II mean. The hypothesis test is used to decide whether this difference is due to sampling variability or to the effects of the independent variable on the dependent variable.

KEY CONCEPTS

research hypothesis

independent variable

dependent variable

independent-samples design

related-samples design

level

control group

blind

double blind

placebo

matched-pairs design

repeated-measures design

counterbalanced design

sampling distribution of the difference between two means

standard error of the difference between two means

level of an independent variable

pooled variance

direct-differences method

SPOTLIGHT 11–a *t*-Tests in Research

The *t*-test for the significance of the difference between two means has been applied to questions in all areas of science, from agronomy to zymurgy. Here is one example from psychology.

Do noisy environments affect human performance? Some people are required to spend their working hours amidst high noise levels. Those who live near airports, railroad tracks, and busy highways may be similarly condemned throughout their leisure hours. It is fairly well established that long exposure to even moderate noise levels leads to some permanent hearing loss. Researchers are less in agreement, however, about the effect of noise on human performance in different situations.

Finkelman and Glass reasoned that *predictability* of noise might be an important factor in determining whether or not performance is impaired by noise when people are working at the limits of their mental ability.[5] To test this idea, they had volunteer subjects carry on mental tasks while being subjected to noise. In the repeated-measures experiment, subjects receiving the "predictable noise" condition were treated to blasts of 80-db noise through earphones, presented for 9 sec each time, with blasts spaced at regular intervals. Under the "unpredictable noise" condition, subjects received 80-db noise at irregular intervals, in blasts of varying duration. All subjects were scored on a dependent variable "number of errors on a digit recall task."

The mean number of errors on the task was 4.0 for the "predictable noise" group and 8.0 for the "unpredictable noise" group. The obtained t for this difference was 2.37, significant at the 0.05 level (degrees of freedom were unspecified). The obtained significant difference suggests that some kinds of noise conditions *do* affect performance.

SPOTLIGHT 11-b Does Party-Going Increase Smoking Behavior?

Many cigarette smokers report that they smoke more at parties or other social gatherings than they do otherwise. Is this true? If so, why?

A recent pair of studies by Brett Silverstein, Lynn Kozlowski and Stanley Schachter was conducted to address some of these questions.[6] In an earlier study, these researchers had demonstrated that manipulation of urinary pH (acidity) can influence the number of cigarettes people choose to smoke. In the studies discussed below, they investigated the possibility that some aspect of social situations raises urinary acidity, thereby producing more smoking.

In one study, eighteen smokers simply recorded the number of cigarettes smoked during the day; from these data, the mean number of cigarettes consumed each waking hour was determined for each person. Subjects also kept track of their social activities, and each smoker's days were individually categorized as "social" or "nonsocial," according to several criteria. For these subjects, the mean number of cigarettes smoked during the social day was 31.23, and the mean number during the nonsocial day was 27.85, a significant difference ($t = 2.71$, $df = 17$, $p < 0.02$). Because it was arguable that this difference arises from the simple

[5]J. M. Finkelman and D. C. Glass, "Reappraisal of the Relationship Between Noise and Human Performance by Means of a Subsidiary Task Measure," *Journal of Applied Psychology* 54 (1970): 211–213. The actual experiment included nine listening conditions and two dependent variables. However, the authors based their conclusions on the comparison presented in this *Spotlight*.

[6]B. Silverstein, L. Kozlowski, and S. Schachter, "Social Life, Cigarette Smoking, and Urinary pH," *Journal of Experimental Psychology: General*, 106 (1977): 20–23.

fact that social days were generally longer than nonsocial days, the mean number of cigarettes smoked per hour on these days was also determined. On social days, subjects averaged 1.85 cigarettes an hour, and nonsocial days they averaged 1.73. This difference between means was almost, but not quite, significant at the 0.05 level ($t = 2.01, df = 17, p < 0.06$). From this result, the experimenters concluded that these people did smoke more when engaged in social activity. The next step was to examine the role of urinary pH in this situation.

For smokers, it was determined that urinary pH readings were lower at the end of social days (mean = 5.86) than at the end of nonsocial days (mean = 6.30); this difference was statistically significant ($t = 3.49, df = 15, p = 0.01$). But does this drop cause more cigarette smoking? In further tests with other subjects, pH readings dropped from a mean of 6.43 before a two-hour party to 6.00 after the party ($t = 3.23, df = 14, p < 0.01$). Moreover, this drop appeared for both smokers and nonsmokers, refuting the possible argument that smoking causes pH changes rather than the reverse. These results, along with the earlier evidence that pH manipulation alters smoking behavior, suggest a partial explanation for increased smoking at parties.

PROBLEMS

1. A farmer wants to find out if two different brands of chicken feed lead to differences in egg production for his flock of 400 chickens. Design a two-sample, independent-samples experiment to examine this question. List all steps in the experiment, identify the independent and dependent variables, and interpret the hypotheses tested in terms of the chicken feed/egg production issue.

2. Repeat problem 1 using a matched-pairs approach; then design a repeated-measures experiment for the same situation.

3. Two normally distributed populations have these parameters:

POPULATION I	POPULATION II
$\mu_1 = 45$ years of age	$\mu_2 = 40$ years of age
$\sigma_1 = 5$ years	$\sigma_2 = 4$ years

a. What is the probability that a randomly selected individual from Population I will be 50 years old or more?

b. What is the probability that a randomly selected individual from Population II will be 50 years old or more?

c. A sample of size $n_1 = 36$, consisting of randomly selected members of Population I is drawn. What is the probability that the sample mean exceeds 47?

   d. A randomly selected sample of size $n_1 = 36$ is drawn from Population I, and a randomly selected sample of size $n_2 = 40$ is drawn from Population II. What is the probability that $\bar{X}_1 - \bar{X}_2$ equals 1 year or more?

4. Compute s_p^2 for the following situations:
   a. $\hat{s}_1 = 250$, $n_1 = 42$, $\hat{s}_2 = 210$, $n_2 = 37$
   b. $\hat{s}_1^2 = 102$, $n_1 = 8$, $\hat{s}_2^2 = 130$, $n_2 = 7$
   c. $\hat{s}_1 = 0.750$, $n_1 = 17$, $\hat{s}_2 = 0.951$, $n_2 = 14$
   d. $\hat{s}_1^2 = 1,450$, $n_1 = 75$, $\hat{s}_2^2 = 1,321$, $n_2 = 64$

5. In an experimental study of two speed reading methods, 30 people are randomly divided into two groups of 15 each. One group receives the training in the "Speed Scan" method for a month, and the other receives training in a method called "Rapid Read" for a month. At the end of that time, the following statistics are computed for each group:

SPEED SCAN READING SPEED	RAPID READ READING SPEED
$\bar{X} = 1,500$ words/min	$\bar{X} = 1,450$ words/min
$\hat{s} = 150$ words/min	$\hat{s} = 170$ words/min

   a. For the study described, what is the independent variable? The dependent variable? If we want to see if the two groups differed significantly from one another, what pair of hypotheses should be tested?
   b. Perform the hypothesis test to see if the groups do indeed differ from one another. Include a verbal interpretation of the hypotheses tested, the decision rule used, and level of significance if it is 0.05 or less.

6. In a two-sample experiment, Group I produced a mean of 530 on the dependent variable and Group II produced a mean of 510. Other statistics obtained in this study were $\hat{s}_1 = 25$, $n_1 = 15$, $\hat{s}_2 = 32$, $n_2 = 13$. Testing the one-tailed hypotheses, where Group I's population mean can only be greater than Group II's, perform the t-test to decide whether or not these means differ significantly from each other.

7. The subjects in the study described in problem 5 also took an examination designed to measure reading comprehension; their individual scores are listed below. Conduct the appropriate t-test to determine if the two groups differed significantly with respect to comprehension scores.

SPEED SCAN COMPREHENSION SCORES			RAPID READ COMPREHENSION SCORES		
85	72	69	66	81	73
58	74	63	88	83	80
71	75	81	77	81	83
64	78	75	89	79	84
69	79	72	81	85	80

8. For the hypothesis tests described in problems 5 and 7 above, describe what a Type I error would be in each case and what a Type II error would be.

9. Is one pain reliever faster than another? Forty people who frequently suffer from headaches were randomly divided into two groups of twenty each. Group I received common aspirin, and Group II received Brand Z pain reliever in the study; people were asked to take the drugs home with them and, at the occurrence of headache, take one tablet and record the amount of time that elapsed before feeling the pain ease. These recorded times are shown below. Perform the hypothesis test to determine whether or not the two drugs differ. (During the experimental test period, five people failed to have headaches; thus you should note carefully that neither sample size is equal to 20.)

GROUP I			GROUP II		
5.7 min	6.8	4.2	5.8	6.1	9.4
7.4	3.1	6.2	9.3	8.2	8.1
8.5	4.7	4.9	7.4	8.5	7.9
5.3	3.2	6.4	9.9	10.4	8.8
7.1	7.6	5.8	7.3	6.2	5.9
6.1	6.9	5.6	6.7	8.5	

10. In problem 4 in Chapter 10, an experiment was described that attempted to assess the effects of alcohol consumption on human reaction time in a driving simulator device. In that study, 100 subjects were tested. A better way to conduct that experiment, however, would be with the use of a repeated-measures design; fewer subjects could be used because this design removes much of the unwanted variability among measurements due to individual differences among subjects. Suppose, instead, that 12 subjects were allowed to practice in the machine until reaction times reached a stable level (reaction times might be the elapsed time between presentation of a simulated road hazard and placement of the subject's foot on the brake pedal). After practice, the subjects were each tested once more for reaction

time (Condition I), given a cocktail containing 50 ml of 80-proof liquor, and tested again (Condition II). These reaction times are listed below in milliseconds (one millisecond, ms, is one one-thousandth of a second).

PERSON	CONDITION I	CONDITION II
1	380 ms	397
2	330	338
3	350	349
4	420	437
5	250	246
6	280	310
7	375	398
8	313	328
9	370	377
10	240	260
11	290	282
12	383	390

Perform the related-samples t-test to determine if indeed alcohol altered reaction times for these subjects.

11. Given below are dependent variable measurements from a matched-pairs experiment. Perform the appropriate two-tailed t-test and decide whether or not the means of the two groups differ significantly from each other.

PAIR	PAIR MEMBER	
	1	2
1	405	620
2	337	341
3	605	710
4	550	442
5	670	381
6	740	831
7	832	841
8	506	521

12. Two sets of measurements are presented in the table. Test the significance of the difference between means (a) using the independent-samples test, and (b) using the related-samples test. Use a two-tailed t-test, and check only at the 0.05 alpha level. What assumption in the related-samples case allows use of the $s_{\bar{D}}$ error term instead of $\sqrt{\dfrac{\hat{s}_p^2}{n_1} + \dfrac{\hat{s}_p^2}{n_2}}$?

(PAIR)[7]	GROUP 1	GROUP 2
1	39	34
2	48	40
3	46	36
4	39	37
5	35	33

13. Two groups of laboratory rats are raised in different size cages. Group I animals are raised in cages with 4 times the floor space in the Group II animal cages. Animals in both groups are treated alike in all other respects. Listed below are body weights for the 10 animals in each group. Perform the two-tailed t-test to decide if cage size had an effect on body weight. Assume that there were no matched pairs (i.e., an independent samples experiment).

GROUP I		GROUP II	
450 g	400	380	420
520	390	360	490
530	470	517	434
410	490	470	398
508	527	462	480

14. The animals in Group II were then transferred to larger cages. The number of revolutions that each animal averaged per day on an activity wheel was assessed before and after the cage transfer. Use the related samples t-test to decide whether activity-wheel behavior was different under different conditions.

ANIMAL	BEFORE TRANSFER	AFTER TRANSFER
1	20	25
2	140	178
3	62	61
4	196	248
5	145	180
6	190	154
7	130	220
8	73	148
9	226	257
10	39	72

[7]"Pair" designations have meaning only in the related-samples case.

12. Parameter Estimation: Educated Guessing

After reading this chapter you should be able to do the following:

1. *Give an example of an application of inferential statistics in which parameter estimation is used instead of hypothesis testing.*

2. *Distinguish between point estimates and interval estimates of parameters.*

3. *Discuss the characteristics of a good estimator.*

4. *Perform point estimates for the mean, standard deviation, and proportion.*

5. *Construct a confidence interval for the mean.*

6. *Explain the effect of sample size, sample standard deviation, and level of confidence on the width of the confidence interval.*

Estimating Parameters

In hypothesis testing, the scientist approaches a study with a research hypothesis in mind — some idea that relations exist between variables. The test either supports or fails to support these hypotheses. Not all research, however, is concerned with differences among groups or with finding cause-and-effect relations. The scientist may simply want to learn unknown characteristics of groups. What was the mean height of Bronze Age man?[1] How much beef does the average adult American consume each year? What proportion of the population may be considered alcoholic? When a researcher selects a sample from these populations and uses sample observations to estimate population characteristics, he or she is engaging in parameter estimation.

Parameters, remember, are characteristics of the entire population. These include the mean, median, standard deviation, proportion in a certain category, correlation between variables in the population, and

---

[1]Bronze Age man's height would be estimated from a sample of available Bronze Age skeletons.

many others. Since the exact value of any parameter can only be determined by making an observation on every member of the population, and since this is usually impossible to do when working with human or animal populations, the behavioral scientist must often be content with **parameter estimates** based on sample observations.

As do all inferential procedures, parameter estimation depends on proper sampling procedures. If a biased sample is selected from the population, a biased parameter estimate will be the likely result. To illustrate parameter-estimation methods, let's assume that publishers of a certain magazine hire a consumer psychologist with a strong background in statistics to find out just who the magazine's readers are (the population is thus "readers of the magazine"). To begin with, she recommends that the following parameters be estimated: mean age of the readers, standard deviation of reader ages, and the proportion of readers who are male (and thus also the proportion who are female). For these parameters, she will provide the publishers with a **point estimate;** for the mean age she will also provide an **interval estimate.**

Point Estimation Versus Interval Estimation

The **point estimate** is simply a single value, based on the sample, that the statistician assumes to be the best estimate of the parameter's value. "Best" means several things to the statistician. First, it means that the estimator statistic is **unbiased.** An unbiased estimator is a statistic whose expected value is equal to the parameter value; that is, the mean of the sampling distribution equals the corresponding parameter value. Second, the statistic provides an estimate that is **consistent.** A consistent estimator is one that has a higher probability of being close to the parameter value when large samples are used than when small samples are used. We will also look at some point estimators that meet this requirement. Another characteristic of an estimator that produces the "best" estimate is that the statistic uses all the available information in the sample; if every sample observation contributes to the value of the estimate, the estimator is said to be **sufficient.** So we will make point estimates using statistics that are unbiased, consistent, and sufficient.

We are already familiar with estimators that meet these criteria.[2] For the population mean μ, the best point estimator is the sample mean $\bar{X}$. For the population proportion, symbolized P, the sample proportion p is the best estimator. The population standard deviation is best estimated with the unbiased sample standard deviation, $\hat{s}$.

The consumer psychologist can thus produce point estimates for these parameters by computing the corresponding sample statistics. Of course,

---

[2]Examples of statistics that do not meet these criteria are s and the median. The sample standard deviation s is a *biased* estimator of σ because the mean of its sampling distribution is *less* than σ. The sample median is not a *sufficient* estimator of the population median since it does not use all the information in the sample (does not use all sample values, just their rank orders and middle values).

she will want to select a representative sample for this purpose, and this could be difficult to obtain. She might use a stratified random sample composed of two strata: subscribers and over-the-counter buyers. Within each stratum she randomly selects the appropriate number. Suppose she begins with an extremely small N of 25. For this sample the following statistics are computed:

$$\bar{X} = 28 \text{ years}$$
$$\hat{s} = 8 \text{ years}$$

Proportion of male readers $= p = 0.30$

Proportion of female readers $= 1 - p = 0.70$

Thus, the best point estimates are:

Estimated mean age $=$ estimated $\mu = 28$ years

Estimated standard
    deviation of ages $=$ estimated $\sigma = 8$ years

Estimated proportion
    of male readers $=$ estimated $P = 0.30$

Estimated proportion
    of female readers $=$ estimated $Q = 0.70$

Point estimates by themselves, however, provide no means of evaluating the precision of those estimates. For this, we must turn to interval estimates. As Figure 12.1 suggests, an interval estimate is a range of values; the statistician reports the upper and lower values of this interval, and the **level of confidence.** The level of confidence is a concept closely related to probability and expresses the statistician's confidence that she has encapsuled the true parameter value in the interval. The width of the specified interval and the stated level of confidence allow the statistical consumer to evaluate the quality of the estimate. We could construct confidence intervals for many different parameters; however, we will simply introduce the concept here by covering only the confidence interval for μ.

Confidence Interval for μ

The mean age of magazine readers in the sample was 28 years. If we knew μ and σ, we could construct a sampling distribution of the mean for $\bar{X}$ like that in Figure 12.2(a) (p. 330). The sampling distribution would have a mean, μ, and a standard error, $\sigma/\sqrt{N}$. To compute a confidence interval we perform a similar operation with our sample statistics. In interval estimation we construct, essentially, our best estimate of the sampling distribution over $\bar{X}$, as shown in Figure 12.2(b). This distribution has a mean of $\bar{X}$ (here equal to 28) and a standard error equal to $\hat{s}/\sqrt{N}$ (here, $8/\sqrt{25} = 8/5 = 1.60$). The 95 percent **confidence interval** for μ is the range of values that lie under the central 95 percent area of this curve. Similarly,

Point Estimation

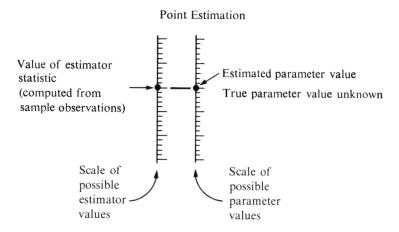

Value of estimator
statistic
(computed from
sample observations)

Estimated parameter value
True parameter value unknown

Scale of
possible
estimator
values

Scale of
possible
parameter
values

Interval Estimation

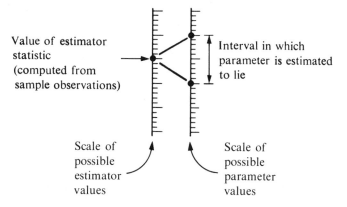

Value of estimator
statistic
(computed from
sample observations)

Interval in which
parameter is estimated
to lie

Scale of
possible
estimator
values

Scale of
possible
parameter
values

Figure 12.1 Comparison of point estimation and interval estimation. In point estimation the estimator statistic produces a single value (a "point"), which is the estimated parameter value. In interval estimation the estimator statistic and its sampling distribution are used to specify a range of values in which the parameter is estimated to lie. The width of the interval and the level of confidence specified allow one to evaluate the precision of the estimate.

the 99 percent **confidence interval** for μ is the range of values that lie under the central 99 percent area of this curve. These intervals are also indicated in Figure 12.2(b). Notice that the higher level of confidence (99 percent) is associated with a wider interval.

Confidence intervals are specified in terms of their **limits.** Since the interval is centered on $\bar{X}$, the upper and lower limits will lie equidistant from $\bar{X}$. Specifically, the amount that is added or subtracted from $\bar{X}$ is known as the **allowance** factor. Thus:

Upper limit = $\bar{X}$ + allowance factor

Lower limit = $\bar{X}$ − allowance factor

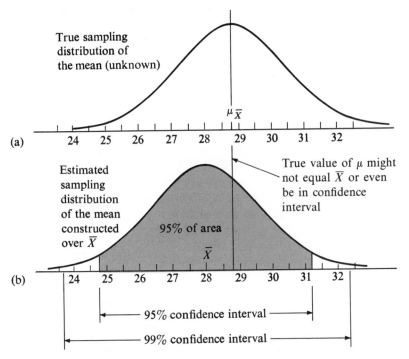

(a)

(b)

Figure 12.2 Two different confidence intervals for the text example statistics. Notice that as the level of confidence increases, the width of the interval also increases.

For the 95 percent confidence interval, the allowance factor is the distance above and below the mean that lies under 95 percent of the area under the distribution. Thus, the upper limit of the interval cuts off $2\frac{1}{2}$ percent of the area under the curve above the mean, and the lower limit cuts off $2\frac{1}{2}$ percent of the area below it. If the shape of the distribution is normal or that of the t-distribution, we can find the t or z values corresponding to these areas and translate them into units of the original measurement scale for the allowance factor.

Here, we are working with an unknown σ, so the sampling distribution is assumed to be shaped as a t-distribution with $N - 1$ degrees of freedom.[3] Then:

$$\text{Allowance factor} = (t\text{-critical})(\text{estimated } \sigma_{\bar{x}})$$

$$= (t\text{-critical})\left(\frac{\hat{s}}{\sqrt{N}}\right)$$

If the confidence level is 95 percent, t-critical is the same as t-critical for a two-tailed hypothesis test at an alpha level of $1 - 0.95$, that is, 0.05. With

---

[3]Remember that this assumption is valid only if the population is normal *or* if $df = 30$ or more.

24 degrees of freedom t-critical is 2.064. Using the expression for the allowance factor, the 95 percent confidence interval is specified by:

$$\text{Upper limit} = \bar{X} + (t\text{-critical})\left(\frac{\hat{s}}{\sqrt{N}}\right)$$

$$= 28 + (2.064)(1.60)$$
$$= 28 + 3.30$$
$$= 31.30$$

$$\text{Lower limit} = \bar{X} - (t\text{-critical})\left(\frac{\hat{s}}{\sqrt{N}}\right)$$

$$= 28 - 3.30$$
$$= 24.70$$

95 percent confidence interval for μ = 24.70–31.30

Changing the level of confidence changes only the value of t-critical. Suppose we want to be even more confident that the value of μ lies within our interval. The 99 percent confidence interval requires a t-critical equal to the t-critical for $\alpha = 0.01$. With 24 degrees of freedom, t-critical is 2.797. Thus, the limits of the 99 percent confidence interval are:

$$\text{Upper limit} = 28 + (2.797)(1.60)$$
$$= 28 + 4.48$$
$$= 32.48$$

$$\text{Lower limit} = 28 - 4.48$$
$$= 23.52$$

99 percent confidence interval for μ = 23.52–32.48

Raising the level of confidence widens the interval; if we are willing to accept a lower level of confidence, the interval becomes narrower.

The width of the confidence interval also depends on the value of $\hat{s}$, since this statistic partially determines the value of the estimated $\sigma_{\bar{x}}$. The value of $\hat{s}$ is, of course, not under control of the statistician, but the other factor involved in computing estimated $\sigma_{\bar{x}}$ is controllable: sample size. The larger the sample size, the smaller the estimated standard error of the mean, and thus the narrower the confidence interval. Suppose that the same study were conducted with a survey of 10,000 magazine readers instead of twenty-five, and the following statistics were obtained:

$\bar{X} = 28$ years

$\hat{s} = 8$ years

$N = 10,000$

Then, the 99 percent confidence interval for μ would become much narrower. The value of t-critical (for $df = 9,999, \alpha = 0.01$) is 2.576. Then:

$$\text{Upper limit} = 28 + (2.576)\left(\frac{8}{\sqrt{10,000}}\right)$$
$$= 28 + (2.576)(0.08)$$
$$= 28 + 0.21$$
$$= 28.21$$
$$\text{Lower limit} = 28 - 0.21$$
$$= 27.79$$

Thus, with 10,000 subjects, the 99 percent confidence interval based on these statistics is 27.79–28.21. With twenty-five subjects, the same statistics produce a 99 percent confidence interval of 24.70–31.30.

Interpretation of the Confidence Interval

What, precisely, does the statistician mean when she says that the 99 percent confidence interval for the mean age of magazine readers is 24.70–31.30? Strictly speaking, she means that if many such intervals were constructed in the same manner, 99 percent of them would enclose the true value of μ between upper and lower limits. Given simply a point estimate of μ, one has no way of evaluating the precision of estimation. For any given confidence interval, the width of the interval and level of confidence make possible such an evaluation.

Notice that this is not the same as saying that the probability that μ lies between 24.70 and 31.30 is 0.99. Such a statement is incorrect, because probability refers to relative frequency of some variable. The parameter μ is assumed fixed; it does not vary and will either always be between 24.70 and 31.30 or never be between those values. However, other intervals constructed in the same manner will have different limits; if all possible intervals, from all possible samples of size $N = 25$ from this population were constructed, 99 percent of them would have limits encapsuling the parameter. Thus, we use the term confidence interval instead of probability interval.

SUMMARY

In hypothesis testing the researcher approaches a study with a research hypothesis in mind, suspecting that there is a relation between variables. Often, however, the scientist simply wants to learn unknown characteristics — parameters — of populations. Since exact values of parameters can be determined only by making an observation on every member of the population, they must usually be estimated from sample observations.

From sample statistics one may produce either point estimates or interval estimates of parameters. The point estimate is an attempt to estimate the value of the parameter. We are already familiar with sample statistics for making point estimates of the mean, standard deviation, and proportion. The statistics used are good estimators because they are unbiased (have an expected value equal to the parameter value), consistent (have a higher probability of being close to the parameter value if larger samples are used), and sufficient (utilize all the information in the sample).

Methods of interval estimation produce two limits between which we estimate the true parameter lies. The width of the interval depends both on the standard error of the sample statistic and upon the level of confidence involved. The level of confidence and width of the interval allow the statistical consumer to evaluate the precision of the estimate.

KEY CONCEPTS

parameter estimates	unbiased estimator	level of confidence
point estimate	consistent estimator	confidence interval
unbiased estimator	sufficient estimator	limits of confidence intervals
interval estimate		allowance factor

SPOTLIGHT 12   Sample Size and Error of Estimate

To repeat a point made in Chapter 1, statistical methods themselves do not automatically reject bad data. Any group of observations, for instance, can be used to compute a confidence interval for the mean at the 95 percent confidence level and, in general, the larger the sample size, the narrower the interval. So, if a sociological researcher knows that 40 interviews provide a good estimate, why shouldn't he obtain 400 interviews to get an even more precise estimate? The problem is that data quality is sometimes affected by the number of observations that must be made. W. A. Wallis and H. V. Roberts addressed the problem of sample size and error in *The Nature of Statistics*:

> A large number of measurements made hurriedly or superficially may not represent as much true information as a small number made carefully. In extreme cases, poor data can be so misleading as to be worse than no information at all. A rather paradoxical example of the effective use of samples is the Bureau of the Census' use of them to check on the accuracy of the census. Although sampling error is almost absent from the census, the nonsampling error is considerable — that is, such errors as those arising from failure to make questions clearly understood, from misrecording replies, from faulty tabulation, from omitting people who should have been interviewed. In the sample census, however, these nonsampling errors may be reduced enough to offset the sampling error, for it is cheaper and easier to select, train, and supervise a few hundred well-qualified interviewers to conduct a few thousand careful interviews than it is to select, train, and supervise 150,000 interviewers to conduct a complete census of the population. Similarly, in measuring the useful life of the equipment in a telephone plant, the practical choice is not between measurements for a sample of the equipment and equally accurate mea-

surements for all the equipment, but between fairly precise measurements of a sample made carefully by competent engineers, and crude measurements of the whole plant made hastily by less skilled people. Even in laboratory experiments in the sciences, the difficulties of precise measurement are often so great that it is better to reduce the number of items measured in order to take more care with the individual measurements. . . .

The reader may wonder why, in view of the advantages of sampling, the entire population of the United States is enumerated completely every ten years. Aside from the overriding fact that the Constitution requires this, perhaps the most important reason is that information is required for very small groups of the population — such as small towns, individual neighborhoods in cities, etc. — as well as for the country as a whole. Even so, however, about half the questions on the 1950 census were asked only of a sample — for some questions a 20 percent sample, and for some questions a $3\frac{1}{3}$ percent sample (namely, a $16\frac{2}{3}$ percent subsample of the 20 percent sample).[4]

PROBLEMS

1. Explain in your own words the difference between
   a. A point estimate and an interval estimate.
   b. A biased estimator and an unbiased estimator.
   c. Confidence and probability.

2. a. What differences might there be between the sample mean and sample median, with regard to *sufficiency* (as discussed on p. 327), when these statistics are used to estimate population parameters?
   b. Consider the sample range as a point estimate of a larger population's range. Is the sample range ever going to be larger than the population range? What is more likely to be the case on every sample? Would this situation tend to make the sample range a biased or unbiased estimator of the population range?

3. How much does the American family pay each year for medical bills? To study this question 20 families are selected at random and asked to report their medical expenses for the past year. The obtained figures were (in dollars):

207	350	120	530	750
55	480	800	260	190
170	320	40	155	333
220	590	120	250	150

Compute point estimates for the following parameters:

a. μ

b. σ

c. Proportion of families that pay more than $500.

[4]W. A. Wallis and H. V. Roberts, *The Nature of Statistics* (New York: The Free Press, 1962), p. 138. Copyright © 1962 by The Free Press of Glencoe, Inc. Reprinted by permission.

4. For the study described in problem 2 above:
   a. Determine the 95 percent confidence interval for the mean;
   b. Determine the 99 percent confidence interval for the mean.

5. Listed below are "stopping distances" obtained in a special test of 25 drivers, randomly selected from people taking their driver's license examination. Each driver was instructed to drive his or her own car on a test track, at 30 miles per hour, and then stop as quickly as possible when a loud warning horn sounded.

250	402	330	220	305
310	150	164	210	187
219	140	310	148	228
215	187	241	179	196
301	202	220	248	105

   From these data, compute point estimates of the following parameters:
   a. μ
   b. σ
   c. Proportion of drivers who stop in 200 feet or less.

6. For the data given in problem 5, determine:
   a. The 80 percent confidence interval for μ.
   b. The 90 percent confidence interval for μ.

7. In constructing a confidence interval for μ, what size allowance factor is used in the following situations?
   a. 95 percent confidence level, $N = 121$, $\hat{s} = 25$.
   b. 80 percent confidence level, $N = 10$, $\hat{s} = 102$.
   c. 99 percent confidence level, $N = 41$, $\hat{s} = 1.45$.

8. Medical expenditures for 10,000 families are surveyed and the (hypothetical) results are $\bar{X} = \$500$; $\hat{s} = \$200$. Construct a confidence interval for the mean at:
   a. The 90 percent confidence level.
   b. The 95 percent confidence level.
   c. The 99 percent confidence level.
   Generalize: What effect does level of confidence have on the width of the interval?

9. Suppose that the statistics reported in problem 3 were obtained on a sample of twenty-five families. Again, construct
   a. The 90 percent confidence interval.
   b. The 95 percent confidence interval.
   c. The 99 percent confidence interval.

Generalize: What effect does sample size have on the width of the interval?

10. Explain why the Type I and Type II error terminology, first discussed for the one-sample hypothesis test in Chapter 10, cannot be applied to the confidence interval situation. What kind of error *can* be made in constructing a confidence interval? What is the probability of making such an error?

13. Nonparametric Tests

After reading this chapter you should be able to do the following:

1. *Describe the differences between parametric hypothesis tests and nonparametric tests.*

2. *Conduct a one-way chi-square test; use this test to perform the goodness-of-fit-test.*

3. *Conduct a two-way chi-square test for association and interpret the results.*

4. *Perform the following tests for differences between two samples: the sign test, the Wilcoxon rank-sum test, and the Wilcoxon matched-pairs signed-ranks test. Explain why these tests are sometimes preferred to their parametric counterparts.*

Parametric Versus Nonparametric Tests

The material in the last three chapters has been concerned almost exclusively with means and variances. Indeed, all of the research questions used as examples were phrased in terms of questions about means. Of course there are many important questions that cannot be related to population means, and this chapter introduces you to a few of the methods used for asking and answering these questions. We will still use the logic of Neyman-Pearson hypothesis testing; only the statistics and sampling distributions will change.

Remember that the hypothesis tests in preceding chapters required us to compute statistics from interval-level data or better; in the behavioral sciences we are not always blessed with such high-quality raw material. Often we would like to ask questions about underlying variables, but the behavioral measures of those variables are probably related in an ordinal relationship at best. (This might be a good time for you to review the material in Chapter 1 on level-of-measurement issues if it has been some time since you covered them.) Thus, the kinds of statements parametric tests allow us to make about underlying variables are often limited.

337

The hypothesis tests about means also required that we use either large samples or samples from normally distributed populations, so that the shape of the sampling distributions would be known. Again, those tests require conditions that cannot always be met.

Tests like the z-test, t-test, and F-test are known as **parametric tests** because the hypotheses involved are stated in terms of population parameters. There is a large family of statistical tests designed for situations that do not meet the above criteria and that do not frame hypotheses in terms of population parameters. These are known as **nonparametric tests** or **distribution-free tests** (because they require no assumptions, or else very few assumptions about the population parameters).

Nonparametric tests, then, will serve much the same purpose as the t-test did in Chapter 11. In a controlled experiment, we want to know whether or not experimental groups differ from each other on some dependent variable. These tests may be used to evaluate all of the two-sample experimental situations we examined in Chapter 11, as well as additional situations where the requirements for at least interval-level data, normal populations, or large samples are not met.

The **chi-square statistic,** symbolized χ^2, is used in many of these. We will begin our coverage of nonparametric statistics by examining some uses of this important statistic.

The One-Way Chi-Square Test

The basic purpose of the **one-way chi-square test** is not very different from that of tests about means. We have a collection of observations on some variable, and we want to know whether the observed distribution can be attributed entirely to sampling variability, or whether some other factor(s) influenced distribution of the observations. Consider, for instance, the following (hypothetical) survey results taken from a random sample of 100 undergraduate students. Each student was asked to indicate on a questionnaire whom he or she would go to first for help with a major personal problem. The students responded as follows:

	f
Campus counseling office	27
A friend	28
Clergyman	24
Parent	11
No one	10

Taking each student's preference is thus making an observation on a nominal-level variable "whom he or she would go to first." Now, the question addressed here is whether or not some categories are (significantly) more popular than others.

Notice that the data for the chi-square test consist of *frequencies*. As you recall, nominal, ordinal, and all other levels of data can be tabulated as frequencies. There is no requirement that these response categories be arranged in a rank order or in any way represent equal units of measurement. We simply want to know whether these frequencies are distributed among categories randomly or nonrandomly. Specifically, the pair of hypotheses under examination is:

$$H_0: O = E$$
$$H_1: O \neq E$$

The null hypothesis says that the observed frequencies (O) may be considered equal to the frequencies expected (E) if the true case is that responses are randomly distributed among categories. The null hypothesis, in other words, says that our observed frequencies do not differ much from the frequencies we would obtain when each category is equally likely to be picked. The alternative hypothesis states that the observed frequencies differ from random frequencies; in other words, H_1 states that some categories are more likely than others to be picked because our data differ substantially from the expected pattern of frequencies one would get under a true H_0. The chi-square hypothesis test, of course, is used to decide between these hypotheses.

The chi-square test always requires *two* frequency table columns, one showing the observed frequencies and the other the **expected frequencies.** Table 13.1 shows these for the present example. The O_i column contains the observed frequencies presented earlier, and the E_i column shows the frequencies that would be most likely to occur if there were no true student differences in preferences among categories. When there are no reasons to expect preferences for any category under the null hypothesis, each expected frequency (E_i) is simply N/k, where k is the number of response categories, and N is the total number of observations — 100 in the above example. If we had made a survey of 200 students instead of 100, each expected frequency would have been 40. The subscript i denotes the specific response category; i will have values ranging from 1 through k.

Do the observed frequencies differ significantly from the expected

Table 13.1 *Observed and Expected Frequencies for Student Choice of Aid Sources*

i		O_i	E_i	$O_i - E_i$
1	Campus counseling office	27	20	7
2	A friend	28	20	8
3	Clergyman	24	20	4
4	Parent	11	20	−9
5	No one	10	20	−10

frequencies? We will approach this question by first computing the chi-square test statistic as follows:

$$\chi^2\text{-observed} = \sum_{i=1}^{k} \frac{(O_i - E_i)^2}{E_i}$$

Before explaining the use of this statistic, let's compute the observed chi-square of these data. Notice that the difference between each O_i and E_i has been listed in the third column of Table 13.1. To compute the obtained statistic, we square the difference between each O_i and its corresponding E_i; this square is then divided by E_i. The sum of these quotients is taken as the chi-square observed. That is:

$$\chi^2\text{-observed} = \frac{(27 - 20)^2}{20} + \frac{(28 - 20)^2}{20} + \frac{(24 - 20)^2}{20}$$

$$+ \frac{(11 - 20)^2}{20} + \frac{(10 - 20)^2}{20}$$

$$= \frac{(7)^2}{20} + \frac{(8)^2}{20} + \frac{(4)^2}{20} + \frac{(-9)^2}{20} + \frac{(-10)^2}{20}$$

$$= \frac{49}{20} + \frac{64}{20} + \frac{16}{20} + \frac{81}{20} + \frac{100}{20}$$

$$= 2.45 + 3.20 + 0.80 + 4.05 + 5.00$$

$$= 15.50$$

This value, 15.50, is assumed under the null hypothesis to represent one possible value from a sampling distribution. Here, the sampling distribution that would appear if we sampled randomly from the same population, an infinite number of times, and if the null hypothesis were true, is approximated by a mathematically determined curve called the **chi-square distribution.** Figure 13.1 shows the shape(s) of the chi-square distribution(s).

As with any other sampling distribution, we can determine probabilities associated with the occurrence of different values of the statistic by finding areas under the curve. Notice that the chi-square distribution, like the t-distribution, is actually a family of distributions whose members differ in shape depending on the number of degrees of freedom. In the one-way chi-square test (which we are doing) $df = k - 1$, where $k =$ the number of response categories. Hence, the degrees of freedom depends on the number of possible observation categories and not on the number of observations. In this example, $k = 5$, so $df = 4$.

Tables of critical values for chi-square are shown in Table C, Appendix 1. An abbreviated chi-square table is given here as Table 13.2 (p. 342). As in the t-distribution table, the left-hand column lists degrees of freedom. Columns labeled "Probability" indicate the relative frequency of occurrence of values of chi-square equal to or higher than the tabled values when H_0 is true; these probabilities are equivalent to alpha levels in a one-tailed test. To test the chi-square hypotheses in this example, at

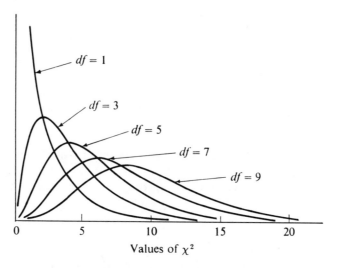

Figure 13.1 The chi-square distribution for various degrees of freedom. Up to about 30 degrees of freedom the shape of the distribution changes drastically with changes in the degrees of freedom.

$\alpha = 0.05$, $df = 4$, we find the appropriate chi-square critical from the table: 9.49. For the indicated df, the observed value of χ^2 exceeds the tabled values with a probability listed at the top of each column in Table 13.2. For 3 df, the value 7.815 is exceeded 0.05 of the time. The test is, in fact, a one-tailed test,[1] and the decision rule is:

If χ^2-observed $\geqslant 9.49$, reject H_0

Thus, we reject H_0 at the 0.05 level. Checking the table further, we discover that this observed statistic is also significant at the 0.01 level.

As with the other hypothesis tests, the selected alpha level represents the probability of making a Type I error. When H_0 is true, values of chi-square greater than the critical value will appear with a relative frequency equal to alpha. Increasing the number of degrees of freedom increases the power of the test; the more response categories possible, the greater the probability of rejecting a false H_0.

What then does this decision say about our data? By rejecting the null hypothesis, we have said that the true distribution of frequencies among categories is not equal; that is, each of the categories is actually *not* an equiprobable choice of students seeking counseling.

Notice that the table only shows chi-square distributions for degrees of freedom up to and including 30. Chi-square distributions for $df > 30$ are similar in shape (although mean and variance of the chi-square distribution changes with increasing degrees of freedom).

---

[1]There is no "lower" region of rejection possible with the chi-square, since the lowest possible value of the obtained statistic is zero, and an occurrence of this value would indicate no difference at all between observed and accepted frequencies, that is, do not reject H_0.

Table 13.2   *Abbreviated Chi-Square Table*

			PROBABILITY		
df	0.20	0.10	0.05	0.02	0.01
3	4.642	6.251	7.815	9.837	11.341
4	5.989	7.779	9.488	11.668	13.277
5	7.289	9.236	11.070	13.388	15.086

The chi-square can even be used when there are only two response categories ($df = 1$). When this is done, however, one should employ **Yates's correction** to the computational process. With Yates's correction:

$$\chi^2\text{-observed} = \sum \frac{(|\ O_i - E_i\ | - 0.5)^2}{E_i}$$

We can use this, for instance, to examine the significance of results advertised by a well-known luxury-car manufacturer. The television advertisement reports that 100 owners of car A were taken for a drive, blindfolded, in car B. After the test, the same individuals were taken for a drive in car A. We hope the order of rides was counterbalanced by taking half the group in A first. Now, the ad reports, 60 out of 100 prefer the ride of B to A. Is this significant? Or, with allowance for random variability, should we say that they chose about equally between cars?

This question can be tested with the chi-square, using the following information:

PREFERENCE CATEGORY	O_i	E_i	$O_i - E_i$	$\|\ O_i - E_i\ \| - 0.5$
Car A	40	50	-10	9.5
Car B	60	50	10	9.5

Notice that an extra column is added for Yates's correction. The expected frequencies of 50 indicate that if there were no preference, drivers would choose equally between categories. The χ^2-critical for $df = 1$, $\alpha = 0.05$, is 3.841. Then:

$$\chi^2\text{-observed} = \frac{(|40 - 50| - 0.5)^2}{50} + \frac{(|60 - 50| - 0.5)^2}{50}$$

$$= \frac{(9.5)^2}{50} + \frac{(9.5)^2}{50}$$

$$= 1.805 + 1.805$$

$$= 3.61$$

This is not greater than χ^2-critical, so we do not reject H_0. We decide (at least with this statistic and level of significance) that there is not sufficient evidence to indicate a preference for one car over the other. Notice that

χ^2-observed computed without Yates's correction is 4.00, a value which *does* call for rejection of H_0.

Yates's correction must also be applied when any *expected* frequency (E_i) is less than 10. The net effect of applying the correction is to raise by 0.5 the difference between O_i and E_i when O_i is less than expected and to lower by 0.5 that difference when O_i is greater than expected. This is an attempt to make the empirically obtained frequencies better approximate the mathematically defined sampling distribution. The obtained chi-square is computed from frequencies — whole numbers. Thus, an empirically determined chi-square distribution would be discrete, while the mathematically determined curve is continuous. The correction lessens the difference between the two.

The difference between empirical and mathematical sampling distributions becomes especially critical when expected frequencies are less than 5; thus many statisticians believe that when E_i is less than 5, Yates's correction is no longer adequate and one must use Fisher's exact method, a technique beyond the scope of this book.[2]

The Goodness-of-Fit Test

The one-way chi-square test just shown allowed us to decide whether or not observations were equally likely to appear in all categories. With the one-way chi-square, however, we do not have to restrict ourselves to questions about equally likely categories. Consider, for example, a situation faced by a sociologist who wishes to determine if the city's three major ethnic groups are making equal use of the municipal swimming pool. She knows that the city's population is 35 percent Afro-American, 20 percent Mexican-American, and 45 percent other Americans. If all these groups used the pool in proportion to their relative part of the city's population, we would expect that 35 out of 100 pool users would be Afro-American, 20 out of 100 Mexican-American, and 45 out of 100 other ethnic groups. Suppose that a random sample of 100 pool users produced the following obtained frequencies (shown with the expected frequencies):

	O	E
Afro-American	25	35
Mexican American	15	20
Other	60	45

Is one group using the pool significantly more than the other groups? Notice that here we do not ask if equal numbers of people are using the

---

[2]Fisher's exact method is discussed in Q. McNemar, *Psychological Statistics*, 3d ed. (New York: Wiley, 1962).

pool, but, rather, if pool usage conforms to a particular *model*: under the null hypothesis, we predict that 35 percent of the pool users will be Afro-American, 20 percent Mexican-American, and 45 percent other-Americans. The one-way chi-square is called the **goodness-of-fit-test** if it is used to determine whether or not our observed frequencies differ significantly from the model. Since none of the expected frequencies are 10 or less, and since $df > 1$, Yates's correction is not needed. Then:

$$\chi^2\text{-observed} = \frac{(25 - 35)^2}{35} + \frac{(15 - 20)^2}{20} + \frac{(60 - 45)^2}{45}$$
$$= 2.86 + 1.25 + 5.00$$
$$= 9.11$$

This is greater than the χ^2-critical for $\alpha = 0.05$, $df = 2$, which is 5.991, so we reject H_0; apparently a disproportionate number of one group is using the pool. (The test is not significant at $\alpha = 0.01$.)

At this point, we can use the goodness-of-fit-test to decide whether a variable is normally distributed or not. We will do the same test that we did with the swimming pool study, except that our model — the basis of our expected-frequency predictions — is the normal curve.

Let us suppose that we have devised an exam to assess basic reading ability of high-school graduates, that we have given an early version of the exam to 500 randomly selected graduating seniors from around the country, and now we want to know if the test produces scores that are normally distributed. We can apply the goodness-of-fit-test to our scores to make that decision. This will take a bit of calculation, but it involves only concepts we have already covered.

With the swimming pool example above, our expected frequencies were different in each of the three categories. Here, with the reading test scores, let us arbitrarily divide the normal distribution into ten intervals that contain equal areas under the normal curve probability density function. If the *areas* over each of ten intervals are the same, then we would expect the relative frequency of observations in each interval to be the same.

Table 13.3 shows the limits of these intervals, along with some other information needed to conduct the test. The z-score limits for the intervals were obtained by dividing the total area under the curve into increments of 0.1000, and finding the z-scores associated with the boundaries of these increments. This much of the process, of course, is done entirely with Table A, Appendix 1, and does not involve the data.

Next, we administer the test to 500 people and obtain a frequency distribution of their scores. Let us assume that this group produced a mean score of 530 and a standard deviation of 110. What we want to know now is whether or not these scores can be considered normally distributed. If they were perfectly normal in distribution. we would expect 50 of them (one-tenth of the total 500) to fall within each of the z-score limits shown in Table 13.3. To see how many actually do fall within each

Table 13.3 *Summary Table for Conducting the Goodness-of-Fit Test for Normality*

INTERVAL	Z-SCORE LIMITS	TEST-SCORE LIMITS	O_i	E_i
1	$-\infty$ to -1.282	0–389	42	50
2	-1.282 to -0.842	389–437	43	50
3	-0.842 to -0.524	437–472	48	50
4	-0.524 to -0.253	472–502	57	50
5	-0.253 to 0.000	502–530	59	50
6	0.000 to 0.253	530–558	62	50
7	0.253 to 0.524	558–588	50	50
8	0.524 to 0.842	588–623	42	50
9	0.842 to 1.282	623–671	48	50
10	1.282 to ∞	671–1,000	49	50

interval, we could find the z-score of each individual score and fill in our O_i column directly. However, we can save some work and translate the z-score limits to test-score points, where we assume a normal distribution with μ equal to our sample $\bar{X}$ and σ equal to our sample $\hat{s}$. (By doing so, incidentally, we are involving some assumptions about parameters. This is one out of many such cases where parameters enter into nonparametric tests; this is still basically a nonparametric test, however, because the tested hypotheses are about frequencies.) This estimation also requires us to alter the involved degrees of freedom as indicated below. Then, all that remains is to calculate χ^2-observed in the usual manner:

$$\chi^2\text{-observed} = \sum_{i=1}^{k} \frac{(O_i - E_i)^2}{E_i}$$

$$= \frac{(42 - 50)^2}{50} + \frac{(43 - 50)^2}{50} + \frac{(48 - 50)^2}{50}$$

$$+ \frac{(57 - 50)^2}{50} + \frac{(59 - 50)^2}{50} + \frac{(62 - 50)^2}{50}$$

$$+ \frac{(50 - 50)^2}{50} + \frac{(42 - 50)^2}{50} + \frac{(48 - 50)^2}{50}$$

$$+ \frac{(49 - 50)^2}{50}$$

$$= 1.28 + 0.98 + 0.08 + 0.98 + 1.62$$
$$+ 2.88 + 0.00 + 1.28 + 0.08 + 0.02$$

$$= 9.20$$

Normally, we would have $df = k - 1$ in a one-way chi-square test with k categories. Here, however, we must subtract an extra 2 df because we have used our sample statistics to estimate two population parameters. Thus:

$$df = k - 3 \qquad \text{Here, } df = 10 - 3 = 7$$

From Table C, Appendix 1, we see that with seven degrees of freedom, the chi-square value that is exceeded 5 percent of the time under a true null hypothesis is 14.067. Our observed value of 9.20 does not exceed the critical value, and we can conclude that our sample does come from a normally distributed population.

The Chi-Square Two-Way Test for Association

The chi-square statistic often appears when researchers want to determine whether or not two variables are independent of one another, that is, whether or not they are associated. "Independence" is used here in the same sense it was in Chapter 8, where we learned that two events A and B are independent *if*:

$$p(A)p(B) = p(A \text{ and } B)$$

As discussed in Chapter 8, the above rule could be used as a "test" for independence, by separately determining $p(A)$, $p(B)$, and $p(A \text{ and } B)$, and then checking to see whether or not the above multiplicative relationship held. We now approach the questions of independence with the chi-square statistic; as you should see, the chi-square is a more useful research tool than the multiplicative rule. Independence between two variables will be tested with the **chi-square two-way test for association.**

Consider, for instance, the example summarized in Table 13.4. In a particular industry the contract between union and management was about to expire. There seemed to be considerable dissension among workers as to whether or not to strike for a substantially better contract; thus, a survey was taken of worker opinion by selecting 600 workers at random from a population of 50,000. Each was asked to indicate his preference in the approaching strike vote as either "Yes," "Uncertain," or "No." The number of workers responding in each category is given as a *column total* in the "Observed Frequencies" tabulation of Table 13.4. Here, it appears that opinion is rather evenly divided, since 200 voted "Yes," 220 voted "Uncertain," and 180 voted "No." A one-way chi-square of these three response categories does not, in fact, indicate that frequencies are significantly different ($\chi^2 = 4.0$, $df = 2$). The one-way chi-square, therefore, simply tells us that, overall, each vote category was about as likely as each other category to be chosen. However, the results become more interesting when they are classified according to age of the worker as well as response type.

Take a minute to inspect Table 13.4's observed frequencies. It appears that a disproportionate number of younger workers favor striking, whereas a larger proportion of the older workers do not favor striking. In other words, it looks like *strike vote preference* (one variable) is not *independent* of age group (another variable). In probabilistic terms, it seems that the probability that a randomly selected worker votes "Yes," for instance, depends on which age category he is in. If we select one

Table 13.4 *Expected and Observed Frequencies for the Two-Way Chi-Square Problem Discussed in the Text*

OBSERVED FREQUENCIES:

AGE	STRIKE VOTE PREFERENCE			TOTAL
	YES	UNCERTAIN	NO	
Workers under 30	110	40	30	180
Workers 31–45	40	100	60	200
Workers 46 and over	50	80	90	220
Total	200	220	180	600 = Grand total

EXPECTED FREQUENCIES:

AGE	STRIKE VOTE PREFERENCE			TOTAL
	YES	UNCERTAIN	NO	
Workers under 30	60	66	54	180
Workers 31–45	66.67	73.33	60	200
Workers 46 and over	73.33	80.67	66	220
Total	200	220	180	600

j = number of rows = 3 $df = (j - 1)(k - 1) = (2)(2) = 4$

k = number of columns = 3

worker at random, and proceed along the lines discussed in Chapter 8, we can establish the following events:

 A = Selection of a worker under 30

 B = Selection of a worker who votes "Yes"

we can determine directly from the table that:

$$p(A) = \frac{180}{600} = 0.300 \quad \text{and} \quad p(B) = \frac{200}{600} = 0.333$$

Furthermore, we see that there are 110 "Yes" votes among the "Under 30" workers, so:

$$p(A \text{ and } B) = \frac{110}{600} = 0.1833$$

According to the Chapter 8 probability operation, A and B are not independent events because:

$$p(A)p(B) \neq p(A \text{ and } B)$$
$$(0.300)(0.333) \neq 0.1833$$

Now, are we justified in stating that different age-group categories have different strike preferences? That is, can we claim that the two variables are associated rather than independent?

On the basis of the multiplicative rule, we cannot yet make such statements. The reason is that we are dealing with a sample of 600 people from a population of 50,000. Application of the multiplicative rule, as discussed in Chapter 8, requires that we restrict our statements about independence to the sample observations that actually went into these a priori probability calculations; we cannot use the rule in this form to make statements about population independence. The two-way chi-square test for association, however, will let us decide whether or not such independence exists in the population, on the basis of our sample observations.

Note that 200 people, overall, responded "Yes"; this is one-third of the total 600. If there is no association (i.e., if there is independence) between variables, then we would *expect* one-third of each age group to respond "Yes." And we would expect similar proportionality for the other two-response categories: The proportion of each age group that falls in a category should be the same for all age groups. The statistical test involved here consists of producing an expected frequency according to the assumption of nonassociation, and using the chi-square in the usual manner to determine if observed frequencies differ significantly from expected frequencies.

We *could* produce an expected frequency for each age-group/response-category cell of our table by using the method suggested in the previous paragraph. That is, since one-third of the overall total voted "Yes," we can calculate that one-third of the 180 workers under 30 are expected to answer "Yes" (that is, 60), one-third of the workers 31–45 should answer "Yes" (66.67), and so on. However there is an easier way to obtain the same expected frequencies using the row totals and column totals for the observed frequencies:

$$\text{Expected frequency for a given cell} = \frac{(\text{Row total})(\text{Column total})}{\text{Grand total}}$$

For the cell in row 1, column 1:

$$E = \frac{(180)(200)}{600}$$

$$= 60.0$$

The same procedure may be used to obtain the other expected frequencies. Notice that in the table of expected frequencies, the row totals and column totals are the same as the observed-frequency row and column totals.

Once the observed frequencies are tabulated, and the expected frequencies are calculated from the observed row and column totals, we can calculate chi-square observed in the same manner as before:

$$\chi^2\text{-observed} = \sum \sum \frac{(O - E)^2}{O}$$

Notice that with the two-way chi-square, the summation sign is doubled. This indicates that we sum across all rows and columns, including each cell of the O and E tables. (To avoid possible confusion with the analysis-of-variance notation that follows in the next two chapters, more comprehensive subscripting will be avoided with chi-square in this chapter.)

Thus, for the example at hand:

$$\chi^2\text{-observed} = \frac{(110 - 60)^2}{60} + \frac{(40 - 66)^2}{66} + \frac{(30 - 54)^2}{54}$$

$$+ \frac{(40 - 66.67)^2}{66.67} + \frac{(100 - 73.33)^2}{73.33}$$

$$+ \frac{(60 - 60)^2}{60} + \frac{(50 - 73.33)^2}{73.33}$$

$$+ \frac{(80 - 80.67)^2}{80.67} + \frac{(90 - 66)^2}{66}$$

$$= 41.67 + 10.24 + 10.67 + 10.67 + 9.70$$

$$+ 0 + 7.42 + 0.01 + 8.73$$

$$= 99.11$$

The degrees of freedom are given by the expression:

$$df = (j - 1)(k - 1) \qquad \text{where } j \text{ is the number of rows and } k \text{ is the number of columns.}$$

Here, with $df = 4$, we see in Table C that a chi-square of 13.277 is exceeded only 0.01 of the time under a true null hypothesis. Our observed chi-square of 99.11 is much greater than this, so we reject the null hypothesis that $O = E$ for the observation categories and decide that workers' strike vote preferences are *not* independent of their ages. Rather, there is an association between the two variables. In this case, the sample data allow us to decide that, in the population the distribution of "Yes," "Uncertain," and "No" votes differs among age-group categories. And, if we accept the above statement, we must also accept the statement that among different vote categories, the distributions of ages differ. If we say that the two variables are associated, *both* statements follow.

The Sign Test for Related Samples

Not all nonparametric tests use the chi-square statistic. We will now look at a nonparametric test that uses the binomial variable r as the sampling statistic, a test that is quite simple to apply (if you are comfortable with the binomial distribution) and that has a wide range of applications. As with many of our previous tests, this one is probably introduced most easily through an example problem.

The **sign test** applies to the same kinds of experimental situations as does the t-test for related samples; that is, experiments where we have either matched pairs or repeated measures. To use the t-test, however, we had to make some assumptions that included (1) normality of the population distribution and (2) interval-level data or better. The sign test does not require these assumptions and is thus applicable in places where the t-test is not.

Suppose that a researcher is investigating the relationship between crowded environments and aggressiveness in rats. He begins an experiment with 10 pairs of litter mates. Each pair will have a different genetic background and several different strains may be represented. However, both members of the pair are from the same litter. One member of each pair is raised in an overcrowded environment; these 10 animals constitute Group I. The other member of each pair is reared in a spacious laboratory environment. So far, you may recognize this as a two-sample experiment with matched pairs where the independent variable is the degree of crowding in the animal's home environment. The dependent variable, however, is not amenable to analysis with the t-test. Aggressiveness here is measured through a test of dominance where the two members of each pair are placed in a narrow pathway leading to a food cup. In such situations, one animal will usually force the other aside. The animal prevailing in the dominance contest will be called the more aggressive of the two.

Before proceeding, it will be useful here to consider the level-of-measurement issue. It is not unreasonable to suggest that "aggressiveness" is a continuously distributed variable: some animals may be very, very aggressive, other animals only slightly less aggressive, and others may represent any other "amount" of aggressiveness. Aggressiveness, however, must be treated as an underlying variable, as discussed in Chapter 1. We can never measure quantities of the underlying variable directly; we must instead specify an observable behavioral variable that we believe tells us something about the underlying variable. In this example, "dominance" is the behavioral variable.

Dominance, however, represents ordinal-level measurement. We can assume only that the dominant animal is more aggressive than the other animal. We do not know *how much* difference there is between the two animals in aggressiveness; we cannot assign interval-level numerical aggressiveness scores because aggressiveness is not directly observable. Nonetheless, our conclusions in the following nonparametric test can be applied to statements about aggressiveness as well as to statements about dominance; to do so we need only assume that the dominant animal is the more aggressive animal. The nonparametric test, in other words, will allow us to make statements about underlying variables that could not properly be made with a parametric test using ordinal-level measurement.

A possible set of data for the sign test is shown below. Assume that the dominance test was carried out for 10 pairs of animals, and each pair is given a + or − sign according to the following system:

Group I rat dominates Group II rat: +

Group II rat dominates Group I rat: −

This system of designating pluses and minuses is purely arbitrary — it is simply a way of listing the results in a single column as shown in the table below.

PAIR	GROUP OF DOMINANT ANIMAL	SIGN
1	Group I	+
2	Group II	−
3	Group I	+
4	Group I	+
5	Group I	+
6	Group II	−
7	Group I	+
8	Group I	+
9	Group I	+
10	Group I	+

Now, it certainly looks like Group I animals tended to dominate Group II animals, but is this apparent difference between groups due to a real difference in dominance or just to sampling variability? To use the sign test in this case, let us assume that the dominance test for each pair constitutes a Bernoulli experiment as discussed in Chapter 8. The two outcomes are, of course, "Group I animal dominant" and "Group II animal dominant"; we have called these + and −, respectively. The null hypothesis is that environment had no effect on dominance, and therefore, the probability of either outcome ought to be 0.50 for each pair. If this (null hypothesis) is true, then, a series of 10 independent Bernoulli trials should constitute a binomial experiment, and the probability of each value of r (r = the number of + events) is described by a binomial probability distribution with parameters $N = 10$ and $p = 0.50$. Such a null hypothesis distribution is shown in Figure 13.2.

The hypothesis test now proceeds in the same way as our other hypothesis tests: if our obtained statistic is very unlikely under a true null hypothesis, we will reject the null hypothesis. Here, we ask, "What is the probability of getting 8 pluses or more when $N = 10$ and $p = 0.50$?" We can calculate that probability directly as:

$$\text{prob}(r = 8) + \text{prob}(r = 9) + \text{prob}(r = 10) = 0.0439$$
$$+ \ 0.0098 + 0.0010$$
$$= 0.0547$$

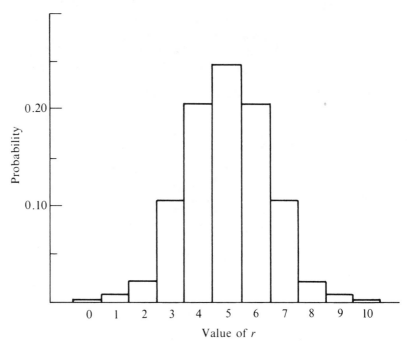

Binomial Distribution with $N = 10$, $p = q = 0.50$

$$\text{prob}(r = 0) = \text{prob}(r = 10) = 0.0010$$
$$\text{prob}(r = 1) = \text{prob}(r = 9) = 0.0098$$
$$\text{prob}(r = 2) = \text{prob}(r = 8) = 0.0439$$
$$\text{prob}(r = 3) = \text{prob}(r = 7) = 0.1172$$
$$\text{prob}(r = 4) = \text{prob}(r = 6) = 0.2051$$
$$\text{prob}(r = 5) = 0.2461$$

Figure 13.2   Binomial distribution used in the sign test example discussed in the text.

This figure, 0.0547 would then be the alpha level obtained in a *one-tailed test*. However, since the experimenter does not hypothesize that animals in Group II cannot be more aggressive than animals in Group I, we must conduct a two-tailed test. This means that our obtained significance level is double 0.0547, that is, 0.1094. This is not a low enough alpha level to decide that the experimental procedure produced a difference between groups, if we stick to our practice of using 0.05 as the division between significant and nonsignificant results.

The sign test can also be used with a numerical dependent variable, and it may, of course, be applied to data at the interval level or better. Consider, for instance, an experiment designed to determine whether human reaction times to a warning device are faster when the warning consists of a loud tone, or a bright light. Reaction times — the amount of time between stimulus onset and initiation of a response from a human observer — represent ratio-level measurement. However, unlike many performance variables, reaction times usually produce skewed distributions because of a floor effect. A reaction time cannot be less than 0, but it has no upper limit. Humans will occasionally take a very long time to respond, and this produces a skewed distribution; this, in turn, may mean that sample means and tests based on them (such as the *t*-test) are inappropriate. In this experiment, each person was tested once with a sound stimulus, once with a light stimulus. You should recognize this as a repeated-measures experiment, of course. The reaction times (in milliseconds) are shown in the table below.

PERSON	REACTION TIME TO LIGHT	REACTION TIME TO SOUND	SIGN
1	200 ms	120	+
2	191	190	+
3	340	320	+
4	255	250	+
5	172	170	+
6	140	890	−
7	230	140	+
8	205	180	+
9	220	220	?
10	210	205	+

There are several very important points to be learned from these data. First, it seems that in most cases, reaction time to sound was less than reaction time to light. When this occurred, a plus sign was assigned to the individual; when reaction time to light was lower, a minus was assigned. The presence of eight + signs indicates that most people responded more quickly to sound. However, the *mean* reaction time to light was 216.3 ms, whereas the *mean* reaction time to sound was 268.5 ms. Close inspection of the data shows that the unexpectedly high sound mean is due largely to one very high observation, person 6's 890 ms reaction time. Because of this rather deviant observation (i.e., because of the presence of substantial skew in the sound-reaction-time distribution), the *t*-test for the difference between means is quite misleading. A related-samples *t*-test shows a *t*-observed of only −0.667; the difference is clearly not significant. And, the difference between means is in the "wrong" direction.

Suppose that you decide to discard observations on person 6, simply because the 890 ms reaction time to sound for that person seems too long to be a "true" reaction time.[3] With the remaining nine individuals, the mean reaction time to light is now 224.8 ms and the mean for sound stimuli is 199.4. Now, the difference between means is in the direction we would expect from looking at the high number of + signs. However, the two-

---

[3]One should discard such data only when *very sure* that they represent spurious measurements.

tailed related-samples t-test for the difference between these means is still not significant at the 0.05 level ($t = 2.17$, $df = 8$). In this case, even if the difference between means is real, there is still too much variability in the data to reject the null hypothesis that reaction-time distributions for light and sound have equal means. When the sign test is applied to these data, however, the variability and the skew introduced by person 6 become less troublesome.

Before performing the test, however, we must decide what to do with person 9, who showed the same reaction time to both stimuli. One way of dealing with such "ties" in the sign test is simply not to count this individual's data. This leaves us with nine individuals who produced + or − signs, and the sign test will be performed for their data.

We then count the number of + and − signs and call this value r. Under the null hypothesis (that reaction times to each stimulus should be about the same), we expect an equal number of + and − signs to appear; however, because sampling variability exists, we realize that the actual value of r will fluctuate from sample to sample, even under a true null hypothesis. We then use the procedures described in Chapter 8 to construct a sampling distribution for r under a true null hypothesis, with N equal to the number of subjects who produced either a + or − sign, and with $p = 0.50$. Here, $r = 8$ and, because we had one subject with tied scores, $N = 9$.[4] We must now ask, "What is the probability that $r = 8$ or more in the null hypothesis distribution?" Computing the probabilities for the binomial variable, we find that:

$$\text{prob } (r = 8) = 0.0176$$
$$\underline{\text{prob } (r = 9) = 0.0020}$$
$$\text{prob } (r = 8 \text{ or } r = 9) = 0.0196$$

Thus, 0.0196 would be the obtained significance level in a one-tailed test. Because we did not hypothesize ahead of time that one sense modality or the other necessarily had to produce lower reaction times, the two-tailed sign test is called for. The obtained significance level, as in the previous example, is found by doubling the above probability. Because, in a symmetrical binomial distribution with $N = 9$, prob ($r = 8$) is equal to prob ($r = 1$), and prob ($r = 9$) is equal to prob ($r = 0$), we determine:

$$\text{prob } (r = 8 \text{ or } r = 1) = 0.0176 + 0.0176 = 0.0352$$
$$\underline{\text{prob } (r = 9 \text{ or } r = 0) = 0.0020 + 0.0020 = 0.0040}$$
$$\text{prob } (r = 9 \text{ or } r = 8 \text{ or } r = 1 \text{ or } r = 0) = 0.0392$$

We can therefore state that for 8 out of 10 people, reaction time to sound was faster than reaction time to light. And, this time advantage for sound stimuli was significant at the 0.039 alpha level. As we have seen,

---

[4]When N is very large, the normal approximation to the binomial distribution can be used (see the discussion on pp. 263–266 in Chapter 9).

significant differences between these groups could not have been demonstrated so easily with a *t*-test.

The Wilcoxon Matched-Pairs, Signed-Ranks Test

The sign test described above is useful in situations where the dependent variable can be summarized with a binomial *r*. However, in many situations calling for a nonparametric test, the sign test is not as *powerful* as the **Wilcoxon matched-pairs, signed-ranks test** (called the signed-ranks test through the rest of this section). We will discuss power and nonparametric tests at the end of this chapter, but for now it will be enough to say that the sign test does not use all of the information available in some data collections. In terms of statistical power, the sign test is often relatively weak in its ability to detect a true alternative hypothesis. The signed-ranks test is more powerful and this means that with it you will be more likely to find a difference between groups if there really is a difference.

Like the sign test, the signed-ranks test is a nonparametric alternative to the related-samples *t*-test. This means that it too applies to two-sample experiments, where either matched pairs or repeated measures are used. Unlike the sign test, however, the signed-ranks test requires data at least at the interval level of measurement. This is because the magnitudes of the differences between scores are considered in the test.

Our signed-ranks example problem is summarized in Table 13.5. We will again use reaction times and an experimental task very similar to the one just covered in the preceding section, so that we may see the added power of the Wilcoxon matched-pairs, signed-ranks test over the ordinary sign test.

Assume that a researcher is trying to determine the best location for a warning light to be installed on aircraft instrument panels. In this experiment, two locations (*A* and *B*) are tested by measuring pilot reaction times

Table 13.5 *Derivation of the Signed Rank Totals W+ and W− for the Wilcoxon Matched-Pairs Signed-Ranks Test*

PILOT	POSITION *A* REACTION TIME (MS)	POSITION *B* REACTION TIME (MS)	*d* DIFFERENCE	RANK	SIGNED RANK +	−
1	750	650	+100	6	+6	
2	830	820	+10	1	+1	
3	1050	610	+440	7	+7	
4	1000	980	+20	2.5	+2.5	
5	660	640	+20	2.5	+2.5	
6	1200	1240	−40	5		−5
7	1170	1140	+30	4	+4	

Total = $W+$ = 23

Total = $W-$ = 5

to light onset. A repeated-measures procedure was used in order to minimize the effects of differences between subjects. Under the position *A* conditions, the seven pilots tested had a mean reaction time of 951.4 milliseconds, and under position *B* conditions the mean reaction time was 868.6 milliseconds. Thus, it appears that the condition *B* location of the light is a better location, and we might be tempted to test the significance of the difference between these two means with a related-samples *t*-test. However, let us suppose that reaction times in this situation are not normally distributed; instead, we know from many previously conducted tests, that reaction times form a frequency distribution that is heavily skewed in the positive direction. This violates one of the assumptions underlying the *t*-test (normality of the population distribution), and it is therefore appropriate that we use the signed-ranks test that does not have that assumption.

Furthermore, if we apply the sign test to these data, the difference between groups is also not significant. Therefore, we decide to use the more powerful Wilcoxon matched-pairs, signed-ranks test.

We will use this test to ask whether or not the differences between groups are due to sampling variability or to the independent variable manipulation — just as we always have with hypothesis tests — but we will *not* concentrate on the sample means. Instead we will look at two statistics called $W+$ and $W-$. By referring to Table 13.5 and following the discussion below you will see that these are easy to derive.

Notice the column headed "*d*", or "Difference." The values here are simply the difference in reaction-time scores between conditions for each pair. When preparing this column, always take care to perform all of the subtractions in the same direction; that is, always take the *B* condition value from the *A* value, or always do the reverse.

Next look at the column headed by "Rank." In this column we assign rank numbers to the *d* differences. This assignment pays no attention to the + and − signs; ranks are assigned only on the basis of the absolute value of *d*. The smallest difference, 10 ms (pilot 2) is assigned a rank of 1, the next largest difference is assigned a rank of 2, and so on through the largest difference, 440 ms (pilot 3). Notice that pilots 4 and 5 had the same size *d* values; when such tied ranks occur, assign each one the (same) average value of the two tied ranks. Since pilots 4 and 5 are tied for ranks 2 and 3, they both receive the average value of those ranks, 2.5.

Finally, consider the last pair of columns labeled + and − under the "Signed Rank" heading. Here we *do* pay attention to the sign of the *d* value. Under the + we list the *ranks* of all *d* values having + signs and under the − sign we list the *ranks* of all *d* values having − signs. The statistics $W+$ and $W-$ are obtained as the sums of these columns.

Under a null hypothesis of "no difference" between conditions (i.e., that location of the light had no effect on reaction time), we would expect $W+$ to about equal $W-$. The *d* ranks should be evenly distributed between plus and minus conditions. However, it appears that here $W+$ is much greater than $W-$. This says that there was a preponderance of cases where the A reaction time was greater than the B reaction time, particularly when the differences were large. But was this difference between $W+$ and $W-$ due to sampling variability or to truly shorter reaction times under the B condition?

To answer this question, we will work out the probabilities of getting the $W+$ and $W-$ values we obtained under a true null hypothesis; if these probabilities turn out to be very low, we will reject the null hypothesis that there is no difference between groups. Then, after showing you how the probabilities are derived, we will show you how to avoid such computations in the future by using Table D, Appendix 1.

To find the probability of obtaining our particular W values under a true null hypothesis, let us briefly consider how many *different* sets of signed ranks are possible when $N = 7$. (N, of course, is the number of pairs.)

We will begin this process by listing *some* of the possible sets of signed ranks and the $W+$ and $W-$ values they produce. Remember that when $N = 7$, the ranks involved are always 1 through 7.

POSSIBLE DISTRIBUTIONS OF + AND − SIGNS	1	2	3	RANK 4	5	6	7	$W+$	$W-$
1	+	+	+	+	+	+	+	28	0
2	+	−	+	+	+	+	+	26	2
3	+	+	−	−	−	+	+	16	12
4	+	−	+	−	+	−	−	9	19
⋮				⋮					
128	+	−	−	−	−	−	+	8	20

Since each rank may be either a plus or minus, there are 2^N possible sequences of signed ranks, that is, 2^N possible pairs of values for $W+$ and $W-$. For $N = 7$, $2^7 = 128$ possible sequences. The sum of $W+$ and $W-$ will always equal $\frac{1}{2}N(N + 1)$; again, with $N = 7$, this sum will equal 28. Since either W value can be zero, this means that the value of either will range from 0 to $\frac{1}{2}N(N + 1)$.

So, with 7 pairs of observations and 7 paired ranks, there are 128 possible sequences of signed ranks. Under the null hypothesis, we assume that all 128 of these are equally likely. Thus, the probability (assuming the null hypothesis) that any one sequence appears is 1/128, or 0.007813.

Now we have to ask a question about these 128 possible sequences: In how many of these do we obtain a $W+$ value of 23 or more? For this example, it is easy to see (by trying out different arrangements of plus and minus signs) that $W+$ can equal 23 in only three such sequences:

			RANK					
1	2	3	4	5	6	7	$W+$	$W-$
+	−	−	+	+	+	+	23	5
−	+	+	−	+	+	+	23	5
+	+	+	+	−	+	+	23	5

Any other sequences produce different values of $W+$ and $W-$. Since the probability of each of these sequences is 0.007813, and since they are mutually exclusive, equally likely events, the probability that $W+ = 23$ is (3)(0.007813), that is, 0.023439.

Similarly, a little experimentation with + and − signs should show you that there are only two sequences possible in which $W+$ equals 24:

			RANK					
1	2	3	4	5	6	7	$W+$	$W-$
+	+	+	−	+	+	+	24	4
−	+	−	+	+	+	+	24	4

The probability that $W+ = 24$ is thus (2)(0.007813), or 0.015626.

There are also only two sequences that yield $W+ = 25$, so the probability that $W+ = 25$ is also 0.015626:

			RANK					
1	2	3	4	5	6	7	$W+$	$W-$
−	−	+	+	+	+	+	25	3
+	+	−	+	+	+	+	25	3

There is only one sequence where $W+$ is equal to 26:

			RANK					
1	2	3	4	5	6	7	$W+$	$W-$
+	−	+	+	+	+	+	26	2

The probability that this sequence appears is 0.007813. Finally, there is only one sequence where $W+$ equals 27 and one sequence where $W+$ equals 28:

			RANK					
1	2	3	4	5	6	7	$W+$	$W-$
−	+	+	+	+	+	+	27	1
+	+	+	+	+	+	+	28	0

Of course the probability of each of these sequences is also 0.007813.

Now we can ask for the probability that $W+$ is equal to or greater than our observed $W+$ value of 2 under the null hypothesis:

$$p(W+ \geq 23) = p(W+ = 23) + p(W+ = 24) + p(W+ = 25)$$
$$+ p(W+ = 26) + p(W+ = 27) + p(W+ = 28)$$
$$= 0.023439 + 0.015626 + 0.015626 + 0.007813$$
$$+ 0.007813 + 0.007813$$
$$= 0.078130$$

In a one-tailed test, this is our observed alpha level. Under a true null hypothesis, the one-tailed test says that we find a value of $W+$ equal to 23 or more only about 0.08 of the time. However, since there is no a priori reason why we know that condition B must be better (produce lower reaction times than condition A), we must perform a two-tailed test. Just as with the two-tailed z- and t-tests, we must ask, "What is the probability that we find a statistic as extreme as our observed statistic *in either direction?* To turn this into a two-tailed test, we simply double the one-tailed probability: $(2)(0.07813) = 0.15626$. This is the observed two-tailed alpha level. If we stick to our practice of using observed significances of 0.05 or less as grounds for rejecting a null hypothesis, we cannot decide on the basis of this evidence that the two positions are different. The best procedure at this point would be to rerun the experiment with a larger number of pilots.

Now that you have suffered through the calculation of $W+$ probabilities, you should know that tables of critical W values are available, and one such table is reproduced as Table D in Appendix 1. To use this table, simply take the *lower* of your two W values and compare this to the tabled values. The row for $N = 7$ is the third row down; our lowest observed W was a $W-$ of 5. If the observed W value is equal to *or lower* than the tabled value at a given alpha level, one may reject the null hypothesis of no difference between groups at that alpha level. Notice in Table D that for $N = 7$, in the two-tailed test, a W value of 2 or less is required to reject the null hypothesis at the 0.05 alpha level. The table thus leads us to the same conclusion as our calculated probabilities.

The Wilcoxon Rank-Sum Test for Independent Samples[5]

Our next test is a nonparametric alternative to the independent samples t-test: the **Wilcoxon rank-sum test.** It is very similar in some ways to the Wilcoxon matched-pairs, signed-ranks test we have just covered.

Again we will consider a situation involving a controlled experiment.

---

[5]This test produces exactly the same results as another test called the Mann-Whitney U-test. The only difference between the U-test and the rank-sum test is the test statistic that has its critical values tabled, R_1 in the rank-sum test and U in the U-test. However, U is just a linear transformation of R_1. Hence, the two tests lead always to the same conclusions. The rank-sum test, however, involves simpler computations.

This time, a researcher is interested in studying the effects of early experience on animals' activity levels as adults. There is a large body of experimental evidence showing that newborn mammals of many kinds grow up to be larger and more active as adults if they receive certain kinds of stimulation during infancy. In the experiment at hand, a researcher had 10 newborn rats to work with, none of them litter mates.[6] These were randomly divided into two groups of five animals each. Group II animals received 10 minutes of physical handling a day during the first months of life; Group I animals received no such experience. In all other ways, the animals in both groups were treated identically. Then, at age 1 year, the animals were placed in a cage with an activity wheel attached; listed below are the number of activity wheel revolutions registered by individual animals in each group during a one-hour period:

GROUP I	GROUP II
78	590
140	170
120	149
105	128
75	220

Clearly, the mean number of revolutions for Group I (103.6) is lower than the mean number of revolutions for Group II (251.4). Yet, inspection of the data suggest that a t-test should not be applied to these data: the high value in Group II (590) produces a heavily skewed distribution of dependent variable measurements. The independent samples t-observed for the difference between these means is -1.70 ($df = 8$) — not significant at the 0.05 level. The extreme value in Group II introduces too much variability in that group's scores for the t-test to show a significant difference. Thus, a nonparametric test that uses ranks rather than dependent variable values should be more appropriate here.

We will begin the Wilcoxon rank-sum test by lining up our ten dependent variable measurements in order, putting both groups in a single rank order from low to high:

Observed Revolutions	75	78	105	120	128	140	149	170	220	590
Rank	1	2	3	4	5	6	7	8	9	10
Group	I	I	I	I	II	I	II	II	II	II

---

[6]If litter mates were included in this group of 10, we might arrange a matched-pairs experiment, as described on p. 360; then, one member of each litter would be assigned to each group; that would create a related samples experiment. The example as described here does not involve litter mates and represents an independent-samples experiment.

Ranking the values from low to high rather than in the reverse order is an arbitrary procedure adopted here so that low ranks are associated with low dependent variable scores — simply an aid to interpreting the results.

It should be noted here that, *if the two groups are unequal in size, it is necessary to designate the smaller group as Group I;* otherwise, the notation used in Table E (Appendix 1) will not be appropriate for the test that follows. If, as here, the two groups are of equal size, either may be called Group I. If the symbols n_1 and n_2 are used to represent sizes of Groups I and II, respectively, then it will always be true that $n_1 \leqslant n_2$. In the following discussion let N represent $n_1 + n_2$.

In ranking values, you may occasionally find two observations with the same value. In such a case, both may be assigned the mean of the two ranks they fill. Two values tied for seventh and eighth ranks, for instance, would each be assigned the (same) mean rank of 7.5.

Our test will focus not on the means of these groups, but rather on the distribution of the ten ranks between groups. We will begin the computation by adding up the ranks for each separate group; call these totals R_1 and R_2.

GROUP I RANKS	GROUP II RANKS
1	5
2	7
3	8
4	9
6	10
$R_1 = 16$	$R_2 = 39$

R_1 is much lower than R_2, suggesting that a disproportionate number of the low ranks appeared in Group II. This, in turn, means that the low activity wheel scores tended to appear in Group I. The null hypothesis in this test is concerned with ranks rather than means. Specifically, H_0 says that there is no effect of early experience on activity wheel scores, and, therefore, the distribution of ranks between the two groups is purely random (each rank has an equal probability of appearing in either group).

If H_0 is true, we would expect R_1 and R_2 to be equal (on the average, if the experiment is run many times), since neither group would be more likely to get high ranks or low ranks. Now, we will want to decide whether our observed difference between R_1 and R_2 is due simply to sampling variability under a true H_0, or whether it is due to a real effect of the independent variable.

Here we will proceed very much as we did with the signed-ranks test in the preceding section. You will be shown how to calculate the exact probabilities involved in the test, and then shown how to avoid those calculations in the future by using tabled values of R_1. Having followed through the calculations, however, you will be in a better position to interpret results of the test.

Consider then our R_1 value. Under H_0, five of the ten ranks available are assigned randomly to the five Group I members. H_0 says that high ranks are as likely as low ranks in Group I. We first ask, How many different sets of five ranks can be selected from a group of ten available ranks? The answer is given by our old friend, the counting rule for combinations. (You may want to review this rule covered in Chapter 8, pp. 220–222). The number of different groups of n_1 ranks that may be selected from N available ranks is:

$$\binom{N}{n_1} = \frac{N!}{n_1! \, (N - n_1)!}$$

For $n_1 = 5$ and $N = 10$:

$$\binom{10}{5} = \frac{10!}{5! \, 5!} = 252$$

Under H_0, each of these 252 ways is equally likely. H_0 says that each possible group of n_1 ranks has a probability of 1/252 of appearing, or 0.003968.

We now ask the familiar question about observed statistics again: How likely is it that we obtain a statistic as deviant as the one we have observed — or more so — when the null hypothesis is true? Under a true H_0, we therefore ask, how likely is it that we obtain $R_1 \le 16$? (We could just as well pick the higher R value, here R_2, and ask how likely it is that $R_2 \ge 39$.)

You can easily verify that, in picking five of the integers 1–10, there are only two sets of five that total 16 or less:

$$1 + 2 + 3 + 4 + 5 = 15$$
$$1 + 2 + 3 + 4 + 6 = 16$$

Any other five integers produce higher totals. Thus, only two of the 252 possible sets of Group I integers produce an R_1 as low as our observed R_1 or lower. The probability under H_0 that either one or the other of these sets of ranks turns up is calculated as the sum of their separate probabilities:

$$p(R_1 \le 16) = 0.003968 + 0.003968$$
$$= 0.007936$$

The above is the obtained alpha for a one-tailed test. Again, however, if there is no reason why one particular group must have higher scores than the other, we must perform a two-tailed test. This is done by doubling the above probability. Thus, our obtained alpha level (significance) is 0.015872, or about 0.016.

The test has shown us, then, that if the null hypothesis is true (early experience has no effect on adult activity wheel scores), then we will observe a discrepancy between R_1 and R_2 as large as our observed difference on about 0.016 of the times this experiment is repeated. This is below our standard criterion of alpha = 0.05, and we reject H_0. Apparently, early experience *did* affect activity scores.

Having now shown the logic involved in the test, we can demonstrate the easy way of doing it. We do not have to calculate probabilities associated with R_1 values because these have been done, and critical values of R_1 for this test have been summarized in tables such as Table E in Appendix 1. The two-tailed test will be described first, using the portion of Table E reproduced as Table 13.6.

To use this table, we must obtain two values from our data: R_1, which we already have, and $2\bar{R}_1 - R_1$. In this latter value, $\bar{R}_1$ is the mean value of R_1 sums that would appear in an infinite number of samples under a true H_0. The value of $\bar{R}_1$ depends only on n_1 and n_2, so the value of $2\bar{R}_1$ can be tabled for each pair of group sizes. In our example problem, $n_1 = 5$, so we select the section of Table E for that Group I size; Table 13.6 shows a portion of that section. Then, we find the row of the table associated with our n_2 size; in our illustration, this too is 5, so we pay attention to the first row of numbers under the upper solid line. From this we find the appropriate value of $2\bar{R}_1$. In this case, $2\bar{R}_1 = 55$. The two values needed for this test, then, are:

$$R_1 = 16$$
$$2\bar{R}_1 - R_1 = 55 - 16 = 39$$

Whichever of these values is smaller is used as the observed statistic for comparison with the tabled values. Here, we will use the value of $R_1 = 16$. (You may notice that $2\bar{R}_1 - R_1$ is equal to our observed value of R_2. This equality exists only when $n_1 = n_2$.)

We are now ready to enter the table. Note first, however, that the alpha levels listed (between n_2 and $2\bar{R}_1$ column headings) are for a one-tailed test. To obtain the two-tailed probabilities, these values must be doubled. We see that for the appropriate n_1 and n_2, our obtained value of R_1 is significant at the one-tailed alpha of 0.010; the obtained value of alpha in

Table 13.6 *Excerpt from Table E, Appendix 1, Showing Critical Values of R_1 for the Wilcoxon Rank-Sum Test*

n_2	0.001	0.005	0.010	$n_1 = 5$ 0.025	0.05	0.10	$2\bar{R}$
5	·	15	16	17	19	20	55
6		16	17	18	20	22	60
7	—	16	18	20	21	23	65
8	15	17	19	21	23	25	70
9	16	18	20	22	24	27	75
			⋮				

the two-tailed test is thus 0.020. If the value of R_1 (or $2\bar{R}_1 - R_1$) is equal to or less than the tabled value, reject the null hypothesis at an alpha level twice that shown in the column heading. (If our R_1 had been 15, we would have rejected H_0 at an alpha level of 0.01 in the two-tailed test, for instance.) Thus, we decide that early experience did have an effect on activity wheel scores.

The one-tailed Wilcoxon rank-sum test is conducted in the same way, except that you must consider the directional predictions specified in the alternative hypothesis. Suppose, for example, that you assume that Group I ranks can only be lower or equal to Group II ranks. Then, you will always enter the table with R_1 (not $2\bar{R}_1 - R_1$, even if it is less than R_1) and you will use the alpha levels just as written in the column headings. Alternatively, if your H_1 states that Group I ranks are greater than Group II ranks, use only $2\bar{R}_1 - R_1$ in the one-tailed test.

Choosing Between Parametric and Nonparametric Tests

What determines whether we use parametric tests or nonparametric tests to evaluate our data? There are a number of factors to be considered in choosing the appropriate statistic, but quite often the issue is not black and white and the statistician is faced with a difficult choice.

Sometimes the nature of the dependent variable makes the choice clear. In the example problem with the two-way chi-square, for instance, where union members preferences on a strike vote were surveyed, there are no appropriate parametric tests. It is hard to imagine how these responses could have been properly translated into interval-level data. They were more appropriately reported as frequencies, data for which the chi-square is ideal. On the other hand, the rat weights used in the early experience example might have been analyzable with either the t-test or the Wilcoxon test that we used in the example. There you were told that a t-test assumption — normality of the population distribution — had been violated. But many times one simply does not know if the dependent variable is normally distributed or not.

One is often inclined to choose t over R as the test statistic because t makes statements about *means,* and these are easier to discuss and interpret than are distributions of ranks. Another consideration that persuades many people to choose the parametric test over its nonparametric counterpart, is statistical power. Very often, when a parametric test and a nonparametric test are both applicable to the same data, the parametric test has more power because it uses more information in the data. That is, the parametric tests are often more likely to find a true alternative hypothesis than are the nonparametric tests, because they make use of intervals on the measurement scale as well as ranks. However, the difference in power is often not as much as some people suppose, and usually the two forms of tests lead to identical conclusions.

The question of just how much more powerful the parametric tests are than their nonparametric alternatives is difficult to answer precisely. Remember that power is the probability of rejecting the null hypothesis when the alternative hypothesis is true; if you refer back to the diagram of probabilities involved in hypothesis testing (Chapter 10, Table 10.3), you can verify that power in a particular test depends on (a) the alpha level chosen, (b) the location of the true H_1 distribution, and (c) the sample size used. Changing any of these changes the power of the test; because the location of the true H_1 distribution is rarely known in the kind of hypothesis testing we have done, power can usually only be estimated.

Nonetheless, there is a rough method of comparing the power of the two kinds of tests, and it makes use of the concept of efficiency, or more specifically, the **relative efficiency** of two tests. To use this concept, we compare the number of individuals required as subjects in test A (N_A), to produce the same power as test B with N_B individuals. That is, suppose that test A required 100 individuals in the sample to produce the same power that test B had with 50 individuals. Then the relative efficiency of the two tests is given by:

$$\frac{N_A}{N_B} = \frac{100}{50} = 2.00$$

Test B is thus twice as efficient as Test A. To evaluate nonparametric tests, the concept **asymptotic relative efficiency** is used. This is an efficiency ratio, where the numerator is the number of subjects required in the t-test and the denominator is the number of subjects required in the nonparametric test, and where the difference between null and (true) alternative hypothesis distributions is very, very small, and where the N's involved are very, very large. (It is called *asymptotic* relative efficiency, because relative efficiency approaches a constant value as the N's approach infinity and the difference between H_0 and H_1 distributions approaches 0.) The asymptotic relative efficiency of the sign test (relative to the t) is 0.637, the efficiency of the Wilcoxon matched-pairs, signed-ranks test is 0.955, and the efficiency of the Wilcoxon rank-sum test is also 0.955 relative to the t-test.

Thus, in those extreme situations, the Wilcoxon rank-sum test and signed-ranks test are nearly as powerful as their corresponding t-tests. However, this measure of efficiency is more helpful in comparing power differences among nonparametrics than it is for comparing nonparametrics with parametrics. We can see that the Wilcoxon test is more powerful than the sign test. The t-test will usually be slightly more powerful than the two Wilcoxon tests, but not always. We have already seen examples of data sets that do not produce significant differences with the t-test, but that do show significant differences with their nonparametric counterparts. With skewed distributions, the assumptions underlying the t-test are no longer met for the small sample situation, and statements about asymptotic relative efficiency become meaningless.

Perhaps the best that can be said is that both kinds of tests have advantages and disadvantages, and that by familiarizing yourself with them, you are better equipped to understand the statistics that other people use and to recognize the limitations of your own.

SUMMARY

Nonparametric statistics (also called distribution-free statistics) are designed for situations where one does not necessarily have interval-level data or above, or where the shape of the population distribution is not known. Hypothesis tests with these statistics are not stated in terms of population parameters.

One highly useful nonparametric statistic is the chi-square. The one-way chi-square hypothesis test requires only that data be collected in the form of frequencies in different observation categories. The null hypothesis states that the observed frequencies are not inconsistent with random distribution of responses among response categories. The alternative hypothesis states that nonrandom factors were involved in determining the observed frequencies. The chi-square table lists values of the chi-square statistic that would occur with a relative frequency of alpha or less if the null hypothesis were true. H_0 is rejected if the obtained chi-square is greater than the tabled value.

The one-way chi-square can also be used in the goodness-of-fit-test. This allows us to decide whether or not observed data come from a distribution with a specified shape, such as the normal distribution. We also examined the two-way chi-square test, which allows us to decide whether or not two variables are related; as with other chi-square tests, the two-way test uses data in the form of frequencies.

This chapter also introduces and demonstrates the sign test and the Wilcoxon matched-pairs, signed-ranks test. These are both nonparametric alternatives to the t-test for related samples. Finally, the Wilcoxon rank-sum test was demonstrated, a nonparametric alternative to the t-test for independent samples.

Usually, the nonparametric tests will be slightly less powerful than their parametric counterparts. However, a number of factors must enter the decision process when one chooses between using a parametric or nonparametric test; if the distribution of dependent variable scores is skewed, for instance, the nonparametric test may be more powerful than the parametric test.

KEY CONCEPTS

parametric tests	**chi-square distribution**	**sign test**
nonparametric tests	**one-way chi-square test**	**Wilcoxon rank-sum test**
distribution-free tests	**Yates's correction**	**Wilcoxon matched-pairs, signed-ranks test**
chi-square statistic	**goodness-of-fit test**	
expected frequencies	**two-way chi-square test**	**relative efficiency**
		asymptotic relative efficiency

SPOTLIGHT 13   **Chi-Square in Research**

The chi-square statistic is very popular in behavioral science research; a glance through any of the psychological or sociological journals, for instance, suggests that a large proportion of the "facts" in these disciplines are based on chi-square evidence. One reason, no doubt, for the popularity of chi-square is that it can be used in many nonparametric tests. This means that the variable under study need not come from a normally distributed population or represent the higher levels of measurement. Following is a short summary of a typical chi-square application from the area of social psychology.

Do Americans support the Bill of Rights? And does it matter who asks them? A number of studies suggest that a majority of adult Americans will not endorse the Bill of Rights (first ten amendments to the U.S. Constitution) when the original text is placed before them but not identified as the Bill of Rights. Dr. William Samuel conducted a study to see if this was true in one area of Sacramento, California, and also whether it made a difference if a "hip" or "straight" canvasser asked for the endorsement.[7]

Thirteen college-age researchers solicited signatures at a number of middle-class homes in a Sacramento suburb. Seven were dressed in "straight" attire and six in "hip" costume. Each researcher carried three different statements: One was a paraphrased version of the real Bill of Rights (which guarantees a number of basic freedoms), one was a negative paraphrase version that urged restriction of these rights, and a third was a "wishy washy" paraphrase that attempted to take a middle ground between the other two versions. At each house a researcher introduced himself or herself as a representative of a student group called Youth for America; the resident was then asked to read *one* of the paraphrases and to sign it if he agreed with it.

There were thus two independent variables: attire of the canvasser ("hip" or "straight") and version of the Bill of Rights (real, negative, and "wishy washy"). All respondents were exposed to one level of each independent variable, and all were measured on the same dependent variable, "signature or no signature."

As Samuel expected, people approached on three days of testing[8] were apparently more ready to sign the negative version than the real or "wishy washy" version: 62 percent of those approached endorsed the negative version, while only 46 percent signed the "wishy washy" text, and 44 percent the real paraphrase. But are these differences significant? Is the difference between 62 percent and 44 percent, for instance, due to the

[7]W. Samuel, "Response to Bill of Rights Paraphrases as Influenced by the Hip or Straight Attire of the Opinion Solicitor," *Journal of Applied Social Psychology*, 2 (1972), 47–62.

[8]The canvassing was actually conducted on four days of testing, but the results were complicated by the fact that one day's canvassing occurred on the Sunday following the Kent State and Jackson State shooting incidents. See Samuel's article for a description of what happened on this day, and the author's theoretical interpretation of these results.

effect of the independent variable "version read," or to chance variability. When the chi-square was computed for the appropriate frequencies, the difference appeared to be real ($\chi^2 = 7.52, df = 2, p \leqslant 0.025$). Thus, the chi-square test supports the conclusion that most of these people would not endorse the Bill of Rights.

However, some other interesting results appeared on closer examination of the data. The above-mentioned preference for the negative version seemed to hold only when the canvasser was dressed as a "straight"; with "hip" canvassers, the frequencies of signatures for different versions was nonsignificant ($\chi^2 = 1.46, df = 2$). Furthermore, "straights" were more likely to obtain signatures than "hips" (for "hip" and "straight" signature rates, $\chi^2 = 4.46, df = 1, p \leqslant 0.05$).

PROBLEMS

1. In your own words, distinguish between parametric and nonparametric hypothesis tests. What advantages are there to each kind of test? What disadvantages?

2. In a recent advertisement, Acme Electronics, Inc., reported the results of a "test." Six television sets were placed side by side in a store display area, and 100 customers were asked to pick the one with the best color picture. Each set was a different brand, and the brand trademarks for each set were covered up. Since more people chose the Acme set than any other, Acme claims that it has the best picture. Use the chi-square test with the data below to examine this claim:

BRAND	NO. OF TIMES CHOSEN AS BEST
Acme	23
B	18
C	19
D	17
E	13
F	10

3. Fred's favorite flipping coin has won the owner a good many dollars in games of chance. One day, after Stan the statistician had lost a friendly contest to Fred, Stan demanded a statistical test of the coin's properties. In 12 flips, the coin landed tails 10 times.

   a. Use the one-way chi-square to evaluate the likelihood that this is an unbiased coin.

   b. Make the same evaluation of the coin's properties using the binomial sign test. To do this you do not have matched pairs, but rather, simply the question of determining the probability that the binomial variable r is equal to 10 or more, when $N = 12$ and the null hypothesis is that $p = 0.50$.

4. A professor suspects that the heaviest library users are sophomores and juniors, perhaps because freshmen haven't yet learned to take advantage of all library resources, and perhaps because seniors have their minds more on postgraduate activities than on getting good grades in their last year. At any rate, he knows that the student body is 50 percent freshmen, 25 percent sophomores, 10 percent juniors, and 15 percent seniors. A random sample of 1,000 library users turns out to be composed of 480 freshmen, 290 sophomores, 180 juniors, and 50 seniors. Use the chi-square goodness-of-fit test to decide whether or not these frequencies support the professor's guess.

5. There are several well-known tests of abstract reasoning ability that yield scores ranging from 0 to 100 and produce a normal distribution of scores when given to the general population of college students. However, we do not know if the same scores would be normally distributed if given to a random sample of engineering students. Suppose that one such test has been given to 100 engineering students, and their scores are as tabulated below. Use the chi-square goodness-of-fit test for normality for a normal distribution divided into *eight* equiprobable regions.

95	56	56	47	67	15	60	61	74	16
61	63	38	88	58	83	59	48	72	40
26	76	20	59	84	56	19	37	50	72
26	24	46	24	22	76	81	57	50	62
66	36	42	34	37	14	60	71	21	48
70	62	55	43	63	78	51	86	43	54
54	42	25	58	34	45	39	55	56	51
44	34	98	45	51	51	63	17	43	22
57	97	52	47	70	50	25	28	26	38
63	71	15	39	63	27	42	57	77	26

6. Do people's political preferences differ from ward to ward in the city? In some cities they certainly do, while in other cities they may not. Use the two-way chi-square test to determine if political preference and ward of residence are independent for the following data. The numbers represent the number of people in each designated ward who voiced a preference or chose to call themselves independents.

| | POLITICAL PREFERENCE | | |
	DEMOCRAT	REPUBLICAN	INDEPENDENT
Ward 1	50	70	30
Ward 2	30	60	20
Ward 3	60	65	15

7. Use the two-way chi-square in another election-related problem to decide if there is a relationship between a voter's sex and his or her choice of mayoral candidate. In a random selection, 100 men and 100 women were asked whom they voted for in the last election. Among the men, 67 voted for candidate A, and 33 for B. Among the women, 22 voted for A, and 78 voted for B.

8. Consider the decision records of two judges who preside in divorce cases in a certain state. The table below summarizes cases where custody of children was contested by both parents. The figures indicate the numbers of cases where custody was awarded to the mother, to the father, and jointly. Assuming that cases are distributed randomly to the two judges, use the two-way chi-square test to decide whether the two judges distribute their decisions differently.

| | NUMBER OF DECISIONS | | |
	TO MOTHER	TO FATHER	JOINT
Judge A	170	62	57
Judge B	310	95	160

9. Scores in gymnastics events are given by judges who evaluate the contestants on a number of criteria. Scores differ among judges, however, even when the same performance is being evaluated. One reason may be that human judges simply cannot attend to all of the details in a rapid maneuver; another possibility is that some subjective bias enters unavoidably into the ratings. Listed below are ratings for ten gymnasts, made by two different judges on the same performance:

GYMNAST	RATINGS	
	JUDGE 1	JUDGE 2
1	9.7	9.6
2	8.2	8.5
3	7.7	7.3
4	7.9	7.3
5	8.0	7.0
6	8.4	8.6
7	8.1	7.3
8	8.9	8.2
9	9.3	8.4
10	9.5	9.0

   a. Use the sign test to decide if there is a significant difference between ratings from the two judges.

   b. Use the Wilcoxon matched-pairs, signed-ranks test to decide if there is a significant difference between judges.

   c. *If* the tests in a and b lead to different statistical decisions, discuss some possible reasons for the difference.

   d. Why would these data not be appropriate for the (parametric) *t*-test for related samples?

10. The Type I and Type II error terminology used with the parametric tests applies to the nonparametric tests as well. For parts a and b of problem 9, indicate what a Type I and a Type II error would be in each case.

11. Another place where subjective judgments appear is when "essay" examinations are graded. While it is true that good instructors can evaluate essays very reliably — that is, use the same criteria for each paper — it is also true that there is often a great deal of variability among instructors in the way they assign grades. Again, we will want to determine whether or not there is a significant difference in the grades assigned by two different instructors. Since essay exam grades are probably not normally distributed, and since they often do not constitute interval-level measurement, a nonparametric test is called for. Use the Wilcoxon matched-pairs, signed-ranks test to decide if scores from the two instructors are really different; the numbers represent exam scores obtained by students who took exams under both instructors.

STUDENT	SCORE FROM INSTRUCTOR 1	2
1	90	89
2	85	76
3	77	72
4	93	96
5	82	74
6	71	67
7	69	65
8	85	78
9	84	82
10	97	91

12. The two groups of numbers below represent reading test scores obtained by 30 fourth graders. Those in Group I had had special instruction in phonetics; those in Group II had only the standard reading course without special instruction. Use the Wilcoxon rank-sum test to decide if the two groups really differ.

GROUP I SCORES					GROUP II SCORES				
75	76	98	87	74	73	61	51	63	60
92	71	77	95	96	49	70	44	67	48
69	62	91	83	72	52	70	59	68	57

13. Evaluate the experimental data in problem 10, Chapter 11 (p. 324) using the sign test.

14. Evaluate the experimental data in problem 10, Chapter 11 (p. 324) using the Wilcoxon matched-pairs, signed-ranks test.

15. Evaluate the experimental data given in problem 7, Chapter 11 (p. 322) using the Wilcoxon rank-sum test.

14. Introduction to the Analysis of Variance

After reading this chapter, you should be able to do the following:

1. *Explain the kinds of research questions addressed by single-factor analysis of variance (ANOVA).*

2. *Explain the general rationale underlying the analysis — why variances are used to answer questions about means.*

3. *Perform a single-factor ANOVA and interpret the results.*

4. *Discuss the assumptions underlying ANOVA, and perform the Hartley F_{max} test to check for homogeneity of variances.*

5. *Discuss the different kinds of independent variables that can be used in ANOVA studies.*

The One-Factor Analysis of Variance

For the final two chapters of this book, we introduce an extremely useful family of parametric statistical tests called the **analysis of variance,** or **ANOVA** for short. ANOVA uses the same Neyman-Pearson hypothesis testing logic we used with the *t*-tests and nonparametric tests, but the computations involved are more complex. The added complexity, however, is often worth the trouble: if our purpose in performing experiments is to learn something about relationships between variables, particularly cause-and-effect relationships, then ANOVA experiments can be designed that are far more informative than those we have covered so far.

The important difference between ANOVA experiments and those we have already learned is that with analysis of variance, we may have more than two experimental groups. Thus, more than two levels of the independent variable may be administered, as suggested in Table 14.1. Except for the additional groups, the experimental procedures shown in the table are exactly the same as those used in the independent samples *t*-test.

Table 14.1   *Single-Factor Analysis-of-Variance Experiment*

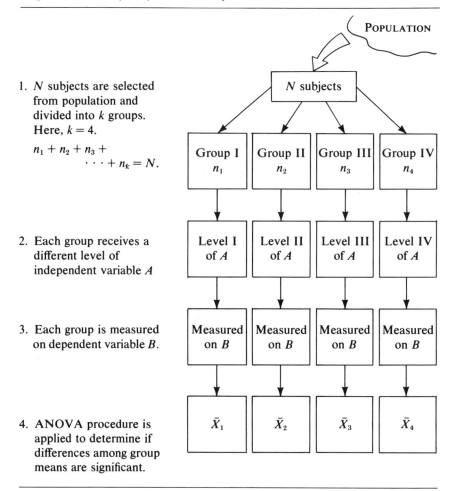

1. N subjects are selected from population and divided into k groups. Here, $k = 4$.
   $$n_1 + n_2 + n_3 + \cdots + n_k = N.$$

2. Each group receives a different level of independent variable A

3. Each group is measured on dependent variable B.

4. ANOVA procedure is applied to determine if differences among group means are significant.

However, using more than two levels allows us to uncover aspects of the relationship between independent and dependent variables that could not be examined with a two-samples experiment. And, in the next chapter, we will see that ANOVA procedures allow us to use two independent variables in the same experiment. We begin our coverage of ANOVA with an example that will help introduce analysis of variance terminology and some symbols.

Let us assume that we are interested in finding out what effect variable A, say, fertilizer strength, has on variable B, adult plant height. We *could* do a two-sample experiment by growing one group of plants with fertilizer and another group without, and then using the independent samples t-test to decide whether or not the two group means differed significantly. Remember, though, that we want to learn more than simply whether or not fertilizer has an effect on plant height, but also as much as possible

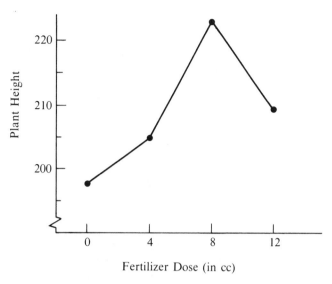

Figure 14.1 Graph of mean plant height for the four groups used in the single-factor experiment.

about the kind of effect. Therefore, we will plan a four-sample experiment such as the one diagrammed in Figure 14.1.

Our population might represent a large bag of seed corn, and from this we randomly sample, say, 40 kernels. Our overall sample size N is thus 40. We then divide these N seeds randomly among k groups of sizes n_1, n_2, n_3, and n_4. The number of groups (here $k = 4$) will also be the number of different *levels*[1] of the independent variable that we will use. In ANOVA terms, an independent variable (such as fertilizer amount) is a **factor.** Since our first kind of ANOVA has only one independent variable, we are dealing with a single-factor analysis of variance; our factor thus has four levels in this study.

To keep the computations simple, we will use the same number of seeds in each group (and assume that all of them grow so that we can get 10 dependent variable measurements for each group). Thus, $n_1 = n_2 = n_3 = n_4 = 10$. We can now refer to each group's size as n.

The next step in the experiment consists of administering the different levels of the independent variable to the different groups. For our example, we will use different amounts (given in cubic centimeters) of the liquid fertilizer per plant. Suppose that these are:

Level I = 0 cc fertilizer per plant
Level II = 4
Level III = 8
Level IV = 12

---

[1]Remember that *level* was a term introduced in Chapter 11 to mean a specific amount or kind of the independent variable administered to a group.

These are administered, of course, while we observe all of the practices necessary to give our controlled experiment meaning. This means that we take great pains to insure that all plants receive the same light, water, temperature, and other conditions that could influence the dependent variable.

At the end of, say, 5 months, all plants are measured on the dependent variable. Suppose that these measurements (in centimeters) are as given in Table 14.2.

The first thing we will want to do is look at the sample means and graph them. This has been done in Figure 14.1 (using a chopped vertical axis to accent differences among groups).

The graph itself begins to give some of the information we sought in performing the experiment. For the 4 cc and 8 cc groups, increasing amounts of fertilizer were associated with increasing mean plant height. Perhaps there is an optimal amount of fertilizer that one does not want to exceed, however, since increasing the dose to 12 cc apparently lowered the mean height. Notice how much more informative this four-sample experiment is than a two-sample experiment would have been. We not only suspect that fertilizer *has* an effect, but we also learn something about the relationship between amounts of fertilizer and plant height, by including more than two levels of the independent variable. If we had included only 4 cc and 12 cc levels, we would probably have concluded that fertilizer has no effect on height.

It is now time to raise the question again that has haunted us in one form or another practically since we began inferential statistics: Is the observed difference in sample means due to the effect of fertilizer on plant

Table 14.2 *Results of the Four-Sample Experiment Investigating the Effect of Fertilizer on Plant Height*

	PLANT HEIGHTS (IN CENTIMETERS)										$\sum\limits_{i=1}^{n} X_{ji}$	$\dfrac{\sum\limits_{i=1}^{n} X_{ji}}{n}$
	PLANT (i)										TOTALS	MEANS
	1	2	3	4	5	6	7	8	9	10		
Group I	190	187	199	220	184	192	191	201	213	195	1,972	197.20
Group II	201	195	214	211	213	202	195	220	206	199	2,056	205.60
Group III	225	230	237	220	219	248	214	201	240	229	2,263	226.30
Group IV	201	210	206	185	209	217	220	213	219	206	2,086	208.60

$$\text{Grand Total} \sum_{j=1}^{k} \sum_{i=1}^{n} X_{ji} = 8,377$$

$$\text{Grand Mean} \frac{\sum\limits_{j=1}^{k} \sum\limits_{i=1}^{n} X_{ji}}{kn} = 209.4250$$

height, or is it due simply to sampling variability? This is the question that ANOVA helps us answer.

The null hypothesis in single-factor ANOVA is much like the null hypothesis in the t-test. It says, simply, that after treatment, all samples still belong to populations with equal means:

$$H_0: \mu_1 = \mu_2 = \mu_3 = \mu_4$$

There are a number of possible alternative hypotheses. For example:

$$\mu_1 = \mu_2 = \mu_3 \neq \mu_4$$
$$\mu_1 \neq \mu_2 = \mu_3 \neq \mu_4$$
$$\mu_1 = \mu_2 \neq \mu_3 = \mu_4$$

and so on.

ANOVA lumps all of these possibilities into a single all-encompassing alternative hypothesis:

$$H_1: \text{Not } H_0$$

Thus, if we reject H_0, we are saying simply that at least one (and maybe more) of our sample means comes from a population with a different μ than the others. In their general form, then, our hypotheses can be stated:

$$H_0: \mu_1 = \mu_2 = \cdots = \mu_k$$
$$H_1: \text{Not } H_0$$

where $k = $ the number of samples.

How shall we test these hypotheses? To answer this, we will return to the concepts of *variation* and *variance*, much as we discussed them in Chapter 4. Before doing this, it will be well worth our time to look carefully at the symbols and notation to be used in ANOVA.

Symbols and Notation in Single-Factor ANOVA

The ANOVA notation is a straightforward extension of the subscripted variable notation we introduced in Chapter 3. To begin with, we will want to identify individual plant heights in such a way as to tell which group they belong to and which member of the group we are dealing with. Thus, a double subscript is used:

$$X_{ji} = \text{dependent variable measurement for individual } ji,$$

where $i = $ the individual's identity within the group; i can be any integer value from 1 through n.

$j = $ the group the individual belongs to; j can be any integer value from 1 through k. (Remember that n stands for the number of individuals in each group, and k stands for the number of groups.)

Now, if we want the total of the measurements *within* Group 1, we symbolize this sum as $\sum_{i=1}^{n} X_{1i}$. The j subscript does not appear on the summation sign because its value (here, 1) stays constant throughout the summation process. These individual group totals are given in Table 14.2.

The mean for each group will be symbolized with a *single* subscript; thus, for Group 1:

$$\bar{X}_1 = \frac{\sum\limits_{i=1}^{n} X_{1i}}{n}$$

The grand total is the sum of all 40 observations, and to introduce its computing formula, we introduce the double summation sign. It is very important that you become comfortable with the use of double summation signs if you are to follow the computations that follow:

$$\text{Grand total} = \sum\limits_{j=1}^{k} \sum\limits_{i=1}^{n} X_{ji}$$

The double summation sign operates as follows. Begin at the outermost (left) $\sum$. Take the subscript value of j that it tells you to begin with — 1. Now, hold j constant at 1 while you perform the operation instructed by the inner $\sum$, that is, sum the individual members of Group 1. Then, when you have used all i values including n, return to the outer $\sum$ momentarily and take the next indicated j value, 2. Holding j constant at 2, proceed to follow the instructions of the inner $\sum$ again, adding up the values of individual group members in Group 2. Continue this process until j has been used at its highest value, k. Here, $k = 4$, of course, so the whole sign asks you to add up the Group I members, and keep right on adding to the same total until you reach the last member of the last group. (If you are unsure of yourself on the use of double summation, perhaps this would be a good time to work the problems numbered 1a–1h at the end of this chapter.) The **grand mean** $\bar{X}_G$ is the sum of all 40 observations divided by 40, and it is:

$$\bar{X}_G = \frac{\sum\limits_{j=1}^{k} \sum\limits_{i=1}^{n} X_{ji}}{N} = \frac{8,377}{40} = 209.4250$$

Some Theory Underlying the Analysis

This section contains some information that will contribute greatly to your ability to interpret the results of an ANOVA experiment, as well as to help you understand why the particular calculations used here help us decide between null and alternative hypotheses. In other words, we are going to look at some theory and assumptions behind the analysis of variance. This presentation will not be very mathematical nor will it present anymore than the most basic components of ANOVA theory. (Those who would like to delve deeper may find Hays, 1973, or Winer, 1971, useful.[2]) Nonetheless, it should help you understand why we can make inferences about means by looking at variances.

Consider first the **structural model** for each observation. Such an expression is sometimes called a *linear model* because it states that each measurement represents a linear combination of the underlying effects. We treat this as a statement telling us where each measurement comes from. For our dependent variable plant height, we assume:

$$X_{ji} = \mu_G + \alpha_j + \epsilon_{ji}$$

Do not let the Greek letters throw you. The model says that our observed value for the ith individual in group j is composed of several components that add up to make the adult plant height. These are:

1. μ_G, which stands for the grand mean of all plants in the *population*.
2. α_j, which stands for the unique effect of receiving level j of the independent variable. This is the **treatment effect.**
3. ϵ_{ji}, which stands for all of the *other* factors that influence plant height. This is the **error effect.**

The value of μ_G is assumed to be the same for every plant in the entire experiment. The value of α_j is assumed to be the same for every plant within group j. Thus we have a possibility of four different α_j values in the study; the α_1 component is the same for every member of Group I, the α_2 component is the same for every member of Group II, and so on. Finally, another component ϵ_{ji} is added to the model to represent all factors not included in the first two components, such as the individual kernel's genetic background, exact soil conditions surrounding it, and anything else that may affect it uniquely. There is the possibility of a different ϵ_{ji} value for every kernel in the study. Values of ϵ_{ji} are assumed to be normally distributed with a mean of 0 and an unknown standard deviation.

These components are population parameters (that is why the Greek letters are used). We are particularly interested in the α_j and the ϵ_{ji} components. We know that there are many differences among our X_{ji} values, but what causes them? If they are due to differences among α_j values, then we can say that fertilizer amount does have an effect on plant height; for this reason, the α_j values are called treatment effects. On the other hand, our differences could be due simply to error effects — other factors besides our treatments. If this is the case, then the differences among X_{ji} values are to be attributed largely to differences among ϵ_{ji} values.

As you know by now, we cannot calculate parameters directly unless we make an observation on every member of the population. Thus, we

[2]W. L. Hays, *Statistics for the Social Sciences*, 2nd ed. (New York: Holt, Rinehart and Winston, 1973); B. J. Winer, *Statistical Principles in Experimental Design*, 2nd ed. (New York: McGraw-Hill, 1971).

must estimate them from our sample statistics. Let us carefully examine the statistics that go into this estimation process.

Concentrate on an individual observation, say, $X_{2,10}$ the tenth plant in Group II. According to our structural model, that individual's value of 199 is composed of several components:

$$X_{2,10} = \mu_G + \alpha_2 + \epsilon_{2,10}$$

We can easily find statistics to serve as estimates of the model's parameters. μ_G, for instance, is best estimated by $\bar{X}_G$, the sample grand mean (of all 40 observations). Treatment effect α_2 is defined as the unique effect of receiving level II treatment. One way of defining this further is to say that α_2 is the difference between the population mean μ_G and the mean that would appear in a large population of corn grown under level II treatment, μ_2. Thus:

$$\alpha_2 = \mu_G - \mu_2$$

We will use $\bar{X}_2$ as an estimator of μ_2 and we can now estimate the treatment effect:

$$\begin{aligned}
\text{Estimated } \alpha_2 &= \bar{X}_2 - \bar{X}_G \\
&= 205.60 - 209.425 \\
&= -3.825
\end{aligned}$$

This means that when compared to the mean value of all different treatment populations (μ_G), we estimate that being specifically in Group II adds a -3.825 cm to plant height.

The final estimate for $X_{2,10}$'s structural model concerns the error component $\epsilon_{2,10}$. The best estimate we have of this is the difference between the individual plant and its group mean. Thus:

$$\text{Estimated } \epsilon_{2,10} = X_{2,10} - \bar{X}_2$$

A very important point to be learned from the above discussion is that the difference between each X_{ji} and the observed sample grand mean, $\bar{X}_G$, can be written as the sum of the statistical estimators for α_j and ϵ_{ji}. For element $X_{2,10}$:

$$X_{2,10} - \bar{X}_G = (X_{2,10} - \bar{X}_2) + (\bar{X}_2 - \bar{X}_G)$$

You can also verify that this is true for element $X_{3,4}$:

$$\begin{aligned}
X_{3,4} - \bar{X}_G &= (X_{3,4} - \bar{X}_3) + (\bar{X}_3 - \bar{X}_G) \\
220 - 209.425 &= (220 - 226.30) + (226.30 - 209.425) \\
&= (-6.30) + 16.875 \\
10.575 &= 10.575
\end{aligned}$$

And, in general:

$$X_{ij} - \bar{X}_G = (X_{ji} - \bar{X}_j) + (\bar{X}_j - \bar{X}_G)$$

This division of the individual deviation scores around the grand mean into two components is the key to the analysis of variance. Remember that the sum of a group of squared deviation scores makes a *variation*, or *sum of squares*. In ANOVA we will compute *three* sums of squares by squaring both sides of the equation above and then taking the sums of those squared values:

$$\sum_{j=1}^{k} \sum_{i=1}^{n} (X_{ji} - \bar{X}_G)^2 = \sum_{j=1}^{k} \sum_{i=1}^{n} [(X_{ji} - \bar{X}_j) + (\bar{X}_j - \bar{X}_G)]^2$$

$$= \sum_{j=1}^{k} \sum_{i=1}^{n} (X_{ji} - \bar{X}_j)^2 + 2 \sum_{j=1}^{k} \sum_{i=1}^{n} (X_{ji} - \bar{X}_j)(\bar{X}_j - \bar{X}_G)$$

$$+ \sum_{j=1}^{k} \sum_{i=1}^{n} (\bar{X}_j - \bar{X}_G)^2$$

The large middle term becomes zero, because each $(X_{ji} - \bar{X}_G)$, when summed across all of the i values in each group, is equal to zero (the sum of the deviation scores about a mean is zero). For the bottom term, the value of $(\bar{X}_j - \bar{X}_G)^2$ is the same for all different i values in a group; since there are n different i values in the group, we may rewrite that last term as $n \sum_{j=1}^{k} (\bar{X}_j - \bar{X}_G)^2$. Thus, our equality becomes:

$$\sum_{j=1}^{k} \sum_{i=1}^{n} (\bar{X}_{ji} - \bar{X}_G)^2 = \sum_{j=1}^{k} \sum_{i=1}^{n} (X_{ji} - \bar{X}_j)^2 + n \sum_{j=1}^{k} (\bar{X}_j - \bar{X}_G)^2$$

$$\qquad (1) \qquad\qquad\qquad (2) \qquad\qquad\qquad (3)$$

Even if you did not follow the derivation of this equation, it is important that you understand that the total variation of individual scores around the grand mean (part 1 of the above equation) is equal to the sum of two components, just as the individual deviation scores around the grand mean can be broken into two components; what we have just done is called **partitioning the variation.** Part 2 of the equation is called the **within groups sum of squares** (SS_{wg}) because it contains the deviation of each score in a group about the group means. Part 3 is called the **between groups sum of squares** (SS_{bg}) because it contains the deviation of each group mean about the grand mean. If we call the total variation (part 1) the **sum of squares total** (SS_{total}) then we can write:

$$SS_{total} = SS_{wg} + SS_{bg}$$

By now you may be wondering what all of this has to do with our four groups of corn: Why do we bother to discuss structural models, parameter estimates, and partitioning of the variation if we are really interested in a hypothesis test about means? The answer is close at hand, and it has to do with two *variances* that we compute from our available *variations*.

Just as we computed an inferential standard deviation $\hat{s}$, we can make an inferential variance (MS) by dividing a variation (SS) by its degree of freedom ($\hat{s}^2$, for instance, is equal to $SS/(N - 1)$, when we are considering only the variance of a single sample). Here we will divide SS_{bg} by its degrees of freedom, which is $k - 1$, and obtain a **between groups mean square:**

$$MS_{bg} = \frac{SS_{bg}}{df_{bg}} = \frac{SS_{bg}}{k - 1}$$

Similarly, we can divide SS_{wg} by its degrees of freedom, which are $k(n - 1)$ and obtain a **within group mean square:**

$$MS_{wg} = \frac{SS_{wg}}{df_{wg}} = \frac{SS_{wg}}{k(n - 1)}$$

The mathematical theory underlying ANOVA (which will not be presented here) says that if our structural model is correct, then we can specify the long-run mean for these statistics — their expected values — as follows:

$$E(MS_{bg}) = \sigma^2_{\epsilon_{ji}} + n\sigma^2_{\alpha_j}$$
$$E(MS_{wg}) = \sigma^2_{\epsilon_{ji}}$$

The mean value of the within group MS will equal the variance of the different ϵ_{ji} components in the overall group, and the mean value of the between group MS will equal that variance *plus* n times the variance of the k different α_j effects.

Remember that our null hypothesis was the following:

$$H_0: \mu_1 = \mu_2 = \mu_3 = \mu_4$$

If this H_0 is true, then the α_j components of each of the k populations will have to be the same, since:

$$\mu_1 = \mu_G + \alpha_1$$
$$\mu_2 = \mu_G + \alpha_2$$
$$\mu_3 = \mu_G + \alpha_3$$
$$\mu_4 = \mu_G + \alpha_4$$

And, if all the α_j components are the same, their variance ($\sigma^2_{\alpha_j}$) will be zero.

Consider then a statistic called **F-observed**[3], that is the ratio of our two variances:

$$F\text{-observed} = \frac{MS_{bg}}{MS_{wg}}$$

---

[3]Named for Sir Ronald Fisher (1890–1962), English statistician and geneticist, who contributed more than anyone else to the development of analysis of variance statistics.

If we make a ratio out of the expected values of the statistics that go into that ratio, we have

$$\frac{E(MS_{bg})}{E(MS_{wg})} = \frac{\sigma^2_{\epsilon_{ji}} + n\sigma^2_{\alpha_j}}{\sigma^2_{\epsilon_{ji}}}$$

What would you expect F-observed to equal if H_0 is true? If H_0 is true, then $\sigma^2_{\alpha_j}$ really does equal zero, and we would expect an F-observed of about 1.0. If H_0 is true, then both MS_{bg} and MS_{wg} have the same expected value; this is because they are both *independent* estimates of the population variance $\sigma^2_{\epsilon_{ji}}$. (They are independent because they are both computed from different components of the total variation.) Any observed difference between the two mean squares — when H_0 is true — would then be due only to sampling variability, and we would not expect an F-observed very different from 1.0 very often. On the other hand, if there is a nonzero $\sigma^2_{\alpha j}$, MS_{bg} will probably be greater than MS_{wg}, and F-observed will be greater than 1.0. In such a case, we would reject H_0.

Here, then, is how we will proceed with the single-factor analysis of variance: we will compute our three variations, use two of them to make variances, and then make an F-observed statistic from those two variances as shown above. We will compare our observed F to a table of critical F values; the critical F values, like our critical t- and z-values, come from a sampling distribution of the statistic that assumes a true H_0. We will learn how to perform the test after showing how we arrive at F-observed for our example problem.

Computations in Single-Factor ANOVA (Definitional Formulas)

As with some of our other statistics, we can obtain the needed sums of squares for ANOVA by using either the definitional formulas — those presented in the preceding section — or the computational formulas. We will cover the definitional formula approach now, because these formulas show you which parts of the deviation scores go into each SS. Later we will present the computational formulas because they are easier to use.

Almost all of the computational work in ANOVA is involved in finding SS_{total}, SS_{wg}, and SS_{bg}. Let us begin the calculations by finding SS_{total}. Even though this variation doesn't ultimately enter our F-observed, it is a good idea to compute SS_{total} so that you can check your computations on the other sums of squares.

SS_{total} is precisely the same variation we learned to compute in Chapter 3. It is the sum of all the squared deviation scores — all 40 of them — around the grand mean:

$$SS_{\text{total}} = \sum_{j=1}^{k} \sum_{i=1}^{n} (X_{ji} - \bar{X}_G)^2$$

Here we start with the first elements in Group I and keep squaring and adding until we include the last element of Group IV:

$$SS_{\text{total}} = (190 - 209.425)^2 + (187 - 209.425)^2 + \cdots$$
$$+ (219 - 209.425)^2 + (206 - 209.425)^2$$
$$= 8,985.775$$

Next we compute our SS_{wg}. This will be the sum of the within group variations for each of our four groups. We can show this by writing:

$$SS_{wg} = \qquad SS_{wg_1} \quad + \quad SS_{wg_2} \quad + \quad SS_{wg_3} \quad + \quad SS_{wg_4}$$

$$\sum_{j=1}^{k} \sum_{i=1}^{n} (X_{ji} - \bar{X}_j)^2 = \sum_{i=1}^{n} (X_{1i} - \bar{X}_1)^2 + \sum_{i=1}^{n} (X_{2i} - \bar{X}_2)^2 + \sum_{i=1}^{n} (X_{3i} - \bar{X}_3)^2 + \sum_{i=1}^{n} (X_{4i} - \bar{X}_4)^2$$

For Group I, where $j = 1$ and $\bar{X}_1 = 197.20$:

$$SS_{wg_1} = (201 - 205.60)^2 + (195 - 205.60)^2 + \cdots$$
$$+ (199 - 205.60)^2$$
$$= 1,187.60$$

Similarly, we find that:

$$SS_{wg_2} = (201 - 205.60)^2 + (195 - 205.60)^2 + \cdots$$
$$+ (199 - 205.60)^2$$
$$= 664.40$$

and

$$SS_{wg_3} = 1,680.10$$
$$SS_{wg_4} = \quad 958.40$$

Thus:

$$SS_{wg} = 1,187.60 + 664.40 + 1,680.10 + 958.40$$
$$= 4,490.50$$

The final variation that we need is the between groups sum of squares:

$$SS_{bg} = n \sum_{j=1}^{k} (\bar{X}_j - \bar{X}_G)^2$$

$$= 10 \, [(197.20 - 209.425)^2 + (205.60 - 209.425)^2$$
$$+ (226.30 - 209.425)^2 + (208.60 - 109.425)^2]$$
$$= 4,495.275$$

We check our computations by insuring that $SS_{\text{total}} - SS_{wg} + SS_{bg}$:

$$8,985.775 = 4,490.50 + 4,495.275$$

Now we are ready to turn SS_{wg} and SS_{bg} into mean squares by dividing them by their degrees of freedom:

$$MS_{bg} = \frac{SS_{bg}}{df_{bg}} = \frac{SS_{bg}}{k - 1} = \frac{4,495.275}{4 - 1} = 1,498.425$$

$$MS_{wg} = \frac{SS_{wg}}{df_{wg}} = \frac{SS_{wg}}{k(n - 1)} = \frac{4,490.500}{36} = 124.736$$

Finally, we compute F-observed:

$$F\text{-observed} = \frac{MS_{bg}}{MS_{wg}} = \frac{1,498.425}{124.736} = 12.0136$$

When you perform an ANOVA, it is convenient to summarize the statistical information in an **ANOVA summary table.** For the single factor situation the summary table might appear as follows:

Summary of Analysis

SOURCE OF VARIATION	SS	df	MS	F
Between Groups Within Groups	4,495.275 4,490.500	3 36	1,498.425 124.736	12.013***
Total	8,985.775	39		

***$p < 0.001$

In the next section we will take up the meaning of the observed F and the asterisks. For the moment, however, you should note two things about the summary table.

First, the left-hand column, titled "Source of Variation" simply lists the parts of the partitioned total variation. Since "Between Groups" variation also corresponds to "Between Experimental Treatments," some researchers would write "Treatments" here; others would write "Fertilizer." The next part of the partitioned variation, "Within Groups" is sometimes called "Error" when it appears in the table, since variation here is assumed to come only from the differences among ϵ_{ji} components of the structural model. Notice that the second column contains the appropriate SS for each source, and the total variation is also given.

Next, notice that the degrees of freedom for each partition add up to the total degrees of freedom. The total df is $N - 1$ (just as it was in Chapter 10, when we computed $\hat{s}$), where $N = kn$, the total number of measurements in the experiment.

$$df_{total} = df_{bg} + df_{wg}$$
$$N - 1 = (k - 1) + k(n - 1)$$
$$40 - 1 = (4 - 1) + 4(10 - 1)$$

The final element in the table, F-observed, is the endpoint of our calculation. It is the statistic that will allow us to decide between our hypotheses.

The F-Test

Remember from the section on ANOVA theory that we should expect an F-observed of about 1.00 if the null hypothesis is true[4], that is, if fertilizer had no effect on plant height. But how much must F differ from 1.0 from before we reject H_0? The answer is given by the sampling distribution of the F-statistic that would appear with a true H_0 and an infinite number of replications of the experiment. All we have to do to perform our hypothesis test is to compare our F-observed with the value of F-critical in the table; F-critical, of course, is a value of F that would be exceeded by a small proportion of the observed F's when H_0 is true.

Like the t-distribution and the χ^2-distribution, F is really a family of distributions. *Which* of these F-distributions we are dealing with depends on the degrees of freedom involved. With the F-distribution, however, we must specify *two* degrees of freedom: the df associated with the numerator variance and the df associated with the denominator variance of the F-observed ratio. In symbols, F for the single-factor ANOVA has $df = k - 1, k(n - 1)$. For our particular example, $df = 3, 36$. These two numbers cannot be combined; if we change either one, we are dealing with a different F-distribution.

Figure 14.2 shows one F sampling distribution. Notice that the lowest possible value of F is 0 (when $MS_{bg} = 0$), and the highest value possible is infinity (when $MS_{wg} = 0$). Table F in Appendix 1 contains the critical values of F that cut off 0.05, 0.01, and 0.001 of the area under the *upper* tail of the sampling distribution. Thus, our F-test will be a one-tailed test, because we assume that F values less than 1.0 (where $MS_{bg} < MS_{wg}$) are due only to sampling variability and not to a true alternative hypothesis. Remember from the ratio of $E(MS)s$ presented in the theory section, that *any* difference among the α_j components of the structural model could only increase the value of $E(MS_{bg})$.

To make the test, we simply compare our F-observed with the appropriate F-critical values in the table. Since there are so many different F-distributions, there is a different page of the table for each alpha level. [Again, we have two symbols standing for the same thing. Alpha (α) stands for the level of significance, and α_j stands for a treatment effect parameter.] Let us begin with $\alpha = 0.05$. Our $df_{numerator}$ (given in the column headings) is 3, and the $df_{denominator}$ (given in the row headings) is 36. For $df = 3, 36$, F-critical at this alpha level is not listed precisely — $df_{denominator}$ values are only given for df of 30 and 40. When this occurs,

---

[4]Actually, the expected value of F, $E(F)$, is greater than 1.0, and it depends on the number of degrees of freedom in the denominator variance, df_{denom}:

$$E(F) = \frac{df_{denom}}{df_{denom} - 2} \qquad (\text{when } df_{denom} > 2)$$

However, 1.0 is a close approximation to $E(F)$ when there are more than a few degrees of freedom in the denominator variance. For our example problem, where $df_{denom} = 36$, $E(F) = 1.059$.

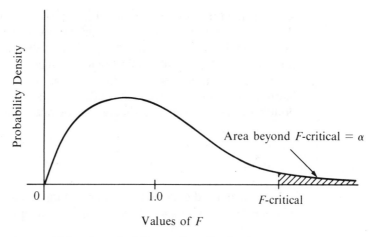

Figure 14.2   A typical F sampling distribution.

one may either interpolate to get the exact F-critical, or use the F-critical for the next lower df listed. If we choose the latter option, that is, select the F-critical for $df = 3, 30$, and our F-observed exceeds the critical F, there is no need to interpolate because the interpolated F-critical would only be lower in value (with higher df) and that too would be exceeded by F-observed. At $\alpha = 0.05$, $df = 3, 30$, F-critical $= 2.92$ and, clearly, we reject H_0 at this significance level. Checking the tables for $\alpha = 0.01$ and then $\alpha = 0.001$, we find that our F-observed is significant at the 0.001 level.

Thus, we have finally decided to reject H_0. Our fertilizer very probably did have an effect on plant height. We note the level of significance obtained in the ANOVA summary table, as shown, by using asterisks. One convention is to use the following symbols: *** $= p < 0.001$, ** $= p < 0.01$, and * $= p < 0.05$. In a printed report, we would very likely want to include a graph of our four group means, such as Figure 14.1, and we would state that the effect of fertilizer on plant height was significant ($F = 12.013$, $df = 3, 36$, $p < 0.001$).

[C]   Computational Formulas for Sums of Squares

In Chapter 4, we learned that the SS for a single group could be computed in either of two ways:

Definitional
Formula

Computational
Formula

$$SS = \sum_{i=1}^{n} (X_i - \bar{X})^2 = \sum X_i^2 - \frac{\left(\sum_{i=1}^{n} X_i\right)^2}{N}$$

The same form of computational formula can be used to compute all of the sums of squares needed for our ANOVA. Here, as elsewhere, the more complex-looking computational formulas are really easier to use than the simpler-looking definitional formulas.

The computational formula for SS_{total} is in fact the same formula we learned in Chapter 4. The only difference is that here it appears with a double summation sign to indicate that we need all values from all groups:

$$SS_{total} = \sum_{j=1}^{k} \sum_{i=1}^{n} X_{ji}^2 - \frac{\left(\sum_{j=1}^{k} \sum_{i=1}^{n} X_{ji}\right)^2}{kn}$$

For our data, the sum of all squared individual X_{ji} values is 1,763,339. The sum of all X_{ji} values is 8,377, and we can compute SS_{total} as:

$$SS_{total} = 1,763,339 - \frac{(8,377)^2}{(4)(10)}$$

$$= 1,763,339 - 1,754,353.225$$

$$= 8,985.775$$

The computational formula for SS_{wg} makes use of a similar process that must be repeated for each group, with the k individual group results being added together. The complete formula for SS_{wg} is:

$$SS_{wg} = \sum_{j=1}^{k} \sum_{i=1}^{n} X_{ji}^2 - \frac{\left(\sum_{i=1}^{n} X_{ji}\right)^2}{n}$$

From the data given in Table 14.2, you can verify that the sum of squared values within the first group is 390,666 and the sum of the n individual X values is 1972. Then:

$$SS_{wg_1} = \sum_{i=1}^{n} X_{1i}^2 - \frac{\left(\sum_{i=1}^{n} X_{1i}\right)^2}{n}$$

$$= 390,066 - \frac{(1,972)^2}{10}$$

$$= 1,187.60$$

Similarly:

$$SS_{wg_2} = \sum_{i=1}^{n} X_{2i}^2 - \frac{\left(\sum_{i=1}^{n} X_{2i}\right)^2}{n}$$

$$= 423,378 - \frac{(2,056)^2}{10}$$

$$= 664.40$$

and

$$SS_{wg_3} = 1,680$$

$$SS_{wg_4} = 958.40$$

When we add these together — that is, add the outer summation sign — we arrive at:

$$SS_{wg} = 1,187.60 + 664.40 + 1,680.10 + 958.40$$
$$= 4,490.500$$

Lastly, our SS_{bg} can be computed from the squared group totals and the grand total:

$$SS_{bg} = \frac{\sum_{j=1}^{k} \left(\sum_{i=1}^{n} X_{ji} \right)^2}{n} - \frac{\left(\sum_{j=1}^{k} \sum_{i=1}^{n} X_{ji} \right)^2}{kn}$$

$$= \frac{(1,972)^2 + (2,056)^2 + (2,263)^2 + (2,086)^2}{10} - \frac{(8,377)^2}{(4)(10)}$$

$$= 1,758,848.50 - 174,353.225$$

$$= 4,495.275$$

These values are identical to those obtained with the definitional formulas. They do not, however, require computation of individual means or all of the individual subtractions that the definitional formula approach does. Thus, they require less computational effort than the definitional approach and they entail less risk of computational error. After the sums of squares are computed and entered in the summary table, there is no difference between computational and definitional formula approaches.

Whenever possible, it is best to use the same number of individuals in each group, such as we did in the example with corn plants. Occasionally, this is not possible. Some human subjects may not show up for an experiment, or animal and plant subjects may die during the course of an experiment. Unequal n's may appear in an experiment for a number of reasons, but when they do, the ANOVA analysis is greatly complicated. Part of the complication arises from the fact that the mathematical theory underlying ANOVA does not state quite so clearly how to proceed with the analysis; and, part of the problem lies with more difficult calculations to carry out. For these reasons, the unequal n situation will not be covered here.

$\boxed{\text{O}}$ The Hartley F_{max} Test for Homogeneity of Variances

Up until this point, little has been said regarding the assumptions underlying ANOVA. Of course, you should have recognized that because we use means and variances in ANOVA, the assumption of interval-level data is necessary. Also, like the t-test, the analysis of variance assumes normality of the population distribution. If you have followed the methods in earlier chapters, you should now be equipped to evaluate your data with regard to level of measurement, and, if sample sizes are large, the problem of insuring a normal population distribution becomes unimportant. The same principle underlying the central limit theorem (which we used with the sampling distributions in Chapters 10 and 11) also applies in the ANOVA situation.

A third assumption underlying ANOVA is that the sample variances (the within group variances for the individual groups) are approximately equal. This is called the assumption of **homogeneity of variances.** There are several readily applied tests for determining whether or not this assumption is met, but we will briefly look at just one: the **Hartley F_{max} test.**

Going back to the example problem involving fertilizer and plant heights, we produced four individual within group variations in the process of computing the overall SS_{wg}. If we divide each of these by *its* degrees of freedom, $(n - 1)$, we have a variance, $\hat{s}^2$, for each group. From that example:

$$\hat{s}_1^2 = \frac{SS_{wg_1}}{n - 1} = \frac{1,187.60}{9} = 131.96$$

Similarly:

$$\hat{s}_2^2 = 73.82$$
$$\hat{s}_3^2 = 186.68$$
$$\hat{s}_4^2 = 106.49$$

The assumption underlying ANOVA is that these four variances are all unbiased estimates of the same population variance, $\sigma^2_{\epsilon_{ij}}$. If that assumption is violated too drastically, then the tabled values of F-critical will no longer be accurate for a true H_0. The Hartley F_{max} test will allow us to decide whether or not we can assume homogeneity.

The F_{max} statistic is taken, simply, as the largest of these sample variances over the smallest of these:

$$F_{max} = \frac{\hat{s}^2_{largest}}{\hat{s}^2_{smallest}}$$

Table G in Appendix 1 gives critical values of F_{max} that, when exceeded, indicate that the assumption of homogeneity of variances is not tenable. For our data:

$$F_{max} = \frac{\hat{s}_3^2}{\hat{s}_2^2} = \frac{186.68}{73.82} = 2.53$$

Table G shows only the critical values at $\alpha = 0.05$ for this test; note that the critical value depends on the overall number of groups in the experiment (k) and the number of elements in each group. Here $k = 3$, and $n - 1 = 9$. The critical value in the table is 5.34, much more than our F_{max}. Hence we do not reject the assumption of homogeneity of variances, and we decide that our data *are* appropriate for ANOVA. However, we should confuse the issue a little here by noting that the reverse is not necessarily true. That is, rejection of the assumption does not necessarily mean that you cannot conduct the ANOVA. If the observed F_{max} is only a little above the critical value, and if you have equal n's, then it is probably still permissible to conduct the ANOVA in the usual manner. Only when both of the above conditions are not present — i.e., when you have unequal n's and an F_{max} much greater than the critical value should the ANOVA be postponed until you have time to seek expert statistical advice. For a little more discussion on these issues, one can begin with the references given in footnote 2 on page 379.

The Independent Variable and Cause and Effect

In the example problem discussed in this chapter, the independent variable consisted of a number of different levels of an experimental treatment that were applied to different groups. This was so that cause-and-effect arguments could be made, in the same way that cause-and-effect arguments were made from the two-sample experiments discussed in Chapter 11. However, there are many uses of ANOVA where the researcher does not control the independent variable, and the resulting analysis is more a study in correlation than in cause and effect.

One such situation would arise, for instance, if we took four groups of children, where the ages in each group were different. We might have, for instance:

Group I   = children aged 5–6
Group II = children aged 7–8
Group III = children aged 9–12
Group IV = children aged 13–15

Each group might be given the same task to work on, say, a survey of vocabulary size. The data from such a test could be subjected to ANOVA calculations in the same way we subjected plant heights to the single-factor ANOVA. Here, however, we could not argue from a significant F-observed that age causes changes in vocabulary size. The reason is that we did not randomly assign children to different age groups — the independent variable, in fact, is something we cannot change for any individual. However, if the different groups produced a significant F, we *could* argue that their scores are truly different. This kind of study is more of a correlational analysis than a controlled experimental manipulation, since we are justified in claiming an association between "independent" variable and dependent variable, but we are not justified in claiming a cause-and-effect relationship between them.

Studies of the kind just described are common in the behavioral sciences. We can characterize them by saying that subjects are assigned to different groups on the basis of **subject variables** — characteristics of the subject that he or she brings to the experiment. Some subject variables commonly used would include age, sex, various kinds of aptitudes, geographical location of home, and many others. When a researcher assigns people to different groups on the basis of some attribute they already possess, the independent variable is a subject variable, and the study is a kind of correlational study.

Qualitative and Quantitative Independent Variables

Another characteristic of our example problem was that it used **quantitative independent variables.** Measurable amounts of fertilizer were chosen for each group's treatment. It is not necessary always to have this kind of independent variable; we may instead design an experiment where a **qualitative independent variable** is used. Suppose, for instance, that instead of giving each group in the fertilizer study different amounts of fertilizer, different brands of fertilizer were used. If the controlled experiment were conducted in the same manner as the example problem we covered, and if a significant F-observed were obtained, we could still argue just as correctly that fertilizer brand caused changes in plant heights.

The only difference between qualitative and quantitative independent variables that we need note here would be in the way we graph the results. Figure 14.1 appropriately has dots representing individual group means connected with lines. If we had used a qualitative independent variable, a bar graph of these results would be more appropriate.

SUMMARY

The single-factor analysis of variance (ANOVA) is a means of analyzing controlled experiments like the two-sample experiments described in Chapter 11, but which may have more than two samples. Different levels of an independent variable are administered to a number of different groups, and the group means are found to differ. ANOVA is designed to help you decide whether the difference is due to sampling variability or to a true effect of the independent variable on the dependent variable.

In ANOVA, inferences about means are made literally from analyses of variances. The total SS in the experimental results is divided into two additive components: the SS due to between-groups differences (SS_{bg}) and the SS due to differences among individuals within groups (SS_{wg}). These sums of squares are then made into variances by dividing each SS by its respective degrees of freedom. Under the null hypothesis of no treatment effect, both MS's should be about the same value, in that both are independent estimates of the population variance. Under a true alternative hypothesis, however, the MS_{bg} will be larger than the MS_{wg}. The actual hypothesis test involves making a ratio of the two variances called F-observed. F-observed is then compared to critical values of F that would appear under a true H_0. If the F-observed is very unlikely under a true H_0, that hypothesis is rejected. Rejection of H_0 is equivalent to deciding that the independent variable does have an effect on the dependent variable.

This chapter discusses some of the assumptions underlying the analysis of variance. A means of testing one of these assumptions, homogeneity of variances, is provided through the Hartley F_{max} test.

Also discussed are some differences among kinds of independent variables in ANOVA studies. When the independent variable is more accurately described as a subject variable, for instance, the ANOVA study is more accurately described as a correlational study than a study of cause and effect.

KEY CONCEPTS

single-factor analysis of variance

factor

double subscript notation

structural model

grand mean

treatment effect

error effect

partitioning of the variation

within groups sum of squares

between groups sum of squares

sum of squares total

within groups mean square

between groups mean square

F-observed

ANOVA summary table

The F-test

homogeneity of variances

Hartley F_{max} test

subject variables

quantitative independent variables

qualitative independent variables

SPOTLIGHT 14   Analysis of Variance in Research

Without question, the analysis of variance is the most popular statistical technique applied to controlled experiments in the behavioral sciences. A conceptual understanding of ANOVA principles is necessary if one hopes to comprehend and evaluate much of the current research literature in psychology, sociology, and education. An introduction to ANOVA, therefore, is an important part of a first course in applied statistics, even though very few students actually go on to perform experiments themselves. If you remember the main concepts presented in this chapter and in Chapter 15, you should be able to make an intelligent, *critical* evaluation of a surprisingly large proportion of the studies reported in professional journals.

A quick glance through publications such as the *Journal of Experimental Psychology*, *Sociometry*, the *Journal of Applied Psychology*, or any of the other current behavioral science journals will reveal that ANOVA results are not always presented in the form described in this chapter and in Chapter 15. Frequently, you will find ANOVA tables that are abbreviated forms of the tables illustrated in this text. For instance, in reporting the results of an experiment designed to examine the effects of "Speaker Credibility" on "Persuasion," the experimenter might tell you only that "The effect associated with credibility was significant ($F = 34.7, df = 3, 76, p < 0.001$)." Consider for a moment the information contained between the parentheses, and you will recognize that it contains all of the important information displayed in larger tables (such as those presented in this text) with SS, MS, df, and "Source of Variation" columns. Because you know that there are three degrees of freedom associated with the F-obtained numerator variance, you will infer there were four experimental groups, each of which received different levels of an independent variable called "Credibility." The error variance has 76 degrees of freedom associated with it. Without even reading the "Methods" section of the article, you should be able to calculate that each group had twenty subjects. ($df_{error} = k(n - 1)$). If $df_{error} = 76$ and $k = 4$, then n must equal 20. What about the individual SS and MS values? It is true that you cannot recover them from the parenthetical information, but then it is unlikely that you would ever want to. It is the ratio of the two MS values as expressed by F-observed that helps you evaluate the experiment, not the separate SS and MS values. In the last decade, authors and editors have begun to realize that printing complete ANOVA summary tables is a waste of space, and current reports are likely to omit all but the essential information.

PROBLEMS

Problems 1–3 are based on the following situation:

In order to study the effects of sleep deprivation on performance of air traffic controllers, you randomly select 24 controllers from the union roster and ask them to participate in your study. All agree, and you randomly assign them to four groups of six individuals each. Members of each group are asked to stay up the indicated number of hours beyond their normal bedtime and then undergo a simulated vigilance task on a radar screen. Listed below are the number of errors made by each:

HOURS OF SLEEP DEPRIVATION	INDIVIDUAL					
	1	2	3	4	5	6
4	37	22	22	25	34	28
12	36	45	47	28	40	37
20	43	70	66	52	58	62
28	76	66	43	62	71	69

1. To practice the use of subscript and summation notation, give the value of

a. $X_{3,2}$

e. $\sum\limits_{i=2}^{n} X_{2,i}$

b. $X_{1,4} + X_{2,3}$

f. $\sum\limits_{j=1}^{2} \sum\limits_{i=1}^{n} X_{ji}$

c. $\sum\limits_{i=1}^{4} X_{1i}$

g. $\left(\sum\limits_{i=1}^{4} X_{3i}\right)^2$

d. $\sum\limits_{i=1}^{n} X_{3i}$

h. $\sum\limits_{i=1}^{n} X_{4i}^2$

2. Perform the analysis of variance on the data above and interpret the results of your analysis. Also graph the sample means.

3. For element X_{11} in the data table above, show that the value of $(X_{ji} - \bar{X}_G)$ is really equal to $(X_{ji} - \bar{X}_j) + (\bar{X}_j - \bar{X}_G)$.

4. Are people of different intelligence more likely to participate in different sports? To answer this question, five athletes from five different sports are given an IQ test. The scores for each group were:

TRACK	HOCKEY	FOOTBALL	WRESTLING	LACROSSE
120	109	108	110	85
110	115	112	115	100
90	100	95	121	121
115	128	130	97	118
115	96	100	140	105

a. Answer the question by performing the single-factor ANOVA.
b. Why is this study more properly called an attempt to find correlation than it is a controlled experiment?

5. Twenty-one high-school seniors decided to take part in an investigation of the special "exam preparation" books that purportedly help one get ready for college entrance examinations. The group divided themselves into three groups of seven each on a purely random basis. Two of the groups used the books, each group selecting a different book. The third group did not use the books. Listed below are the obtained entrance exam scores. Perform the ANOVA and interpret your results.

Exam Scores

NO BOOK	BOOK I	BOOK II
380	532	540
470	455	570
441	440	520
487	620	620
420	560	660
390	522	605
450	517	602

6. Test all three of the data sets above (problems 2, 4, and 5) for homogeneity of variances using Hartley's F_{max} test.

7. Perform an ANOVA on the dependent variable measurements below to decide whether or not group means differ significantly from each other.

$$\text{Group I:} \quad 75, 64, 73, 84, 71$$
$$\text{Group II:} \quad 42, 68, 55, 71, 59$$
$$\text{Group III:} \quad 71, 81, 89, 67, 74$$

8. Thirty typists take part in a test of typing speed. The typists are randomly divided into three groups of 10 each, and each group uses a different brand of typewriter. Is there a difference in the mean speeds of the various groups?

Typing Speed Scores:

GROUP I	GROUP II	GROUP III
75	81	75
80	74	58
91	92	64
74	83	71
72	78	65
83	86	59
71	91	62
82	72	64
65	64	54
74	82	71

9. The data below are dependent variable measurements from a four-sample experiment. Perform the ANOVA to decide whether changes in the independent variable produce changes in the dependent variable.

$$\text{Group I:} \quad 2.10, 3.44, 2.70, 1.92, 1.85$$
$$\text{Group II:} \quad 3.01, 2.50, 3.44, 1.40, 2.66$$

Group III: 3.00, 2.54, 2.62, 2.48, 3.71

Group IV: 4.00, 3.01, 2.47, 3.58, 4.40

10. When there are only two groups in the independent samples experiment, you may perform either the t-test or the ANOVA to test the significance of the difference between means. Show that the two tests lead to identical conclusions by performing both for the following data set. Show also that your observed F-ratio, with $df = 1, 18$ is equal to the squared t-observed with $df = 18$. (In general, with two groups, F-observed with $df = 1, k(n - 1)$ will equal the squared independent samples t-observed with $df = n_1 + n_2 - 2$, if $n_1 = n_2$ and $k = 2$.)

GROUP I		GROUP II	
550	620	370	280
430	580	290	365
610	490	489	237
573	579	420	394
602	639	441	402

15. Analysis of Variance With Two Factors

After reading this chapter, you should be able to do the following:

1. *Explain some kinds of research questions addressed by two-factor analysis of variance (ANOVA).*

2. *Explain the separate meanings of main effect and interaction effect, and show how to interpret such effects from graphs.*

3. *Perform a two-factor ANOVA.*

4. *Show how tables of expected mean squares are used in constructing F-values for the analysis.*

The Two-Factor Analysis of Variance

Scientists of all kinds recognize simplicity as a virtue. If there are two different theories to account for a phenomenon, other things being equal, the simpler of the two is considered the more likely explanation *because* it is simpler. This principle is known as Occam's Razor, and it makes very good sense because nature almost always accomplishes things in the simplest way possible. In designing experiments, too, if there are two ways to approach a question in the laboratory, the simpler experiment is usually the better experiment. The simpler the experimental design, the fewer things that can go wrong during the study and the fewer potential weaknesses there are in arguments made from the results. An experiment may require very complex technical equipment — like some of the studies of subatomic particles in physics — and yet be simple in design. And, the simpler the design, the simpler and clearer the statistical analyses that are called for. Indeed, many of the truly important experiments in the history of science have required little or no statistical analysis; many have involved only one independent and one dependent variable; many have required only the statistical methods we have already covered. Highly

398

complex statistical methods have their places, but complex experiments and complex statistics are not inherently superior to simple statistics.

With these things said, we turn now to some more complex statistics. We will extend the principles developed in the last chapter to include experiments with *two* independent variables. Complexities should be added to an experiment only when doing so adds to the information one can obtain from it. In this chapter we will see how using two independent variables more than doubles the information we can extract from an experiment with a single independent variable. In particular, we will see how the two-factor analysis of variance (**two-factor ANOVA,** for short) allows us to separately evaluate the effect of each independent variable in the experiment, and how two-way ANOVA lets us ask a question impossible to ask with a pair of single-factor experiments: What unique effects do various *combinations* of these independent-variable levels have? As usual, we will introduce the statistics by way of an example problem.

A Two-Factor Experiment

Our example concerns a (hypothetical) study of memory performed by an experimental psychologist. He is particularly interested in examining some factors that influence memory of vocabulary words in a foreign language. Suppose that he obtains the help of 30 students who are studying elementary German. They agree to participate in his study, and they are randomly divided into six groups of five each.

From Table 15.1, we can learn the structure of the experiment and the reason for six different groups. The figure is called a **design matrix** and it shows both independent variables and how they are combined.

One independent variable (or factor) in this study is "Method," by which new vocabulary words are presented. This factor has two levels, and half of the subjects receive one level, half receive the other. Students in the "Verbal" level of that factor are simply given 100 new German vocabulary words printed next to their closest English equivalents. The other 15 students are in the "Pictorial" level, and are presented with the same new words printed in German and accompanied by a picture or a cartoon strip that portrays the meaning of the new word without resorting to using English translations.

The second independent variable is "Practice Sessions" and it has three levels. Ten students are in each level (five in the "Verbal Level" of factor 1 and 5 in the "Pictorial Level" of factor 1); each level of factor 2 receives the indicated number of practice material on as many successive days. The matrix shows how the combination of 3 levels from one factor and 2 levels from the other factor produce 6 different experimental **combinations.** Each student, of course, is assigned to only one combination and there are five students in each combination.

All combinations, or groups, are scheduled to complete treatment on the same day. After the last session, a vocabulary test is given to all

Table 15.1 *Design Matrix and Notation for the Two-Factor Experiment*

| | | PRACTICE SESSIONS | | | ROW TOTALS | ROW MEANS |
		2	5	10		
Method of Presentation	Verbal	X_{111} X_{112} X_{113} X_{114} X_{115}	X_{121} X_{122} X_{123} X_{124} X_{125}	X_{131} X_{132} X_{133} X_{134} X_{135}	$\sum_{c=1}^{m}\sum_{i=1}^{n} X_{1ci}$	$\bar{X}_{1..}$
	Pictorial	X_{211} X_{212} X_{213} X_{214} X_{215}	X_{221} X_{222} X_{223} X_{224} X_{225}	X_{231} X_{232} X_{233} X_{234} X_{235}	$\sum_{c=1}^{m}\sum_{i=1}^{n} X_{2ci}$	$\bar{X}_{2..}$
Column Totals		$\sum_{r=1}^{k}\sum_{i=1}^{n} X_{r1i}$	$\sum_{r=1}^{k}\sum_{i=1}^{n} X_{r2i}$	$\sum_{r=1}^{k}\sum_{i=1}^{n} X_{r3i}$	Grand Total $= \sum_{r=1}^{k}\sum_{c=1}^{m}\sum_{i=1}^{n} X_{rci}$	
Column Means		$\bar{X}_{.1.}$	$\bar{X}_{.2.}$	$\bar{X}_{.3.}$	Grand Mean $= \bar{X}_{...}$	

X_{rci} = observation on an individual, where

r = row $(1 \leq r \leq k)$

c = column $(1 \leq c \leq m)$

i = individual within cell $(1 \leq i \leq n)$

students; the dependent variable is the number of words correctly identified by each student.

Of course, there are many factors besides the experimental factors included in the study that could conceivably affect the dependent variable. A careful experimenter would want to insure that all of these were the same for all members of all combinations, and this might be quite difficult to achieve in practice. He would want to be sure, for instance, that the 100 words were previously unknown to all participants in the study; that some people did not engage in extra study or extra practice between their experimental sessions; and so on. All of these added precautions help to make the study well controlled. However, let us assume that these controls have been satisfactorily achieved and proceed with the analysis.

There are several questions we can ask about results from such a study. What effect did the teaching method have on vocabulary scores? What effect did the number of practice sessions have? Was the effect of practice sessions different for "Verbal" students than it was for "Pictorial" students? As with single-factor ANOVA, we should not try to answer these questions until we discuss the symbols and notation used in two-factor ANOVA.

Symbols and Notation in Two-Factor ANOVA

Referring back to Table 15.1, you will notice that we must now label each X observation with three subscripts. The multiple subscripts make summation signs look complex and cumbersome, but they are no more difficult to use than the double subscript system we used in Chapter 14. There is one unfortunate development in the subscript situation, however. We must change symbols slightly when we advance to two-factor ANOVA.[1] If you understand all of the symbols presented in Table 15.1, though, you should have no problem in conducting the analysis.

The general symbol for a single observation will be X_{rci}. As with single-factor ANOVA, the i stands for the individual within a particular cell. Since there are n individuals in each cell, i can have any value from 1 to n. The mathematician's way of writing this is $1 \leq i \leq n$.

The symbol r stands for the individual's row, that is, the level of factor 1, and the symbol k will stand for the number of rows in the design matrix. For our study, $k = 2$, and r must be either 1 or 2.

Finally, note that c stands for the column an individual is placed in and, of course, determines the level of factor 2 that we are dealing with. If the total number of columns in the study is m, then c must be between 1 and m; here, $m = 3$.

Using the triple summation sign, such as you do when finding the grand total, is a process just like using a double summation sign. Start at the outermost $\sum$, and assume the first subscript value it indicates. Then move to the second $\sum$, assume the first subscript value it indicates, and then carry out the instructions given by the inner $\sum$. For our example problem, that would involve adding up the values of all n individuals in the first row, first column cell. After doing this, retreat to the second $\sum$, change its subscript to the next value indicated, and return to the inner $\sum$ and keep adding. Only when you have gone through all subscripts on the middle $\sum$ in this manner do you return to the outermost summation sign and increment the subscript indicated by it; then, of course, you repeat the addition process with the inner two $\sum$'s. This continues until all subscripts on the outer summation sign have been used. It is all just a symbolic way of saying that you add up all individuals in the group, for all groups in each row and column of the design matrix.

One more item having to do with notation should be mentioned. Notice the special symbol to represent the various means. A dot after a $\bar{X}$ indicates that summation has been carried out across all values of the subscript that occupy that position. For instance, $\bar{X}_{.2.}$ represents the mean

---

[1] If you survey a few statistical textbooks — either undergraduate or graduate — you will quickly discover that practically all authors have their own unique system of notation. This is an unfortunate situation that shows no sign of improving. However, the differences among notation systems largely concern which letters of the alphabet to use in representing each level, number of levels, and so on, and to a lesser extent, the order in which subscripts appear. Thus, if you learn one system, you should be able to transfer to another system because they are almost all based on the same principles. The systems used in this book for one-factor and two-factor ANOVA are very close to those used in a majority of current textbooks.

of the second column. The dots replace the r and i subscripts because all values of those subscripts are included in the total and mean. Similarly, $\bar{X}_{2..}$ represents the mean of the second row; c and i subscripts are replaced by dots because all their values (that appear in the second row) are part of the mean. The grand mean is represented by $\bar{X}_{...}$ because all values of all subscripts go into it. Finally, the individual cell means will have a single dot in their subscripts. For example, $\bar{X}_{23.}$ represents the mean of all i individuals in row 2, column 3.

Graphing Main Effects

A good way to begin analysis of any experimental data is with a graph. The graph by itself will not tell you whether the various effects were significant or not — only the hypothesis tests can do that — but it will show you the direction of the effect and suggest to you which effects may be significant. The advantage of ANOVA is that it allows us to examine all of our effects independently of one another; there is probably no better way to define the effects than to turn to their graphs.

Table 15.2 contains the results of the study, and the main effects associated with each independent variable are graphed in Figure 15.1. Consider first the independent variable "Method of Presentation," our first factor. We want to know if students in one level scored (significantly) differently from students in the other level. The nm students in the "Verbal" level has $\bar{X}_{1..}$ equal to 36.40, and the nm students in the "Pictorial" level had $\bar{X}_{2..}$ of 77.07. The difference between these means is called the **main effect** associated with factor 1. Clearly there is a difference of some magnitude between the methods, but we will have to wait until performing the appropriate F-test before knowing if it is statistically significant.

Imagine what the upper graph in Figure 15.1 would look like if there were little or no difference between levels of factor 1. The two endpoints of the line would be at about the same height; any difference between these means shows up as a difference in height of different parts of the line.

With this in mind, consider then the lower graph in Figure 15.1. This shows the column means, that is, the mean of each level of the "Practice Session" factor. Are the means here more or less alike? Or, are they different? Since the line is not horizontal, we know that there is some main effect associated with factor 2 as well. Of course, we will have to perform a separate F-test on this main effect to determine whether or not it is statistically significant, but for the present it looks like we can say that increasing practice sessions led to an increase in vocabulary scores.

Incidentally, notice that we have a *qualitative* independent variable ("Methods"), and a *quantitative* independent variable ("Practice Sessions"). To make the graphs from all factors comparable, the qualitative variable was graphed as shown rather than with bars; the line connecting

Table 15.2 *Results of the Two-Factor Example Problem*

		PRACTICE SESSIONS 2	5	10	ROW TOTALS	ROW MEANS
Method of Presentation — Verbal		25	28	55		
		18	40	42		
		31	25	61	546	36.40
		24	37	47		
		36	34	43		
Method of Presentation — Pictorial		82	78	78		
		69	74	76		
		73	89	81	1156	77.07
		64	85	71		
		78	69	89		
Column Totals		500	559	643	Grand Total = 1,702	
Column Means		50.00	55.90	64.30	Grand Mean = 56.73	

INDIVIDUAL CELL TOTALS

	2	5	10
Verbal	134	164	248
Pictorial	366	395	395

INDIVIDUAL CELL MEANS

	2	5	10
Verbal	26.80	32.80	49.60
Pictorial	73.20	79.00	79.00

the means of the two levels of "Methods" is permissible because its purpose is to make clear the difference in height between them.

Graphing Interaction Effects

The remaining effect we can analyze with this study is called the **interaction effect,** or more specifically, the *Methods × Practice Sessions interaction.* What we will want to ask here is whether or not the "Practice" factor had the same effect on level 1 of "Methods" as it did on level 2 of "Methods." The psychologist conducting this study may have deliberately combined these two independent variables in a single study because he suspected that specific numbers of practice sessions have different effects on memory when used with different teaching methods.

Main effect associated with factor I

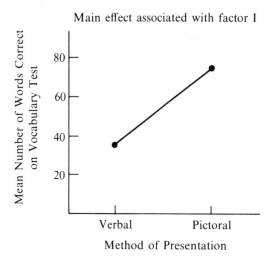

Main effect associated with factor II

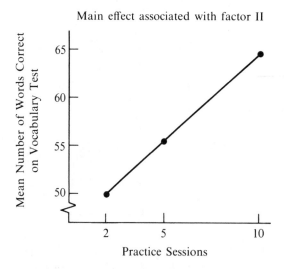

Figure 15.1 Main effects obtained in the example study.

Consider then the upper graph in Figure 15.2. The axes of this graph are the same axes we used in looking at the "Practice Sessions" main effect (Figure 15.1). There are two lines on this graph, however, one representing each level of the "Methods" factor. The lower one, for instance, shows the "Verbal Method" means for each level of the "Practice" factor. Therefore, each of the three means that defines this (lower) line is based on a single combination of levels from the two factors. In this study, that is equivalent to saying that each mean is a cell mean. It may be a good idea now to turn back to Table 15.2 and be sure that you see how the cell

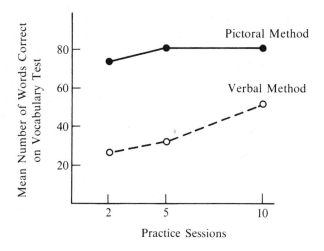

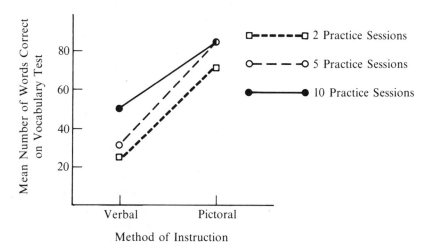

Figure 15.2   Interaction effect obtained in the example study. Either graph alone is sufficient to portray the interaction.

means given at the bottom of the table produce the upper graph in Figure 15.2.

Staying with this graph for the moment, some important items of information can be gleaned from it. First, the "Pictorial" line is much higher overall than the "Verbal" line. This altitude difference is a consequence of the factor 1 main effect — just another way of showing that there was a clear difference between the levels of factor 1. It is also possible to deduce from this graph the size of the main effect associated with "Practice Sessions." Take a pencil and lightly place a dot between the pair of dots over each practice-session level. Now connect these dots and you will have re-created the main-effect graph shown at the bottom of Figure 15.1. More to the point, however, is the fact that the lines for "Pictorial" and "Verbal" levels are not parallel to each other. Non-parallel lines are strong evidence that we might be in the presence of a real interaction effect. One way to describe what we see in this figure is to say that for the "Pictorial" method, increasing practice sessions from 2 to 5 increased memory performance, while increasing practice sessions from 5 to 10 was of no added benefit. With the "Verbal" method, however, increasing sessions from 5 to 10 produced a dramatic improvement in memory. Thus, the effect of "Practice Sessions" does seem to differ from level to level of "Methods." Whether we decide to believe that, or, alternatively, attribute that characteristic of the graph to sampling variability will depend on a third F-test in this ANOVA. This will be the test for the significance of the interaction effect.

Before proceeding to the computations, however, it will be wise to insure that you really understand the interaction concept. And, since that concept is probably best discussed graphically, let us return to the figures.

You should know that every two-factor interaction has two sides to it, two ways of describing it. The *same* interaction in the upper part of Figure 15.2 also appears in the bottom part of the figure. The horizontal axis represents "Method of Instruction," and there is a separate graph line for each level of the "Practice Sessions" factor. An alternative definition of our interaction — if indeed we really have an interaction effect — would be to say that the effect of "Methods" depends on how many practice sessions are involved. There is some small support for this statement in the lower graph, in that there is a (slightly) bigger difference between methods for 5 practice sessions than there is for 2 or 10 practice sessions. If, in our forthcoming F-test for the interaction effect, we decide that the interaction is real, that decision necessarily entails both kinds of verbal interpretations we have presented. In this case, perhaps it is easier to discuss the separate effects of practice on the kind of method used than it is to discuss the alternative definition. In writing reports of experiments with interaction effects, it is not necessary to phrase the definition in both ways, nor is it necessary to include both graphs as in Figure 15.2. If you say that the effect of factor A depends on the level of factor B, then it has to be true that the effect of factor B depends on the level of factor A.

Figure 15.3 illustrates another important characteristic of interaction effects and main effects: you can have one kind of effect without the other. All of the small graphs are some possible experimental results in a two-factor experiment having 3 levels of factor A and 2 levels of factor B. All of the vertical axes represent (combination) means on the dependent variable, and all of the horizontal axes represent levels of factor A. Consider first graphs a, b, and c. In each case, there is probably an A main effect, because the mean of the two B level lines (if it were drawn) would be a line that is not horizontal (you can pencil in the A main effect line by drawing a third line midway between the two given lines). Figure c also probably has a significant B main effect because, overall, the B level 1 line

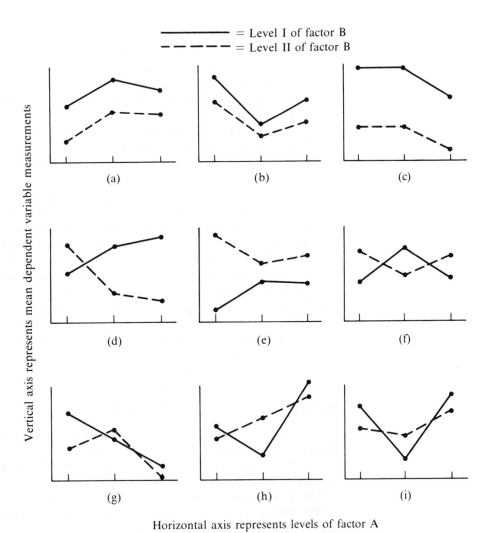

Figure 15.3 Graphs illustrating different combinations of main effects and interactions. In graphs a, b, and c, the main effect associated with factor A is probably significant, but the $A \times B$ interaction is probably not significant. The reverse is true in figures d, e, and f; in these, the interaction effect is probably significant but the A main effect is probably not. Graphs g, h, and i probably show both a significant A main effect and a significant $A \times B$ interaction.

is much higher than the overall B level 2 line. Conversely, there is probably no B main effect in b because the B level 1 and B level 2 lines, overall, are at about the same height. In none of the first-row graphs, however, is it likely that the $A \times B$ interaction would turn out to be significant because each of the three pairs of lines is nearly parallel. Remember that the strength of the interaction is reflected in the amount that the lines depart from being parallel — it has nothing to do with how far apart or how close they are.

In the middle-row graphs d, e, and f, it is very likely that we have a significant interaction. Here, the effect of A does seem to depend on the level of B, since all pairs of lines are decidedly nonparallel. Notice, however, that the mean value between these lines is nearly horizontal in all three cases. Hence, one would not expect a significant A main effect from the d, e, or f data.

In graphs g, h, and i, we have pairs of lines that are not parallel and that produce nonhorizontal main-effect graphs for factor A. We could expect both a significant A main effect and an $A \times B$ interaction in the bottom-row graphs.

The graphical analysis above is important because it shows you what the various effects mean in terms of the two factors in the study. All effects (both main effects and the interaction effect) are defined as differences among means; in any collection of data, there will almost certainly be differences among the various combination and level means, and hence some main effects and an interaction effect that are not zero. However, when experimenters say that they "have" a main effect or that they "have" an interaction effect, they mean of course that the effect has been deemed statistically significant. And, while the graphs by themselves can never answer the question of significance, they can usually give an indication of what the F-tests will show and they are a useful aid in interpreting those effects.

Basis of the Two-Factor Analysis

The same principles that we applied to the single-factor experiment in Chapter 14 will be extended to apply to the two-factor case. Again, your ability to interpret the F-tests will be enhanced if you grasp the fundamental ideas underlying the analysis.

The theory underlying the analysis begins with another structural model. This time, we assume that each X_{rci} data point is the sum of *five* independent components:

$$X_{rci} = \mu_{...} + \alpha_r + \beta_c + \alpha\beta_{rc} + \epsilon_{rci}$$

These parameters are defined as follows:

$\mu_{...}$ = The grand mean of all treatment populations combined.

α_r = The effect due specifically to receiving level r of factor 1. For all members of the same row, this parameter value is the same. Differences in α values for different rows lead to a significant main effect associated with factor 1.

β_c = The effect due specifically to receiving level c of factor 2. For all members of the same column, this parameter is the same value. Differences in β_c values for different columns lead to a significant main effect associated with factor 2.

$\alpha\beta_{rc}$ = The effect due specifically to receiving level r of factor 1 *and* level c of factor 2, that is, the effect due specifically to receiving that combination of levels. It is *not* the product of the individual α and β values. All members of the same rc combination are assumed to receive the same $\alpha\beta$ effect. Differences in $\alpha\beta$ values for different rows and columns lead to a significant factor 1 $\times$ factor 2 interaction effect.

ϵ_{rci} = All other effects that contribute to an individual's X value. Each ϵ value is unique to subject rci, and all ϵ values in the study are assumed to be independent of each other. This is sometimes called the error component of the model, because it includes errors of measurement and anything else that helps take the X value away from $\mu_{...}$. These values are assumed to be normally distributed in the population, with a mean of 0.

Each of these components can be estimated with a statistic, and the best way to show how this is done is to take an individual X_{rci} and divide its deviation from the (observed) grand mean into additive components.

The observed grand mean $\overline{X}_{...}$ is 56.73 and our example individual will be X_{224}. This individual scored 85 on the vocabulary test (it will be helpful during the following discussion if you refer back to Table 15.2 to see where the various values come from); we can compute X_{224}'s deviation score:

$$X_{224} - \overline{X}_{...} = 85.00 - 56.73$$
$$= 28.27$$

We will now find the components of that deviation score — components that estimate the various parameters of the specific structural model:

$$X_{224} = \mu_{...} + \alpha_2 + \beta_2 + \alpha\beta_{22} + \epsilon_{224}$$

The α_2 parameter is the specific effect of receiving level 2 of factor 1 and is equivalent to $\mu_{2..} - \mu_{...}$. We estimate this effect by taking the difference between the observed row mean and the grand mean:

$$\text{estimated } \alpha_2 = \overline{X}_{2..} - \overline{X}_{...}$$
$$= 77.07 - 56.73$$
$$= 20.34$$

Similarly, the β_2 applies to everyone in the second level of factor 2 and it is equivalent to $\mu_{.2.} - \mu_{...}$. Its estimate is the difference between column and grand means:

$$\text{estimated } \beta_2 = 55.90 - 56.73$$
$$= -0.83$$

Next we turn to the interaction parameter $\alpha\beta_{22}$. This is the unique effect of being in level 2 of factor 1 *and* level 2 of factor 2, *after the α_2 effect and*

β_2 *effect have been subtracted*. This is easier to define in symbols than it is in words:

$$\alpha\beta_{22} = \mu_{22.} - \mu_{...} - \alpha_2 - \beta_2$$

That is the way to visualize the effect: as the difference between the combination mean and the grand mean, minus the individual row and column effects. However, to see clearly where our statistical estimator for $\alpha\beta_{22}$ comes from, take the above equation and substitute $(\mu_{2..} - \mu_{...})$ for α_2 and $(\mu_{.2.} - \mu_{...})$ for β_2:

$$\alpha\beta_{22} = \mu_{22.} - \mu_{...} - (\mu_{2..} - \mu_{...}) - (\mu_{.2.} - \mu_{...})$$

In this form, $\mu_{...}$ appears several times. When we remove the parentheses and add all the $+$ and $-$ values of $\mu_{...}$, we have:

$$\alpha\beta_{22} = \mu_{22.} - \mu_{2..} - \mu_{.2.} + \mu_{...}$$

The parameter is then estimated:

$$\begin{aligned}
\text{estimated } \alpha\beta_{22} &= \bar{X}_{22.} - \bar{X}_{2..} - \bar{X}_{.2.} - \bar{X}_{...} \\
&= 79.00 - 77.07 - 55.90 + 56.73 \\
&= 2.76
\end{aligned}$$

The final component of the deviation score is the estimated error parameter ϵ_{224}, the "catch all" parameter that includes everything not accounted for by the other effects. Its estimate is the difference between the individual and his or her cell mean:

$$\begin{aligned}
\text{estimated } \epsilon_{224} &= X_{224} - \bar{X}_{22.} \\
&= 85.00 - 79.00 \\
&= 6.00
\end{aligned}$$

All of these components (parameter estimates) should add up to make the original deviation score:

$$\begin{aligned}
(X_{224} - \bar{X}_{...}) &= (\bar{X}_{2..} - \bar{X}_{...}) + (\bar{X}_{.2.} - \bar{X}_{...}) \\
&\quad + (\bar{X}_{22.} - \bar{X}_{2..} - \bar{X}_{.2.} + \bar{X}_{...}) + (X_{224} - \bar{X}_{22.})
\end{aligned}$$

$$\begin{aligned}
(85 - 56.73) &= (77.07 - 56.73) + (55.90 - 56.73) \\
&\quad + (79.00 - 77.07 - 55.90 + 56.73) + (85 - 79.00)
\end{aligned}$$

Remember now the process we followed in the single-factor case, dividing the total variation into two components. In Chapter 14 we showed that both sides of our deviation score equation, when squared and then summed, produced:

$$\sum_{j=1}^{k} \sum_{i=1}^{n} (X_{ji} - \bar{X}_G)^2 = \sum_{j=1}^{k} \sum_{i=1}^{n} (X_{ji} - \bar{X}_j)^2 + n \sum_{j=1}^{k} (\bar{X}_j - \bar{X}_G)^2$$

or

$$SS_{\text{total}} = \qquad SS_{wg} \qquad + SS_{bg}$$

We can take our two-factor deviation-score equation and perform a similar squaring and adding process. The computational steps will not be given here, but you may wish to work them out on your own if you are comfortable with summation notation. This, then, is the partition of the total variation:

$$\sum_{r=1}^{k} \sum_{c=1}^{m} \sum_{i=1}^{n} (X_{rci} - \bar{X}_{...})^2 = nm \sum_{r=1}^{k} (\bar{X}_{r..} - \bar{X}_{...})^2 + nk \sum_{c=1}^{m} (\bar{X}_{.c.} - \bar{X}_{...})^2$$

$$+ n \sum_{r=1}^{k} \sum_{c=1}^{m} (\bar{X}_{rc.} - \bar{X}_{r..} - \bar{X}_{c..} + \bar{X}_{...})^2$$

$$+ \sum_{r=1}^{k} \sum_{c=1}^{m} \sum_{i=1}^{n} (X_{rci} - \bar{X}_{rc.})^2$$

Each term on the right side of this equation is an independent SS based on different components of the total deviation score; if we substitute the names of each SS for their formulas:

$$SS_{\text{total}} = SS_{\text{rows}} + SS_{\text{columns}} + SS_{\text{interaction}} + SS_{\text{within groups}}$$

Notice that there is a different variation for each effect we wish to test.

Now we can return to the experimental example and look at the hypotheses to be tested. First, we will want to know if the difference between levels of factor 1 is significant. Our hypotheses to be tested are:

H_0: $\mu_{1..} = \mu_{2..}$

H_1: Not H_0

As in the single-factor case, this kind of hypothesis test can be conducted by looking at the variation associated with different row means. In particular, we recognize that the above H_0 is equivalent to asserting that $\alpha_1 = \alpha_2$, which is equivalent to saying $\sigma_{\alpha_r}^2 = 0$. We will, therefore, compute the estimated variance for "rows," by dividing SS_{rows} (henceforth called SS_r) by its degrees of freedom, $k - 1$.

$$MS_r = \frac{SS_r}{df_r} = \frac{SS_r}{k - 1}$$

The next three sections will show how this MS is calculated and how it is used in the F-test.

Similarly, our hypotheses in the test on the main effect associated with factor 2 are:

H_0: $\mu_{.1.} = \mu_{.2.} = \mu_{.3.}$

H_1: Not H_0

Because this H_0 is equivalent to saying that $\sigma_{\beta_c}^2 = 0$, we will also want to look at the estimated variance between the columns in order to assess the factor 2 main effect. This MS is computed by dividing SS_{columns} (SS_c for short) by its degrees of freedom, $m - 1$. Thus:

$$MS_c = \frac{SS_c}{df_c} = \frac{SS_c}{m - 1}$$

The third pair of hypotheses to be tested, of course, involve the interaction effect:

H_0: $\alpha\beta_{11} = \alpha\beta_{12} = \alpha\beta_{13} = \alpha\beta_{21} = \alpha\beta_{22} = \alpha\beta_{23}$

H_0: Not H_0

As you can probably guess by now, this null hypothesis is equivalent to the assertion that $\sigma^2_{\alpha\beta_{rc}} = 0$, and we will want an estimated variance associated with the interaction effect. $SS_{\text{interaction}}$ ($SS_{r \times c}$ for short) will be divided by its degrees of freedom $(k - 1)(m - 1)$ to obtain $MS_{r \times c}$.

$$MS_{r \times c} = \frac{SS_{r \times c}}{df_{r \times c}} = \frac{SS_{r \times c}}{(k - 1)(m - 1)}$$

The final MS necessary for our analysis is a variance computed from $SS_{\text{within groups}}$ (i.e., SS_{wg}); the within groups df in the two-factor experiment is given by $km(n - 1)$. Our MS_{wg} is then:

$$MS_{wg} = \frac{SS_{wg}}{df_{wg}} \frac{SS_{wg}}{km(n - 1)}$$

Computations in the Two-Factor ANOVA (Definitional Formulas)

As with the single-factor ANOVA, the bulk of the computational effort in the analysis involves finding the various sums of squares. Once this is done, the MS values and the F values are obtained easily.

The definitional formulas for the five different SS that we need have, in fact, already been presented on page 411 in the complex equation near the top of the page. The term on the left side of the equal sign is, of course, SS_{total}, and each term on the right represents one of the other sums of squares we will need. We will compute them one by one, again drawing all the values we need from Table 15.2.

SS_{total} uses the grand mean and every observation in the experiment — as indicated by the tripled summation sign:

$$
\begin{aligned}
SS_{\text{total}} &= \sum_{r=1}^{k} \sum_{c=1}^{m} \sum_{i=1}^{n} (X_{rci} - \bar{X}_{..})^2 \\
&= (25.00 - 56.73)^2 + (18.00 - 56.73)^2 \\
&\quad + \cdots + (71.00 - 56.73)^2 \\
&= 15{,}167.87
\end{aligned}
$$

Our SS_r uses the row means and the grand mean. Notice that the sum of these squared deviations is multiplied by the number of observations that go into each row mean, nm. (A row has n observations per cell and m cells — one cell for each column).

$$SS_r = nm \sum_{r=1}^{k} (\bar{X}_{r..} - \bar{X}_{...})^2$$
$$= (5)(3) [(36.40 - 56.73)^2 + (77.07 - 56.73)^2]$$
$$= 12,403.33$$

(Note: If you performed all calculations according to the definitional formula, and used values carried to only two decimal places, you will obtain an SS_r of 12,399.27. To obtain the (correct) answer given above, you must carry more decimal places in your calculations, or use the computational formulas in the upcoming computational section. The same practice will be followed in the rest of the SS definitional formula examples: the correct SS will be shown, but to obtain it, you may have to carry more than the indicated two decimal places through your calculations.)

In the same manner, SS_c is computed from column means and the grand mean. Here the sum of squared deviation scores is multiplied by the number of observations in each column mean, nk.

$$SS_c = nk \sum_{c=1}^{m} (\bar{X}_{.c.} - \bar{X}_{...})^2$$
$$= (5)(2) [(50.00 - 56.73)^2 + (55.90 - 56.73)^2 + (64.30 - 56.73)^2]$$
$$= 1,032.87$$

The $SS_{r \times c}$ is obtained from the combination (cell) means, row means, column means, and the grand mean. Here, notice that the subtraction for each combination involves a different pair of row and column means; be sure to subtract each combination's own row and column mean correctly.

$$SS_{r \times c} = n \sum_{r=1}^{k} \sum_{c=1}^{m} (\bar{X}_{rc.} - \bar{X}_{r..} - \bar{X}_{.c.} + \bar{X}_{...})^2$$
$$= 5 [(26.80 - 36.40 - 50.00 + 56.73)^2$$
$$+ (32.80 - 36.40 - 55.90 + 56.73)^2$$
$$+ (49.60 - 36.40 - 64.30 + 56.73)^2$$
$$+ (73.20 - 77.07 - 50.00 + 56.73)^2$$
$$+ (79.00 - 77.07 - 55.90 + 56.73)^2$$
$$+ (79.00 - 77.07 - 64.30 + 56.73)^2]$$
$$= 476.07$$

The final SS needed is our SS_{wg}. Remember that this computation entails finding the variation within each of the km groups (i.e., combinations, or cells).

$$SS_{wg} = \sum_{r=1}^{k} \sum_{c=1}^{m} \sum_{i=1}^{n} (X_{rci} - \bar{X}_{rc.})^2$$

For Group 1, 1 (the combination in row 1, column 1), we find $SS_{wg_{1,1}}$ by dealing with the innermost summation sign:

$$SS_{wg_{1,1}} = \sum_{i=1}^{n} (X_{11i} - \bar{X}_{11.})^2$$

$$= (25.00 - 26.80)^2 + (18.00 - 26.80)^2 + (31.00 - 26.80)^2$$
$$+ (24.00 - 26.80)^2 + (36.00 - 26.80)^2$$
$$= 190.80$$

The outer two summation signs direct us to repeat the above process for all of the other groups, and in doing so we find:

$$SS_{wg_{1,2}} = 154.80$$
$$SS_{wg_{1,3}} = 267.20$$
$$SS_{wg_{2,1}} = 202.80$$
$$SS_{wg_{2,2}} = 262.00$$
$$SS_{wg_{2,3}} = 178.00$$

The overall SS_{wg} is the sum of these:

$$SS_{wg} = 190.80 + 154.80 + 267.20 + 202.80 + 262.00 + 278.00$$
$$= 1,255.60$$

We should now check the computations to insure that all of the components of variation add up to SS_{total}:

$$SS_{\text{total}} = SS_r + SS_c + SS_{r \times c} + SS_{wg}$$
$$15,167.87 = 12,403.33 + 1,032.87 + 476.07 + 1,255.60$$

This completes the heavy computational labor; the above SS can be fitted into the two-factor summary table, and the MS for each calculated by dividing each SS by its respective degrees of freedom. Once again, the df for each source of variation, that is, each SS, will be simply presented.

SOURCE OF VARIATION	df	df FOR THE EXAMPLE
Rows (factor 1)	$k - 1$	1
Columns (factor 2)	$m - 1$	2
$r \times c$ interaction	$(k - 1)(m - 1)$	2
Within groups	$km (n - 1)$	24
Total	$N - 1$, i.e., $nkm - 1$	29

Therefore:

$$MS_r = \frac{SS_r}{k - 1} = \frac{12,403.33}{1} = 12,403.33 \quad df = 1$$

$$MS_c = \frac{SS_c}{m - 1} = \frac{1,032.87}{2} = 516.43 \quad df = 2$$

$$MS_{r \times c} = \frac{SS_{r \times c}}{(k-1)(m-1)} = \frac{476.07}{2} = 238.04 \qquad df = 2$$

$$MS_{wg} = \frac{SS_{wg}}{km(n-1)} = \frac{1,255.60}{24} = 52.32 \qquad df = 24$$

In this kind of experiment we can go ahead and compute F-observed values for each of our three effects by dividing the appropriate MS by the MS_{wg}. To complete the example problem, we will do this; however, it is very important to note that it is not always appropriate to use MS_{wg} in the denominator of our F-observed ratio. The next section is included to clarify this point.

For our factor 1 main effect:

$$F\text{-observed} = \frac{MS_r}{MS_{wg}} = \frac{12,403.33}{52.32} = 237.07 \qquad df = 1, 24$$

For our factor 2 main effect:

$$F\text{-observed} = \frac{MS_c}{MS_{wg}} = \frac{516.43}{52.32} = 9.87 \qquad df = 2, 24$$

For our factor 1 × factor 2 interaction:

$$F\text{-observed} = \frac{MS_{r \times c}}{MS_{wg}} = \frac{238.04}{52.32} = 4.55 \qquad df = 2, 24$$

All of the preceding calculations can be summarized in the two-factor summary table:

Summary of Analysis

SOURCE OF VARIATION	SS	df	MS	F
Method of Presentation	12,403.33	1	12,403.33	237.07***
Practice Sessions	1,032.87	2	516.43	9.87**
Interaction	476.07	2	238.04	4.55*
Within groups	1,255.60	24	52.32	
Total	15,167.87	29		

*$p < 0.05$
**$p < 0.01$
***$p < 0.001$

Of course, the table could have listed "Rows" instead of "Method of Presentation," and "Columns" instead of "Practice Sessions," but it is easier to interpret in the form shown.

All three effects exceed tabled critical F values, though at different alpha levels. Thus, we decide that the method of instruction does affect vocabulary scores ($F = 237.07$, $df = 1, 24$, $p < 0.001$), the number of practice sessions also affects scores ($F = 9.87$, $df = 2, 24$, $p < 0.01$), and that the effect of practice is different for each of the two methods used; that is, we have a significant interaction ($F = 4.55$, $df = 2, 24$, $p < 0.05$).

One more point before leaving this example: the size and direction of an effect are determined entirely by differences between means. These characteristics can be seen clearly in graphs. The lower graph in Figure 15.1, for instance, shows that 2, 5, or 10 practice sessions led to mean scores of 50.00, 55.90, and 64.30, respectively. The difference between these means constitutes the size of the main effect, and the fact that they increase with increasing practice is its direction. The significance of these effects is entirely another matter; as we have learned throughout this book, significance is assessed in the hypothesis test, and it refers only to the probability that our observed results would appear under a true null hypothesis. All of this means that significance should not be confused with importance. Unfortunately, it is quite common for advertisers to report research findings in which their products were found to be "significantly more effective" than competing brands. Such ads represent misleading use of a technical term that has broader connotations in general usage, and should be punished by banishment to the South Pole.

The Problem of Fixed vs. Random Factors

All three F-observed values in the example problem were obtained by using MS_{wg} in the denominator. In Chapter 14, we showed why it was appropriate to do this in the single-factor case by making a ratio of $E(MS)$ values. When we move to the two-factor case, however, the $E(MS)$ situation becomes a little more complicated and it is not universally true that F values should be computed by using MS_{wg} in the denominator.

The problem has to do with the way we select levels of each factor to use in the experiment. There are two types of independent variables (factors), and the way that we choose the levels of the factors determines which type we have. And, the type of factors we have determines the way F-observed should be computed.

The two kinds of factors are known as fixed factors and random factors. We have a **fixed factor** if the experimenter arbitrarily picks the levels to be used; if statistical generalizations about observed effects involving the factor are made only to the levels used in the experiment, the factor is fixed. An independent variable is a **random factor** when the levels used in the experiment are picked randomly from a very large population of potential levels, and when generalizations about effects involving the factor are to be made to levels other than those used in the study.

Most factors used in experimental studies are appropriately treated as fixed. The number of practice sessions were picked arbitrarily for the example study (perhaps on the basis of some previous research that suggested these levels should be tried); strict generalizations about the "Practice Sessions" main effect and the interaction effect will be made only to 2, 5, and 10 sessions. One rather absurd way to have made this factor random would have been to put, say, the numbers 1–100 in a hat and to draw three of them randomly, letting each drawn number designate the number of sessions in a level. Our "Method of Instruction" factor is also fixed, because the two levels used are the only two we were interested in.

Random factors might appear, for instance, if we tried the memory experiment in a slightly different way. Suppose we took our "Verbal" and "Pictorial" method levels (which make a fixed factor), and then tried them out at several different schools. The design matrix for this study might appear as follows:

	SCHOOL A	SCHOOL B	SCHOOL C
Verbal Method			
Pictorial Method			

Now, if the schools were picked randomly from a long list of potential schools, "Schools" would be a random factor. Strict generalizations about the effect of differences among schools could be extended beyond the three used in the study.

When random factors appear in an experiment, the analysis differs slightly from the ANOVA used in the example study, in that one or both of the main-effect F-ratios may be computed differently from their counterparts in the experiment with two fixed factors.

The different forms of F-ratio computation are dictated by our $E(MS)$ values again. Remember from Chapter 14 that in the single-factor case, the long-run average $E(MS)$ for MS_{bg} was $\sigma^2_{\epsilon_{ji}} + n \sigma^2_{\alpha_j}$, and the $E(MS_{wg})$ was $\sigma^2_{\epsilon_{ji}}$. We made a ratio of the $E(MS)$'s to show that, when H_0 is true (and $\sigma^2_{\alpha_j} = 0$), the two expected values are equal, and therefore we should expect an F-observed of about 1.0; that an observed F much greater than 1.0 was evidence that $\sigma^2_{\alpha_j} \neq 0$.

The same principle will apply here. Without showing the derivation, the $E(MS)$ values for a two-factor experiment are presented in Table 15.3. We will have to leave the derivation of these values to more advanced texts and advanced courses. However, it will be necessary for us to learn how to use them.

Believe it or not, this rather large and rather Greek-looking table is really quite easy to use in construction of ratios for F-observed. Let us begin with the table's simplest points.

Notice first that $E(MS_{wg})$ and $E(MS_{r \times c})$ are always the same regardless of the fixed or random nature of the factors. This means that in testing for interaction effect significance, F-observed always equals $MS_{\text{interaction}}$ divided by MS_{wg}. Consider the ratio of expected values:

$$\frac{E(MS_{r \times c})}{E(MS_{wg})} = \frac{\sigma^2_{\epsilon_{rci}} + n\sigma^2_{\alpha\beta_{rc}}}{\sigma^2_{\epsilon_{rci}}}$$

Table 15.3 *Expected Mean Squares in the Two-Factor Experiment*

	BOTH FACTORS FIXED	FACTOR 1 FIXED FACTOR 2 RANDOM	FACTOR 1 RANDOM FACTOR 2 FIXED	BOTH FACTORS RANDOM
$E(MS_r)$	$\sigma^2_{\epsilon_{rci}} + nm\sigma^2_{\alpha_r}$	$\sigma^2_{\epsilon_{rci}} + n\sigma^2_{\alpha\beta_{rc}} + nm\sigma^2_{\alpha_r}$	$\sigma^2_{\epsilon_{rci}} + nm\sigma^2_{\alpha_r}$	$\sigma^2_{\epsilon_{rci}} + n\sigma^2_{\alpha\beta_{rc}} + nm\sigma^2_{\alpha_r}$
$E(MS_c)$	$\sigma^2_{\epsilon_{rci}} + nk\sigma^2_{\beta_c}$	$\sigma^2_{\epsilon_{rci}} + nk\sigma^2_{\beta_c}$	$\sigma^2_{\epsilon_{rci}} + n\sigma^2_{\alpha\beta_{rc}} + nk\sigma^2_{\beta_c}$	$\sigma^2_{\epsilon_{rci}} + n\sigma^2_{\alpha\beta_{rc}} + nk\sigma^2_{\beta_c}$
$E(MS_{r\times c})$	$\sigma^2_{\epsilon_{rci}} + n\sigma^2_{\alpha\beta_{rc}}$	$\sigma^2_{\epsilon_{rci}} + n\sigma^2_{\alpha\beta_{rc}}$	$\sigma^2_{\epsilon_{rci}} + n\sigma^2_{\alpha\beta_{rc}}$	$\sigma^2_{\epsilon_{rci}} + n\sigma^2_{\alpha\beta_{rc}}$
$E(MS_{wg})$	$\sigma^2_{\epsilon_{rci}}$	$\sigma^2_{\epsilon_{rci}}$	$\sigma^2_{\epsilon_{rci}}$	$\sigma^2_{\epsilon_{rci}}$

where r = row identity, k = the number of rows (levels of factor 1);
c = column identity, m = the number of columns (levels of factor 2);
i = individual within group identity, n = the number of individuals per group.

When the null hypothesis that all interaction effects are the same, i.e., that $\sigma^2_{\alpha\beta_{rc}} = 0$, the expected values become:

$$\frac{\sigma^2_{\epsilon_{rci}}}{\sigma^2_{\epsilon_{rci}}}$$

and the expected value of the statistic $F = MS_{\text{interaction}}/MS_{wg}$ is about 1.0. The degrees of freedom for this F-ratio are, of course, $(k - 1)(m - 1)$, $km(n - 1)$.

The next "simple" aspect of Table 15.3 involves the main effects when both factors are fixed. This was the situation in the example problem of the preceding section, where F-observed values were calculated:

$$\text{Factor 1 main effect: } F = \frac{MS_r}{MS_{wg}} \qquad df = (k - 1), km(n - 1)$$

$$\text{Factor 2 main effect: } F = \frac{MS_c}{MS_{wg}} \qquad df = (m - 1), km(n - 1)$$

The reason we use these values is also clear from a ratio of $E(MS)$'s. For instance:

$$\frac{E(MS_r)}{E(MS_{wg})} = \frac{\sigma^2_{\epsilon_{rci}} + nm\sigma^2_{\alpha_r}}{\sigma^2_{\epsilon_{rci}}}$$

Under an H_0 that $\sigma^2_{\alpha_r} = 0$, this too is a ratio of expected values that should equal about 1.0.

Now consider the cases where random factors appear. Introduction of random factors alters some $E(MS)$ associated with main effects. If, for instance, factor 1 is fixed and factor 2 is random (as in the design matrix on page 417), then:

$$E(MS_r) = \sigma^2_{\epsilon_{rci}} + n\sigma^2_{\alpha\beta_{rc}} + nm\sigma^2_{\alpha_r}$$

It may seem paradoxical that the *fixed* factor of the mixed pair acquires the more complicated $E(MS)$. Be careful not to get this situation reversed.

In this case, F-observed requires a different denominator, or "error" term. Here:

$$F\text{-observed} = \frac{MS_r}{MS_{r\times c}}$$

The reason can again be seen by making a ratio of expected values:

$$\frac{E(MS_r)}{E(MS_{r\times c})} = \frac{\sigma^2_{\epsilon_{rci}} + n\sigma^2_{\alpha\beta_{rc}} + nm\sigma^2_{\alpha_r}}{\sigma^2_{\epsilon_{rci}} + n\sigma^2_{\alpha\beta_{rc}}}$$

Because $E(MS_r)$ has the extra term, we must find a denominator MS whose expected value also has that term. For the main effect F-test, it doesn't matter what value $\sigma^2_{\alpha\beta_{rc}}$ has, since that term appears in both numerator and denominator of the F-observed ratio. Only when the null hypothesis $\sigma^2_{\alpha_r} = 0$ is true do we expect this fraction to have a value of about 1.0. Using MS_{wg} in the denominator, as before, would be incorrect because it would give us a main-effect test contaminated by the interaction effect.

These same principles apply then to produce the following rules for testing main effects:

1. When both factors are fixed, both main effects are tested by using MS_{wg} in the denominator of each F-observed.
2. When both factors are random, both main effects are tested by using $MS_{r\times c}$ in the denominator of each F-observed.
3. When one factor is fixed and one factor is random, the main effect F-ratios use different denominator terms:
   a. F-observed for the fixed-factor main effect is computed using $MS_{r\times c}$ in the denominator.
   b. F-observed for the random-factor main effect is computed using MS_{wg} in the denominator.

This is why one cannot always tell simply by looking at the data how to compute F-ratios. Prewritten computer programs for the two-factor ANOVA (or more complicated ANOVAS not covered in this book) sometimes leave the F part of the summary table blank, because computation of F depends on the way levels are chosen for the factors, and the data themselves give indication of that. Alternatively, other programs have you specify which factors are fixed and which are random, and the computer can then produce the appropriate F. This latter kind of program is a real blessing when one is working with four, five, or six or more factors in a single design, because $E(MS)$ tables for these experiments are — to use an understatement — unwieldy. A third approach taken by some programs is simply to assume that all factors are fixed and compute F-observed using MS_{wg} in the denominator for all cases. This is a situation where it pays to know some ANOVA theory, even if the computer does do all of the computational labor for you.

ⓒ Computational Formulas for the Two-Factor ANOVA

The following computational formulas present alternative means of obtaining the needed sums of squares for the two-factor ANOVA. Even though the formulas given are more complex and the computations involve larger numbers than the definition formulas called for, they are easier to use and less susceptible to errors. Notice that the main difference between definitional formulas and their computational counterparts is that the former use means and the latter use totals. Again, it will be helpful to refer to Table 15.2 throughout the following, so that you can be certain you know where each total comes from.

SS_{total} is, again, practically the same formula we used in the previous chapter and in Chapter 3. Here, the triple summation signs show that all observations in all groups are included.

$$SS_{total} = \sum_{r=1}^{k} \sum_{c=1}^{m} \sum_{i=1}^{n} X_{rci}^2 - \frac{\overbrace{\left(\sum_{r=1}^{k} \sum_{c=1}^{m} \sum_{i=1}^{n} X_{rci} \right)^2}^{\text{Grand total}}}{nkm} \leftarrow N$$

$$= (25)^2 + (18)^2 + \cdots + (89)^2 - \frac{(1{,}702)^2}{(5)(2)(3)}$$

$$= 111{,}728.00 - 96{,}569.13$$

$$= 15{,}167.87$$

SS_r uses row totals and the grand total as follows:

$$SS_r = \frac{\overbrace{\sum_{r=1}^{k} \left(\sum_{c=1}^{m} \sum_{i=1}^{n} X_{rci} \right)^2}^{\text{Row } r \text{ totals}}}{nm} - \frac{\overbrace{\left(\sum_{r=1}^{k} \sum_{c=1}^{m} \sum_{i=1}^{n} X_{rci} \right)^2}^{\text{Grand total}}}{nkm}$$

Number of observations per row

$$= \frac{(546)^2 + (1{,}156)^2}{(5)(3)} - \frac{(1{,}702)^2}{(5)(2)(3)}$$

$$= 12{,}403.33$$

Similarly, the SS_c is obtained from column totals and the grand total:

$$SS_c = \frac{\overbrace{\sum_{c=1}^{m} \left(\sum_{r=1}^{k} \sum_{i=1}^{n} X_{rci} \right)^2}^{\text{Column } c \text{ totals}}}{nk} - \frac{\left(\sum_{r=1}^{k} \sum_{c=1}^{m} \sum_{i=1}^{n} X_{rci} \right)^2}{nkm}$$

Number of observations per column

$$= \frac{(500)^2 + (559)^2 + (643)^2}{(5)(2)} - \frac{(1,702)^2}{(5)(2)(3)}$$

$$= 1,032.87$$

$SS_{r\times c}$ requires all of the totals we have used so far, as well as the combination totals:

Combination
rc total

$$SS_{r\times c} = \frac{\overbrace{\sum\limits_{r=1}^{k}\sum\limits_{c=1}^{m}\left(\sum\limits_{i=1}^{n} X_{rci}\right)^2}}{n} - \frac{\sum\limits_{r=1}^{k}\left(\sum\limits_{c=1}^{m}\sum\limits_{i=1}^{n} X_{rci}\right)^2}{nm}$$

Number of observations per combination

$$- \frac{\sum\limits_{c=1}^{m}\left(\sum\limits_{r=1}^{k}\sum\limits_{i=1}^{n} X_{rci}\right)^2}{nk} + \frac{\left(\sum\limits_{r=1}^{k}\sum\limits_{c=1}^{m}\sum\limits_{i=1}^{n} X_{rci}\right)^2}{nkm}$$

$$= \frac{(134)^2 + (164)^2 + (248)^2 + (366)^2 + (395)^2 + (395)^2}{5}$$

$$- \frac{(546)^2 + (1,156)^2}{(5)(3)} - \frac{(500)^2 + (559)^2 + (643)^2}{(5)(2)}$$

$$+ \frac{(1,702)^2}{(5)(2)(3)}$$

$$= 476.07$$

Finally, the SS_{wg} computational formula requires us to compute SS within each group, and then add these values for all groups:

$$SS_{wg} = \sum\limits_{r=1}^{k}\sum\limits_{c=1}^{m}\left(\sum\limits_{i=1}^{n} X_{rci}^2 - \frac{\left(\sum\limits_{i=1}^{n} X_{rci}\right)^2}{n}\right)$$

For group 1,1:

$$SS_{wg_{1,1}} = (25)^2 + (18)^2 + (31)^2 + (24)^2 + (36)^2 - \frac{(134)^2}{5}$$

The individual with group SS's are found for each of the other groups (combinations) in the same way. These values, of course, turn out to be the same values we found with the definitional formulas earlier:

$$SS_{wg_{1,2}} = 154.80 \qquad SS_{wg_{2,1}} = 202.80 \qquad SS_{wg_{2,3}} = 178.00$$
$$SS_{wg_{1,3}} = 267.20 \qquad SS_{wg_{2,2}} = 262.00$$

SS_{wg} is the sum of these:

$$SS_{wg} = 190.80 + 154.80 + 267.20 + 202.80 + 262.00 + 278.00$$
$$= 1,255.60$$

The rest of the analysis is, of course, identical to the analysis we performed earlier with the definitional formulas.

SUMMARY

This chapter extended the principles developed in Chapter 14, to show how analysis of variance methods can be applied to experiments with two independent variables. Such experiments yield more information than two individual experiments with a single independent variable, because they make possible a test of the following *three* effects: (1) the effect of factor 1 on the dependent variable (called the factor 1 main effect), (2) the effect of factor 2 on the dependent variable (factor 2 main effect), and (3) the effect of the interaction between factors 1 and 2 (the interaction effect); this is the extent to which the effect of factor 1 depends on the level of factor 2, and vice versa.

A rough estimation of main and interaction effects can be obtained from graphs of the various effects, but their significance must be determined in *F*-tests. This chapter shows how the variances for these tests are computed, and how *F*-ratios are constructed for each test. As with the single-factor situation, the total variation among observations is partitioned into a number of additive components; each component is associated with a different source of variation. Each variation (*SS*) is divided by its degrees of freedom to produce a variance, and each variance participates in an *F*-test.

KEY CONCEPTS

two-factor ANOVA	interaction effects	two-factor mean squares
design matrix	interpreting effects from graphs	fixed factor
combinations of levels		random factor
main effects	two-factor structural model	

SPOTLIGHT 15   Interactions of Significance

Progress in some fields is so rapid that many college science courses cover material that was unknown even ten years ago. Visible products of progress — increasingly sophisticated pocket computers or space travel, for instance — have clearly shown everyone that the rate of scientific advancement is accelerating. Understandably, many people accept spectacular achievements in the natural sciences and assume that the behavioral sciences, too, have advanced dramatically in the last few decades. The emergence of several popular magazines dealing with psychological topics and an ever-growing market for guides to self-analysis and applied social psychology are just a few of many signs showing the general public's great expectations from behavioral scientists.

Unfortunately, however, progress in understanding human behavior has been slower than most people realize. True, we know considerably more about some aspects of human psychology than our great-grandparents did, but we are a long way from being able to use much of our new knowledge to improve or explain the human condition. Most people would be surprised to learn that psychologists still can barely define intelligence, much less specify its determinants or measure it satisfactorily. If we try to explain why some people become schizophrenics and some do not, or why some teachers are more effective than others, our explanations today are little better than those offered decades ago. Progress in other behavioral sciences, such as economics, also comes at a very slow pace: investing in the stock market or advising the President on anti-inflation measures were risky activities forty years ago, and they remain so today, despite the application of new computer systems and new data analysis techniques to economic problems. These examples do not suggest that behavioral research is unimportant or a waste of time. They merely emphasize that the scientific method, which has been applied so successfully in medicine, astrophysics, and electronics, has not yielded results so readily in the behavioral sciences. Why not?

Through some of the important concepts in experimental design that we have now covered, we should be able to understand some of the special problems involved in psychological and other behavioral research. We have seen that underlying psychological processes can only be measured indirectly, by making inferences from overt behavior. In addition, they are highly complex phenomena: the status of any one psychological variable is determined by simultaneous interactions between it and many other variables. As we have learned, however, controlled experiments are contrived situations designed to reduce complexity. All important variables except the independent variable(s) are held constant. Indeed, the logic of statistical inference demands this if we are to make cause-and-effect arguments based on the results confidently. The same simplification process that strengthens conclusions about laboratory effects, however, often prevents us from generalizing these conclusions to the complex world outside the laboratory. As a result, the one-hundred-year history of experimental psychological research has taught us a great deal about what happens in the laboratory, but not as much as we would like about everyday psychological processes in the real world.

Consider, for instance, problems faced by experimental psychologists in trying to design informative experiments. Behavior (the term refers to both overt actions and covert mental activity) is certainly determined by complex interactions among many genetic and environmental factors; "environmental factors," moreover, include a lifetime of events. How can all of these factors be held constant, or equated among subjects?

Most psychologists who study learning and motivation achieve such control by experimenting with animal subjects. Laboratory rats and pigeons can be raised under carefully controlled conditions so that each animal's life history is practically identical to every other animal's. (Any college student who takes a psychology course called "Learning," quickly discovers that the basic research literature of this field is concerned almost entirely with these creatures.) "Motivation" (with respect to food or water rewards) can be manipulated by depriving the subject of food or water; experimenters arbitrarily select convenient behaviors to condition, such as having the rat depress a lever, or conditioning the pigeon to peck a circular "key" in an experimental chamber. In such ways, rigid experimental control is achieved. But do experiments of this kind have meaning for human psychology?

If we generalize the results of such research to human behavior, several assumptions are necessary. We must assume that laws of behavior are the same for different species (i.e., there is no interaction between species type and the variables that govern learning), that the laws of behavior are the same, regardless of which specific behaviors are conditioned and which forms of motivation are used (i.e., there is no interaction between type of task conditioned and type of motivation), and, most importantly, that behavioral processes do not interact significantly with variables not manipulated in the laboratory, but are present in the everyday experience of the animal. If these assumptions were completely untrue, animal behavioral research would be a complete waste of time. That principles of conditioning *have* been applied with some success to therapeutic and educational settings indicates, however, that there is some validity to them. On the other hand, if they were completely true, we would by now understand almost everything interesting about the causes of human behavior.

Psychologists who experiment with human subjects face different kinds of generalization problems. A chronic problem in social psychology, for example, is that experimental effects demonstrated in one laboratory are not easily replicated under slightly different conditions in other laboratories. In these cases, it is likely that some unknown aspects of the specific laboratory setting interact with the behavior under study. The psychologist has no way of discovering this unless he or she repeats the experiment in a number of different settings — a costly and ambitious undertaking.

Understanding the problematic nature of these interactions, then, should at least show us why behavioral research proceeds slowly. If we simplify them out of the experiments we design, we run the risk of getting results that have meaning only for, say, pigeons in small boxes. If, on the other hand, we do not simplify the experimental setting and equate many important variables, we are likely to obtain a disorderly set of experimental results having no discernible meaning. Our understanding of human behavior advances only when intelligent people design studies that fall somewhere between these extremes.

PROBLEMS

1. It is very well known in drug research that two drugs, when administered together, may produce an effect on humans or animals that neither drug produces alone. This is called a *drug interaction*, and it is not unlike the kind of statistical interaction we covered in this chapter. Consider an experiment where one factor consists of different dosage levels of drug 1 (0.5 cc, 1.0 cc and 1.5 cc), and the other factor consists of different dosage levels of drug 2 (2 cc, 4 cc and 8 cc). The experimental subjects are rats, and the dependent variable is "activity," measured by rotations of an exercise wheel. The numbers in the table below represent the results of that study:

| DRUG 2 | DRUG 1 | | |
	0.5 cc	1.0 cc	1.5 cc
2 cc	121	147	161
	139	163	184
	126	152	165
	134	158	170
4 cc	143	141	147
	171	164	166
	161	162	168
	153	145	159
8 cc	168	147	117
	193	144	130
	176	164	122
	183	149	139

a. Graph the main effects and interaction effects obtained in this study; from the graphs, interpret the interaction of these two drugs.

b. Perform the two-factor analysis of variance, and present the results in a summary table. Interpret the ANOVA results in terms of the independent and dependent variables in this study.

2. How many different Type I and Type II errors are possible in analyzing the data in problem 1? Describe briefly what each of these would consist of, in terms of the experiment described.

3. Complete the following summary tables. For each *F-value*, indicate whether or not the null hypothesis should be rejected. (Assume both factors fixed.)

a.

SOURCE OF VARIATION	df	SS	MS	F
Factor A	3	3,413	_____	_____
Factor B	2	_____	_____	_____
Interaction	_____	2,914	_____	_____
Within Groups	108	10,421	_____	
Total	119	17,506		

b.

SOURCE OF VARIATION	df	SS	MS	F
Factor A	5	8,765	_____	_____
Factor B	2	11,432	_____	_____
Interaction	10	_____	61	_____
Within Groups	_____	144,900	1,150	
Total	_____	165,707		

4. Perform a two-factor ANOVA on the dependent variable measurements given in the table below. (Assume both factors fixed.)

		FACTOR B LEVELS				
		1		2		3
FACTOR A LEVELS	1	68 47	38 39		21 27	
		42 57	27 34		35 24	
	2	47 41	38 30		31 29	
		75 63	41 42		23 20	
	3	81 49	51 57		38 24	
		52 74	33 44		29 22	

5. The following data are from a two-factor study; however, the factors and levels will remain unnamed so that you can examine the effects of changing a variable's designation from *fixed* to *random*. Perform all parts of the analysis except computing the *F*-observed values; then

   a. Conduct the three *F*-tests assuming that both factors are fixed.

   b. Conduct the three *F*-tests assuming that both factors are random.

   c. Conduct the three *F*-tests assuming that factor 1 is fixed and factor 2 is random.

FACTOR 2 LEVELS

	1	2	3
1	27	34	62
	35	51	47
	46	43	54
2	22	28	30
	28	46	44
	35	36	52

FACTOR 1 LEVELS

6. Perform a two-factor ANOVA on the dependent variable measurements below. (Assume both factors fixed.)

FACTOR B LEVELS

	1		2	
1	485	520	290	240
	410	462	320	313
2	270	206	490	540
	310	258	431	501

FACTOR A LEVELS

7. Perform the Hartley F_{max} test for each of the following two-factor ANOVA situations. In each case, the table shows the within-group variance ($\hat{s}^2$) for each cell (group). The test is performed here in the same manner as it was for the one-factor ANOVA (Chapter 14, pp. 389–391). In each case, $n = 10$.

Within cell variances:

a.

FACTOR B LEVELS

	1	2
1	121	102
2	278	98

FACTOR A LEVELS

b.

FACTOR B LEVELS

	1	2
1	15.14	27.17
2	2.03	47.15

FACTOR A LEVELS

c.

FACTOR B LEVELS

	1	2
1	317	983
2	821	1308

FACTOR A LEVELS

8. Explain in your own words how to predict, roughly, whether or not main effects and interaction effects will be significant from inspection of a graph showing group means in the two-factor experiment. Explain also, however, why you cannot determine with certainty whether or not these effects are really significant simply by inspecting graphs of group means.

9. Thirty men and thirty women agreed to take part in an experimental study of several teaching methods used by the Army to train technicians in radio repair. One-third of the members of each sex level were assigned to a different teaching method (level) of the "Method of Instruction" factor. Their scores on an exam designed to measure technical proficiency were:

METHOD OF
INSTRUCTION

	1	2	3
MEN	57	59	50
	56	74	37
	51	79	32
	64	63	49
	70	69	48
	69	75	36
	59	71	41
	68	74	47
	64	78	44
	52	68	36
WOMEN	95	42	39
	81	49	49
	79	51	47
	76	54	51
	84	39	42
	87	38	34
	92	46	31
	81	36	45
	89	32	40
	76	53	37

SEX OF SUBJECT

a. Graph the main effects and interaction effect from this study.

b. Perform the ANOVA and interpret the results.

10. In problem 9 above, "Sex" is a subject factor and "Method of Instruction" is an experimentally manipulated factor. Discuss the different kinds of interpretations that must be placed on significant effects involving subject factors or factors that represent experimental manipulations.

Appendixes

Appendix 1: Tables

Table A *Areas Under the Unit Normal Distribution*

z	Area Between Mean and z	Area Beyond z	z	Area Between Mean and z	Area Beyond z	z	Area Between Mean and z	Area Beyond z
0.00	.0000	.5000	0.45	.1736	.3264	0.90	.3159	.1841
0.01	.0040	.4960	0.46	.1772	.3228	0.91	.3186	.1814
0.02	.0080	.4920	0.47	.1808	.3192	0.92	.3212	.1788
0.03	.0120	.4880	0.48	.1844	.3156	0.93	.3238	.1762
0.04	.0160	.4840	0.49	.1879	.3121	0.94	.3264	.1736
0.05	.0199	.4801	0.50	.1915	.3085	0.95	.3289	.1711
0.06	.0239	.4761	0.51	.1950	.3050	0.96	.3315	.1685
0.07	.0279	.4721	0.52	.1985	.3015	0.97	.3340	.1660
0.08	.0319	.4681	0.53	.2019	.2981	0.98	.3365	.1635
0.09	.0359	.4641	0.54	.2054	.2946	0.99	.3389	.1611
0.10	.0398	.4602	0.55	.2088	.2912	1.00	.3413	.1587
0.11	.0438	.4562	0.56	.2123	.2877	1.01	.3438	.1562
0.12	.0478	.4522	0.57	.2157	.2843	1.02	.3461	.1539
0.13	.0517	.4483	0.58	.2190	.2810	1.03	.3485	.1515
0.14	.0557	.4443	0.59	.2224	.2776	1.04	.3508	.1492
0.15	.0596	.4404	0.60	.2257	.2743	1.05	.3531	.1469
0.16	.0636	.4364	0.61	.2291	.2709	1.06	.3554	.1446
0.17	.0675	.4325	0.62	.2324	.2676	1.07	.3577	.1423
0.18	.0714	.4286	0.63	.2357	.2643	1.08	.3599	.1401
0.19	.0753	.4247	0.64	.2389	.2611	1.09	.3621	.1379
0.20	.0793	.4207	0.65	.2422	.2578	1.10	.3643	.1357
0.21	.0832	.4168	0.66	.2454	.2546	1.11	.3665	.1335
0.22	.0871	.4129	0.67	.2486	.2514	1.12	.3686	.1314
0.23	.0910	.4090	0.68	.2517	.2483	1.13	.3708	.1292
0.24	.0948	.4052	0.69	.2549	.2451	1.14	.3729	.1271
0.25	.0987	.4013	0.70	.2580	.2420	1.15	.3749	.1251
0.26	.1026	.3974	0.71	.2611	.2389	1.16	.3770	.1230
0.27	.1064	.3936	0.72	.2642	.2358	1.17	.3790	.1210
0.28	.1103	.3897	0.73	.2673	.2327	1.18	.3810	.1190
0.29	.1141	.3859	0.74	.2704	.2296	1.19	.3830	.1170
0.30	.1179	.3821	0.75	.2734	.2266	1.20	.3849	.1151
0.31	.1217	.3783	0.76	.2764	.2236	1.21	.3869	.1131
0.32	.1255	.3745	0.77	.2794	.2206	1.22	.3888	.1112
0.33	.1293	.3707	0.78	.2823	.2177	1.23	.3907	.1093
0.34	.1331	.3669	0.79	.2852	.2148	1.24	.3925	.1075
0.35	.1368	.3632	0.80	.2881	.2119	1.25	.3944	.1056
0.36	.1406	.3594	0.81	.2910	.2090	1.26	.3962	.1038
0.37	.1443	.3557	0.82	.2939	.2061	1.27	.3980	.1020
0.38	.1480	.3520	0.83	.2967	.2033	1.28	.3997	.1003
0.39	.1517	.3483	0.84	.2995	.2005	1.29	.4015	.0985
0.40	.1554	.3446	0.85	.3023	.1977	1.30	.4032	.0968
0.41	.1591	.3409	0.86	.3051	.1949	1.31	.4049	.0951
0.42	.1628	.3372	0.87	.3078	.1922	1.32	.4066	.0934
0.43	.1664	.3336	0.88	.3106	.1894	1.33	.4082	.0918
0.44	.1700	.3300	0.89	.3133	.1867	1.34	.4099	.0901

Tables A, B, and C are taken, respectively, from Tables IIi, III, and IV of R. Fisher and F. Yates, *Statistical Tables for Biological, Agricultural and Medical Research,* published by Longman Group Ltd., London. (Previously published by Oliver & Boyd, Edinburgh.) Reprinted and modified by permission of the authors and publishers.

Table A *Areas Under the Unit Normal Distribution (Cont.)*

z	Area Between Mean and z	Area Beyond z	z	Area Between Mean and z	Area Beyond z	z	Area Between Mean and z	Area Beyond z
1.35	.4115	.0885	1.90	.4713	.0287	2.45	.4929	.0071
1.36	.4131	.0869	1.91	.4719	.0281	2.46	.4931	.0069
1.37	.4147	.0853	1.92	.4726	.0274	2.47	.4932	.0068
1.38	.4162	.0838	1.93	.4732	.0268	2.48	.4934	.0066
1.39	.4177	.0823	1.94	.4738	.0262	2.49	.4936	.0064
1.40	.4192	.0808	1.95	.4744	.0256	2.50	.4938	.0062
1.41	.4207	.0793	1.96	.4750	.0250	2.51	.4940	.0060
1.42	.4222	.0778	1.97	.4756	.0244	2.52	.4941	.0059
1.43	.4236	.0764	1.98	.4761	.0239	2.53	.4943	.0057
1.44	.4251	.0749	1.99	.4767	.0233	2.54	.4945	.0055
1.45	.4265	.0735	2.00	.4772	.0228	2.55	.4946	.0054
1.46	.4279	.0721	2.01	.4778	.0222	2.56	.4948	.0052
1.47	.4292	.0708	2.02	.4783	.0217	2.57	.4949	.0051
1.48	.4306	.0694	2.03	.4788	.0212	2.576	.4950	.0050
1.49	.4319	.0681	2.04	.4793	.0207	2.58	.4951	.0049
						2.59	.4952	.0048
1.50	.4332	.0668	2.05	.4798	.0202			
1.51	.4345	.0655	2.06	.4803	.0197	2.60	.4953	.0047
1.52	.4357	.0643	2.07	.4808	.0192	2.61	.4955	.0045
1.53	.4370	.0630	2.08	.4812	.0188	2.62	.4956	.0044
1.54	.4382	.0618	2.09	.4817	.0183	2.63	.4957	.0043
						2.64	.4959	.0041
1.55	.4394	.0606	2.10	.4821	.0179			
1.56	.4406	.0594	2.11	.4826	.0174	2.65	.4960	.0040
1.57	.4418	.0582	2.12	.4830	.0170	2.66	.4961	.0039
1.58	.4429	.0571	2.13	.4834	.0166	2.67	.4962	.0038
1.59	.4441	.0559	2.14	.4838	.0162	2.68	.4963	.0037
						2.69	.4964	.0036
1.60	.4452	.0548	2.15	.4842	.0158			
1.61	.4463	.0537	2.16	.4846	.0154	2.70	.4965	.0035
1.62	.4474	.0526	2.17	.4850	.0150	2.71	.4966	.0034
1.63	.4484	.0516	2.18	.4854	.0146	2.72	.4967	.0033
1.64	.4495	.0505	2.19	.4857	.0143	2.73	.4968	.0032
1.645	.4500	.0500				2.74	.4969	.0031
1.65	.4505	.0495	2.20	.4861	.0139			
1.66	.4515	.0485	2.21	.4864	.0136	2.75	.4970	.0030
1.67	.4525	.0475	2.22	.4868	.0132	2.76	.4971	.0029
1.68	.4535	.0465	2.23	.4871	.0129	2.77	.4972	.0028
1.69	.4545	.0455	2.24	.4875	.0125	2.78	.4973	.0027
						2.79	.4974	.0026
1.70	.4554	.0446	2.25	.4878	.0122			
1.71	.4564	.0436	2.26	.4881	.0119	2.80	.4974	.0026
1.72	.4573	.0427	2.27	.4884	.0116	2.81	.4975	.0025
1.73	.4582	.0418	2.28	.4887	.0113	2.82	.4976	.0024
1.74	.4591	.0409	2.29	.4890	.0110	2.83	.4977	.0023
						2.84	.4977	.0023
1.75	.4599	.0401	2.30	.4893	.0107			
1.76	.4608	.0392	2.31	.4896	.0104	2.85	.4978	.0022
1.77	.4616	.0384	2.32	.4898	.0102	2.86	.4979	.0021
1.78	.4625	.0375	2.326	.4900	.0100	2.87	.4979	.0021
1.79	.4633	.0367	2.33	.4901	.0099	2.88	.4980	.0020
			2.34	.4904	.0096	2.89	.4981	.0019
1.80	.4641	.0359	2.35	.4906	.0094	2.90	.4981	.0019
1.81	.4649	.0351	2.36	.4909	.0091	2.91	.4982	.0018
1.82	.4656	.0344	2.37	.4911	.0089	2.92	.4982	.0018
1.83	.4664	.0336	2.38	.4913	.0087	2.93	.4983	.0017
1.84	.4671	.0329	2.39	.4916	.0084	2.94	.4984	.0016
1.85	.4678	.0322	2.40	.4918	.0082	2.95	.4984	.0016
1.86	.4686	.0314	2.41	.4920	.0080	2.96	.4985	.0015
1.87	.4693	.0307	2.42	.4922	.0078	2.97	.4985	.0015
1.88	.4699	.0301	2.43	.4925	.0075	2.98	.4986	.0014
1.89	.4706	.0294	2.44	.4927	.0073	2.99	.4986	.0014

Table A *Areas Under the Unit Normal Distribution (Cont.)*

z	Area Between Mean and z	Area Beyond z	z	Area Between Mean and z	Area Beyond z	z	Area Between Mean and z	Area Beyond z
3.00	.4987	.0013	3.15	.4992	.0008	3.50	.4998	.0002
3.01	.4987	.0013	3.16	.4992	.0008	3.60	.4998	.0002
3.02	.4987	.0013	3.17	.4992	.0008	3.70	.4999	.0001
3.03	.4988	.0012	3.18	.4993	.0007	3.80	.4999	.0001
3.04	.4988	.0012	3.19	.4993	.0007	3.90	.49995	.00005
3.05	.4989	.0011	3.20	.4993	.0007	4.00	.49997	.00003
3.06	.4989	.0011	3.21	.4993	.0007			
3.07	.4989	.0011	3.22	.4994	.0006			
3.08	.4990	.0010	3.23	.4994	.0006			
3.09	.4990	.0010	3.24	.4994	.0006			
3.10	.4990	.0010	3.25	.4994	.0006			
3.11	.4991	.0009	3.30	.4995	.0005			
3.12	.4991	.0009	3.35	.4996	.0004			
3.13	.4991	.0009	3.40	.4997	.0003			
3.14	.4992	.0008	3.45	.4997	.0003			

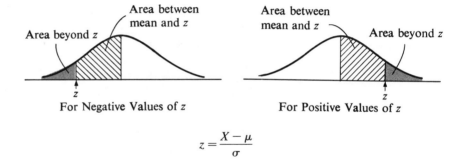

For Negative Values of z For Positive Values of z

$$z = \frac{X - \mu}{\sigma}$$

Table B *Critical Values of Student's t-Distribution*

	Level of Significance for One-Tailed Test (α)					
df	0.10	0.05	0.025	0.01	0.005	0.0005
	Level of Significance for Two-Tailed Test (α)					
	0.20	0.10	0.05	0.02	0.01	0.001
1	3.078	6.314	12.706	31.821	63.657	636.619
2	1.886	2.920	4.303	6.965	9.925	31.598
3	1.638	2.353	3.182	4.541	5.841	12.941
4	1.533	2.132	2.776	3.747	4.604	8.610
5	1.476	2.015	2.571	3.365	4.032	6.859
6	1.440	1.943	2.447	3.143	3.707	5.959
7	1.415	1.895	2.365	2.998	3.499	5.405
8	1.397	1.860	2.306	2.896	3.355	5.041
9	1.383	1.833	2.262	2.821	3.250	4.781
10	1.372	1.812	2.228	2.764	3.169	4.587
11	1.363	1.796	2.201	2.718	3.106	4.437
12	1.356	1.782	2.179	2.681	3.055	4.318
13	1.350	1.771	2.160	2.650	3.012	4.221
14	1.345	1.761	2.145	2.624	2.977	4.140
15	1.341	1.753	2.131	2.602	2.947	4.073
16	1.337	1.746	2.120	2.583	2.921	4.015
17	1.333	1.740	2.110	2.567	2.898	3.965
18	1.330	1.734	2.101	2.552	2.878	3.992
19	1.328	1.729	2.093	2.539	2.861	3.883
20	1.325	1.725	2.086	2.528	2.845	3.850
21	1.323	1.721	2.080	2.518	2.831	3.819
22	1.321	1.717	2.074	2.508	2.819	3.792
23	1.319	1.714	2.069	2.500	2.807	3.767
24	1.318	1.711	2.064	2.492	2.797	3.745
25	1.316	1.708	2.060	2.485	2.787	3.725
26	1.315	1.706	2.056	2.479	2.779	3.707
27	1.314	1.703	2.052	2.473	2.771	3.690
28	1.313	1.701	2.048	2.467	2.763	3.674
29	1.311	1.699	2.045	2.462	2.756	3.659
30	1.310	1.697	2.042	2.457	2.750	3.646
40	1.303	1.684	2.021	2.423	2.704	3.551
60	1.296	1.671	2.000	2.390	2.660	3.460
120	1.289	1.658	1.980	2.358	2.617	3.373
∞	1.282	1.645	1.960	2.326	2.576	3.291

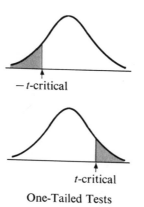

One-Tailed Tests

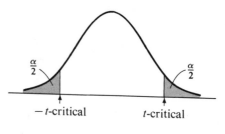

Two-Tailed Tests

Table C   *Chi-Square Probabilities*

	Probability						
	0.50	0.30	0.20	0.10	0.05	0.02	0.01
1	0.455	1.074	1.642	2.706	3.841	5.412	6.635
2	1.386	2.408	3.219	4.605	5.991	7.824	9.210
3	2.366	3.665	4.642	6.251	7.815	9.837	11.341
4	3.357	4.878	5.989	7.779	9.488	11.668	13.277
5	4.351	6.064	7.289	9.236	11.070	13.388	15.086
6	5.348	7.231	8.558	10.645	12.592	15.033	16.812
7	6.346	8.383	9.803	12.017	14.067	16.622	18.475
8	7.344	9.524	11.030	13.362	15.507	18.168	20.090
9	8.343	10.656	12.242	14.684	16.919	19.679	21.666
10	9.342	11.781	13.442	15.987	18.307	21.161	23.209
11	10.341	12.899	14.631	17.275	19.675	22.618	24.725
12	11.340	14.011	15.812	18.549	21.026	24.054	26.217
13	12.340	15.119	16.985	19.812	22.362	25.472	27.688
14	13.339	16.222	18.151	21.064	23.685	26.873	29.141
15	14.339	17.322	19.311	22.307	24.996	28.259	30.578
16	15.338	18.418	20.465	23.542	26.296	29.633	32.000
17	16.338	19.511	21.615	24.769	27.587	30.995	33.409
18	17.338	20.601	22.760	25.989	28.869	32.346	34.805
19	18.338	21.689	23.900	27.204	30.144	33.687	36.191
20	19.337	22.775	25.038	28.412	31.410	35.020	37.566
21	20.337	23.858	26.171	29.615	32.671	36.343	38.932
22	21.337	24.939	27.301	30.813	33.924	37.659	40.289
23	22.337	26.018	28.429	32.007	35.172	38.968	41.638
24	23.337	27.096	29.553	33.196	36.415	40.270	42.980
25	24.337	28.172	30.675	34.382	37.652	41.566	44.314
26	25.336	29.246	31.795	35.563	38.885	42.856	45.642
27	26.336	30.319	32.912	36.741	40.113	44.140	46.963
28	27.336	31.391	34.027	37.916	41.337	45.419	48.278
29	28.336	32.461	35.139	39.087	42.557	46.693	49.588
30	29.336	33.530	36.250	40.256	43.773	47.962	50.892

Degrees of Freedom

The probability that an obtained chi-square equals or exceeds the tabled values is shown at the top of each column. With $df = 16$, for instance, the probability that χ^2-observed ≥ 26.296 is 0.05, when H_0 is true.

Table D *Critical values of W in the Wilcoxon Matched Pairs, Signed Ranks Test*

N	$\frac{1}{2}N(N + 1)$	.05 .10	.025 .05	.01 .02	.005 .01	one-tail two-tail
				alpha level		
5	15	1				
6	21	2	1			
7	28	4	2	0		
8	36	6	4	2	0	
9	45	8	6	3	2	
10	55	11	8	5	3	
11	66	14	11	7	5	
12	78	17	14	10	7	
13	91	21	17	13	10	
14	105	26	21	16	13	
15	120	30	25	20	16	
16	136	36	30	24	19	
17	153	41	35	28	23	
18	171	47	40	33	28	
19	190	54	46	38	32	
20	210	60	52	43	37	
21	231	68	59	49	43	
22	253	75	66	56	49	
23	276	83	73	62	55	
24	300	92	81	69	61	
25	325	101	90	77	68	
26	351	110	98	85	76	
27	378	120	107	93	84	
28	406	130	117	102	92	
29	435	141	127	111	100	
30	465	152	137	120	109	
31	496	163	148	130	118	
32	528	175	159	141	128	
33	561	188	171	151	138	
34	595	201	183	162	149	
35	630	214	195	174	160	
36	666	228	208	186	171	
37	703	242	222	198	183	
38	741	256	235	211	195	
39	780	271	250	224	208	
40	820	287	264	238	221	
41	861	303	279	252	234	
42	903	319	295	267	248	
43	946	336	311	281	262	
44	990	353	327	297	277	
45	1035	371	344	313	292	
46	1081	389	361	329	307	
47	1128	408	379	345	323	
48	1176	427	397	362	339	
49	1225	446	415	380	356	
50	1275	466	434	398	373	

The probability that a given value of W is equal to or less than tabled values is the alpha level listed at the top of a column. Table D is taken from Table 23.B of E. S. Pearson and H. O. Hartley (eds.), *Biometrika Tables for Statisticians, Vol. II* (Cambridge: Cambridge University Press, 1972). Reprinted by permission of the editors and the trustees of *Biometrika*.

Table E   *Critical Values of R_1 for the Wilcoxon Rank Sum Test*

Reject H_0 if R_1 or $2\bar{R} - R_1$ is equal to or less than the tabled value.

n_1 = size of smaller group (if group sizes are unequal)

Alpha levels for a one-tailed test are given between n_2 and $2\bar{R}_1$; double these alpha values for the two-tailed test.

| n_2 | $n_1 = 1$ | | | | | | | $n_1 = 2$ | | | | | | | n_2 |
	0.001	0.005	0.010	0.025	0.05	0.10	$2\bar{R}_1$	0.001	0.005	0.010	0.025	0.05	0.10	$2\bar{R}_1$	
2							4						—	10	2
3							5						3	12	3
4							6					—	3	14	4
5							7					3	4	16	5
6							8					3	4	18	6
7							9				—	3	4	20	7
8					—		10				3	4	5	22	8
9						1	11				3	4	5	24	9
10						1	12				3	4	6	26	10
11						1	13				3	4	6	28	11
12						1	14			—	4	5	7	30	12
13						1	15			3	4	5	7	32	13
14						1	16			3	4	6	8	34	14
15						1	17			3	4	6	8	36	15
16						1	18			3	4	6	8	38	16
17						1	19			3	5	6	9	40	17
18					—	1	20		—	3	5	7	9	42	18
19					1	2	21		3	4	5	7	10	44	19
20					1	2	22		3	4	5	7	10	46	20
21					1	2	23		3	4	6	8	11	48	21
22					1	2	24		3	4	6	8	11	50	22
23					1	2	25		3	4	6	8	12	52	23
24					1	2	26		3	4	6	9	12	54	24
25	—	—	—	—	1	2	27	—	3	4	6	9	12	56	25

| n_2 | $n_1 = 3$ | | | | | | | $n_1 = 4$ | | | | | | | n_2 |
	0.001	0.005	0.010	0.025	0.05	0.10	$2\bar{R}_1$	0.001	0.005	0.010	0.025	0.05	0.10	$2\bar{R}_1$	
3					6	7	21								
4				—	6	7	24			—	10	11	13	36	4
5				6	7	8	27		—	10	11	12	14	40	5
6			—	7	8	9	30		10	11	12	13	15	44	6
7			6	7	8	10	33		10	11	13	14	16	48	7
8		—	6	8	9	11	36		11	12	14	15	17	52	8
9		6	7	8	10	11	39	—	11	13	14	16	19	56	9
10		6	7	9	10	12	42	10	12	13	15	17	20	60	10
11		6	7	9	11	13	45	10	12	14	16	18	21	64	11
12		7	8	10	11	14	48	10	13	15	17	19	22	68	12
13		7	8	10	12	15	51	11	13	15	18	20	23	72	13
14		7	8	11	13	16	54	11	14	16	19	21	25	76	14
15		8	9	11	13	16	57	11	15	17	20	22	26	80	15
16	—	8	9	12	14	17	60	12	15	17	21	24	27	84	16
17	6	8	10	12	15	18	63	12	16	18	21	25	28	88	17
18	6	8	10	13	15	19	66	13	16	19	22	26	30	92	18
19	6	9	10	13	16	20	69	13	17	19	23	27	31	96	19
20	6	9	11	14	17	21	72	13	18	20	24	28	32	100	20
21	7	9	11	14	17	21	75	14	18	21	25	29	33	104	21
22	7	10	12	15	18	22	78	14	19	21	26	30	35	108	22
23	7	10	12	15	19	23	81	14	19	22	27	31	36	112	23
24	7	10	12	16	19	24	84	15	20	23	27	32	38	116	24
25	7	11	13	16	20	25	87	15	20	23	28	33	38	120	25

Table E is taken, with changes in notation, from L. R. Verdooren, "Extended Tables of Critical Values for Wilcoxon's Test Statistic." *Biometrika*, 50 (1963): 177–185. Reprinted by permission of the author and the trustees of *Biometrika*.

Table E *Critical Values of R_1 (Cont.)*

n_2	0.001	0.005	$n_1 = 5$ 0.010	0.025	0.05	0.10	$2\bar{R}_1$	0.001	0.005	$n_1 = 6$ 0.010	0.025	0.05	0.10	$2\bar{R}_1$	n_2
5		15	16	17	19	20	55								
6		16	17	18	20	22	60	—	23	24	26	28	30	78	6
7	—	16	18	20	21	23	65	21	24	25	27	29	32	84	7
8	15	17	19	21	23	25	70	22	25	27	29	31	34	90	8
9	16	18	20	22	24	27	75	23	26	28	31	33	36	96	9
10	16	19	21	23	26	28	80	24	27	29	32	35	38	102	10
11	17	20	22	24	27	30	85	25	28	30	34	37	40	108	11
12	17	21	23	26	28	32	90	25	30	32	35	38	42	114	12
13	18	22	24	27	30	33	95	26	31	33	37	40	44	120	13
14	18	22	25	28	31	35	100	27	32	34	38	42	46	126	14
15	19	23	26	29	33	37	105	28	33	36	40	44	48	132	15
16	20	24	27	30	34	38	110	29	34	37	42	46	50	138	16
17	20	25	28	32	35	40	115	30	36	39	43	47	52	144	17
18	21	26	29	33	37	42	120	31	37	40	45	49	55	150	18
19	22	27	30	34	38	43	125	32	38	41	46	51	57	156	19
20	22	28	31	35	40	45	130	33	39	43	48	53	59	162	20
21	23	29	32	37	41	47	135	33	40	44	50	55	61	168	21
22	23	29	33	38	43	48	140	34	42	45	51	57	63	174	22
23	24	30	34	39	44	50	145	35	43	47	53	58	65	180	23
24	25	31	35	40	45	51	150	36	44	48	54	60	67	186	24
25	25	32	36	42	47	53	155	37	45	50	56	62	69	192	25

n_2	0.001	0.005	$n_1 = 7$ 0.010	0.025	0.05	0.10	$2\bar{R}_1$	0.001	0.005	$n_1 = 8$ 0.010	0.025	0.05	0.10	$2\bar{R}_1$	n_2
7	29	32	34	36	39	41	105								
8	30	34	35	38	41	44	112	40	43	45	49	51	55	136	8
9	31	35	37	40	43	46	119	41	45	47	51	54	58	144	9
10	33	37	39	42	45	49	126	42	47	49	53	56	60	152	10
11	34	38	40	44	47	51	133	44	49	51	55	59	63	160	11
12	35	40	42	46	49	54	140	45	51	53	58	62	66	168	12
13	36	41	44	48	52	56	147	47	53	56	60	64	69	176	13
14	37	43	45	50	54	59	154	48	54	58	62	67	72	184	14
15	38	44	47	52	56	61	161	50	56	60	65	69	75	192	15
16	39	46	49	54	58	64	168	51	58	62	67	72	78	200	16
17	41	47	51	56	61	66	175	53	60	64	70	75	81	208	17
18	42	49	52	58	63	69	182	54	62	66	72	77	84	216	18
19	43	50	54	60	65	71	189	56	64	68	74	80	87	224	19
20	44	52	56	62	67	74	196	57	66	70	77	83	90	232	20
21	46	53	58	64	69	76	203	59	68	72	79	85	92	240	21
22	47	55	59	66	72	79	210	60	70	74	81	88	95	248	22
23	48	57	61	68	74	81	217	62	71	76	84	90	98	256	23
24	49	58	63	70	76	84	224	64	73	78	86	93	101	264	24
25	50	60	64	72	78	86	231	65	75	81	89	96	104	272	25

n_1	0.001	0.005	$n_1 = 9$ 0.010	0.025	0.05	0.10	$2\bar{R}_1$	0.001	0.005	$n_1 = 10$ 0.010	0.025	0.05	0.10	$2\bar{R}_1$	n_2
9	52	56	59	62	66	70	171								
10	53	58	61	65	69	73	180	65	71	74	78	82	87	210	10
11	55	61	63	68	72	76	189	67	73	77	81	86	91	220	11
12	57	63	66	71	75	80	198	69	76	79	84	89	94	230	12
13	59	65	68	73	78	83	207	72	79	82	88	92	98	240	13
14	60	67	71	76	81	86	216	74	81	85	91	96	102	250	14
15	62	69	73	79	84	90	225	76	84	88	94	99	106	260	15
16	64	72	76	82	87	93	234	78	86	91	97	103	109	270	16
17	66	74	78	84	90	97	243	80	89	93	100	103	113	280	17
18	68	76	81	87	93	100	252	82	92	96	103	110	117	290	18
19	70	78	83	90	96	103	261	84	94	99	107	113	121	300	19
20	71	81	85	93	99	107	270	87	97	102	110	117	125	310	20
21	73	83	88	95	102	110	279	89	99	105	113	120	128	320	21
22	75	85	90	98	105	113	288	91	102	108	116	123	132	330	22
23	77	88	93	101	108	117	297	93	105	110	119	127	136	340	23
24	79	90	95	104	111	120	306	95	107	113	122	130	140	350	24
25	81	92	98	107	114	123	315	98	110	116	126	134	144	360	25

Table E   *Critical Values of R_1 (Cont.)*

			$n_1 = 11$								$n_1 = 12$				
n_2	0.001	0.005	0.010	0.025	0.05	0.10	$2\bar{R}_1$	0.001	0.005	0.010	0.025	0.05	0.10	$2\bar{R}_1$	n_2
11	81	87	91	96	100	106	253								
12	83	90	94	99	104	110	264	98	105	109	115	120	127	300	12
13	86	93	97	103	108	114	275	101	109	113	119	125	131	312	13
14	88	96	100	106	112	118	286	103	112	116	123	129	136	324	14
15	90	99	103	110	116	123	297	106	115	120	127	133	141	336	15
16	93	102	107	113	120	127	308	109	119	124	131	138	145	348	16
17	95	105	110	117	123	131	319	112	122	127	135	142	150	360	17
18	98	108	113	121	127	135	330	115	125	131	139	146	155	372	18
19	100	111	116	124	131	139	341	118	129	134	143	150	159	384	19
20	103	114	119	128	135	144	352	120	132	138	147	155	164	396	20
21	106	117	123	131	139	148	363	123	136	142	151	159	169	408	21
22	108	120	126	135	143	152	374	126	139	145	155	163	173	420	22
23	111	123	129	139	147	156	385	129	142	149	159	168	178	432	23
24	113	126	132	142	151	161	396	132	146	153	163	172	183	444	24
25	116	129	136	146	155	165	407	135	149	156	167	176	187	456	25

			$n_1 = 13$								$n_1 = 14$				
n_2	0.001	0.005	0.010	0.025	0.05	0.10	$2\bar{R}_1$	0.001	0.005	0.010	0.025	0.05	0.10	$2\bar{R}_1$	n_2
13	117	125	130	136	142	149	351								
14	120	129	134	141	147	154	364	137	147	152	160	166	174	406	14
15	123	133	138	145	152	159	377	141	151	156	164	171	179	420	15
16	126	136	142	150	156	165	390	144	155	161	169	176	185	434	16
17	129	140	146	154	161	170	403	148	159	165	174	182	190	448	17
18	133	144	150	158	166	175	416	151	163	170	179	187	196	462	18
19	136	148	154	163	171	180	429	155	168	174	183	192	202	476	19
20	139	151	158	167	175	185	442	159	172	178	188	197	207	490	20
21	142	155	162	171	180	190	455	162	176	183	193	202	213	504	21
22	145	159	166	176	185	195	468	166	180	187	198	207	218	518	22
23	149	163	170	180	189	200	481	169	184	192	203	212	224	532	23
24	152	166	174	185	194	205	494	173	188	196	207	218	229	546	24
25	155	170	178	189	199	211	507	177	192	200	212	223	235	560	25

			$n_1 = 15$								$n_1 = 16$				
n_2	0.001	0.005	0.010	0.025	0.05	0.10	$2\bar{R}_1$	0.001	0.005	0.010	0.025	0.05	0.10	$2\bar{R}_1$	n_2
15	160	171	176	184	192	200	465								
16	163	175	181	190	197	206	480	184	196	202	211	219	229	528	16
17	167	180	186	195	203	212	495	188	201	207	217	225	235	544	17
18	171	184	190	200	208	218	510	192	206	212	222	231	242	560	18
19	175	189	195	205	214	224	525	196	210	218	228	237	248	576	19
20	179	193	200	210	220	230	540	201	215	223	234	243	255	592	20
21	183	198	205	216	225	236	555	205	220	228	239	249	261	608	21
22	187	202	210	221	231	242	570	209	225	233	245	255	267	624	22
23	191	207	214	226	236	248	585	214	230	238	251	261	274	640	23
24	195	211	219	231	242	254	600	218	235	244	256	267	280	656	24
25	199	216	224	237	248	260	615	222	240	249	262	273	287	672	25

			$n_1 = 17$								$n_1 = 18$				
n_2	0.001	0.005	0.010	0.025	0.05	0.10	$2\bar{R}_1$	0.001	0.005	0.010	0.025	0.05	0.10	$2\bar{R}_1$	n_2
17	210	223	230	240	249	259	595								
18	214	228	235	246	255	266	612	237	252	259	270	280	291	666	18
19	219	234	241	252	262	273	629	242	258	265	277	287	299	684	19
20	223	239	246	258	268	280	646	247	263	271	283	294	306	702	20
21	228	244	252	264	274	287	663	252	269	277	290	301	313	720	21
22	233	249	258	270	281	294	680	257	275	283	296	307	321	738	22
23	238	255	263	276	287	300	697	262	280	289	303	314	328	756	23
24	242	260	269	282	294	307	714	267	286	295	309	321	335	774	24
25	247	265	275	288	300	314	731	273	292	301	316	328	343	792	25

Table E *Critical Values of R_1 (Cont.)*

n_2	0.001	0.005	$n_1 = 19$ 0.010	0.025	0.05	0.10	$2\bar{R}_1$	0.001	0.005	$n_1 = 20$ 0.010	0.025	0.05	0.10	$2\bar{R}_1$	n_2
19	267	283	291	303	313	325	741								
20	272	289	297	309	320	333	760	298	315	324	337	348	361	820	20
21	277	295	303	316	328	341	779	304	322	331	344	356	370	840	21
22	283	301	310	323	335	349	798	309	328	337	351	364	378	860	22
23	288	307	316	330	342	357	817	315	335	344	359	371	386	880	23
24	294	313	323	337	350	364	836	321	341	351	366	379	394	900	24
25	299	319	329	344	357	372	855	327	348	358	373	387	403	920	25

n_2	0.001	0.005	$n_1 = 21$ 0.010	0.025	0.05	0.10	$2\bar{R}_1$	0.001	0.005	$n_1 = 22$ 0.010	0.025	0.05	0.10	$2\bar{R}_1$	n_2
21	331	349	359	373	385	399	903								
22	337	356	366	381	393	408	924	365	386	396	411	424	439	990	22
23	343	363	373	388	401	417	945	372	393	403	419	432	448	1012	23
24	349	370	381	396	410	425	966	379	400	411	427	441	457	1034	24
25	356	377	388	404	418	434	987	385	408	419	435	450	467	1056	25

n_2	0.001	0.005	$n_1 = 23$ 0.010	0.025	0.05	0.10	$2\bar{R}_1$	0.001	0.005	$n_1 = 24$ 0.010	0.025	0.05	0.10	$2\bar{R}_1$	n_2
23	402	424	434	451	465	481	1081								
24	409	431	443	459	474	491	1104	440	464	475	492	507	525	1176	24
25	416	439	451	468	483	500	1127	448	472	484	501	517	535	1200	25

n_2	0.001	0.005	$n_1 = 25$ 0.010	0.025	0.05	0.10	$2\bar{R}_1$
25	480	505	517	536	552	570	1275

440

Table F *Critical Values of the F-distribution* ($\alpha = 0.05$)

df numerator

df denom	1	2	3	4	5	6	7	8	9	10	12	15	20	24	30	40	60	120	∞
1	161.4	199.5	215.7	224.6	230.2	234.0	236.8	238.9	240.5	241.9	243.9	245.9	248.0	249.1	250.1	251.1	252.2	253.3	254.3
2	18.51	19.00	19.16	19.25	19.30	19.33	19.35	19.37	19.38	19.40	19.41	19.43	19.45	19.45	19.46	19.47	19.48	19.49	19.50
3	10.13	9.55	9.28	9.12	9.01	8.94	8.89	8.85	8.81	8.79	8.74	8.70	8.66	8.64	8.62	8.59	8.57	8.55	8.53
4	7.71	6.94	6.59	6.39	6.26	6.16	6.09	6.04	6.00	5.96	5.91	5.86	5.80	5.77	5.75	5.72	5.69	5.66	5.63
5	6.61	5.79	5.41	5.19	5.05	4.95	4.88	4.82	4.77	4.74	4.68	4.62	4.56	4.53	4.50	4.46	4.43	4.40	4.36
6	5.99	5.14	4.76	4.53	4.39	4.28	4.21	4.15	4.10	4.06	4.00	3.94	3.87	3.84	3.81	3.77	3.74	3.70	3.67
7	5.59	4.74	4.35	4.12	3.97	3.87	3.79	3.73	3.68	3.64	3.57	3.51	3.44	3.41	3.38	3.34	3.30	3.27	3.23
8	5.32	4.46	4.07	3.84	3.69	3.58	3.50	3.44	3.39	3.35	3.28	3.22	3.15	3.12	3.08	3.04	3.01	2.97	2.93
9	5.12	4.26	3.86	3.63	3.48	3.37	3.29	3.23	3.18	3.14	3.07	3.01	2.94	2.90	2.86	2.83	2.79	2.75	2.71
10	4.96	4.10	3.71	3.48	3.33	3.22	3.14	3.07	3.02	2.98	2.91	2.85	2.77	2.74	2.70	2.66	2.62	2.58	2.54
11	4.84	3.98	3.59	3.36	3.20	3.09	3.01	2.95	2.90	2.85	2.79	2.72	2.65	2.61	2.57	2.53	2.49	2.45	2.40
12	4.75	3.89	3.49	3.26	3.11	3.00	2.91	2.85	2.80	2.75	2.69	2.62	2.54	2.51	2.47	2.43	2.38	2.34	2.30
13	4.67	3.81	3.41	3.18	3.03	2.92	2.83	2.77	2.71	2.67	2.60	2.53	2.46	2.42	2.38	2.34	2.30	2.25	2.21
14	4.60	3.74	3.34	3.11	2.96	2.85	2.76	2.70	2.65	2.60	2.53	2.46	2.39	2.35	2.31	2.27	2.22	2.18	2.13
15	4.54	3.68	3.29	3.06	2.90	2.79	2.71	2.64	2.59	2.54	2.48	2.40	2.33	2.29	2.25	2.20	2.16	2.11	2.07
16	4.49	3.63	3.24	3.01	2.85	2.74	2.66	2.59	2.54	2.49	2.42	2.35	2.28	2.24	2.19	2.15	2.11	2.06	2.01
17	4.45	3.59	3.20	2.96	2.81	2.70	2.61	2.55	2.49	2.45	2.38	2.31	2.23	2.19	2.15	2.10	2.06	2.01	1.96
18	4.41	3.55	3.16	2.93	2.77	2.66	2.58	2.51	2.46	2.41	2.34	2.27	2.19	2.15	2.11	2.06	2.02	1.97	1.92
19	4.38	3.52	3.13	2.90	2.74	2.63	2.54	2.48	2.42	2.38	2.31	2.23	2.16	2.11	2.07	2.03	1.98	1.93	1.88
20	4.35	3.49	3.10	2.87	2.71	2.60	2.51	2.45	2.39	2.35	2.28	2.20	2.12	2.08	2.04	1.99	1.95	1.90	1.84
21	4.32	3.47	3.07	2.84	2.68	2.57	2.49	2.42	2.37	2.32	2.25	2.18	2.10	2.05	2.01	1.96	1.92	1.87	1.81
22	4.30	3.44	3.05	2.82	2.66	2.55	2.46	2.40	2.34	2.30	2.23	2.15	2.07	2.03	1.98	1.94	1.89	1.84	1.78
23	4.28	3.42	3.03	2.80	2.64	2.53	2.44	2.37	2.32	2.27	2.20	2.13	2.05	2.01	1.96	1.91	1.86	1.81	1.76
24	4.26	3.40	3.01	2.78	2.62	2.51	2.42	2.36	2.30	2.25	2.18	2.11	2.03	1.98	1.94	1.89	1.84	1.79	1.73
25	4.24	3.39	2.99	2.76	2.60	2.49	2.40	2.34	2.28	2.24	2.16	2.09	2.01	1.96	1.92	1.87	1.82	1.77	1.71
26	4.23	3.37	2.98	2.74	2.59	2.47	2.39	2.32	2.27	2.22	2.15	2.07	1.99	1.95	1.90	1.85	1.80	1.75	1.69
27	4.21	3.35	2.96	2.73	2.57	2.46	2.37	2.31	2.25	2.20	2.13	2.06	1.97	1.93	1.88	1.84	1.79	1.73	1.67
28	4.20	3.34	2.95	2.71	2.56	2.45	2.36	2.29	2.24	2.19	2.12	2.04	1.96	1.91	1.87	1.82	1.77	1.71	1.65
29	4.18	3.33	2.93	2.70	2.55	2.43	2.35	2.28	2.22	2.18	2.10	2.03	1.94	1.90	1.85	1.81	1.75	1.70	1.64
30	4.17	3.32	2.92	2.69	2.53	2.42	2.33	2.27	2.21	2.16	2.09	2.01	1.93	1.89	1.84	1.79	1.74	1.68	1.62
40	4.08	3.23	2.84	2.61	2.45	2.34	2.25	2.18	2.12	2.08	2.00	1.92	1.84	1.79	1.74	1.69	1.64	1.58	1.51
60	4.00	3.15	2.76	2.53	2.37	2.25	2.17	2.10	2.04	1.99	1.92	1.84	1.75	1.70	1.65	1.59	1.53	1.47	1.39
120	3.92	3.07	2.68	2.45	2.29	2.17	2.09	2.02	1.96	1.91	1.83	1.75	1.66	1.61	1.55	1.50	1.43	1.35	1.25
∞	3.84	3.00	2.60	2.37	2.21	2.10	2.01	1.94	1.88	1.83	1.75	1.67	1.57	1.52	1.46	1.39	1.32	1.22	1.00

df denominator

Table F *(Cont.) Critical Values of the F-distribution ($\alpha = 0.01$)*

df denominator	\ df numerator → 1	2	3	4	5	6	7	8	9	10	12	15	20	24	30	40	60	120	∞
1	4052	4999.5	5403	5625	5764	5859	5928	5981	6022	6056	6106	6157	6209	6235	6261	6287	6313	6339	6366
2	98.50	99.00	99.17	99.25	99.30	99.33	99.36	99.37	99.39	99.40	99.42	99.43	99.45	99.46	99.47	99.47	99.48	99.49	99.50
3	34.12	30.82	29.46	28.71	28.24	27.91	27.67	27.49	27.35	27.23	27.05	26.87	26.69	26.60	26.50	26.41	26.32	26.22	26.13
4	21.20	18.00	16.69	15.98	15.52	15.21	14.98	14.80	14.66	14.55	14.37	14.20	14.02	13.93	13.84	13.75	13.65	13.56	13.46
5	16.26	13.27	12.06	11.39	10.97	10.67	10.46	10.29	10.16	10.05	9.89	9.72	9.55	9.47	9.38	9.29	9.20	9.11	9.02
6	13.75	10.92	9.78	9.15	8.75	8.47	8.26	8.10	7.98	7.87	7.72	7.56	7.40	7.31	7.23	7.14	7.06	6.97	6.88
7	12.25	9.55	8.45	7.85	7.46	7.19	6.99	6.84	6.72	6.62	6.47	6.31	6.16	6.07	5.99	5.91	5.82	5.74	5.65
8	11.26	8.65	7.59	7.01	6.63	6.37	6.18	6.03	5.91	5.81	5.67	5.52	5.36	5.28	5.20	5.12	5.03	4.95	4.86
9	10.56	8.02	6.99	6.42	6.06	5.80	5.61	5.47	5.35	5.26	5.11	4.96	4.81	4.73	4.65	4.57	4.48	4.40	4.31
10	10.04	7.56	6.55	5.99	5.64	5.39	5.20	5.06	4.94	4.85	4.71	4.56	4.41	4.33	4.25	4.17	4.08	4.00	3.91
11	9.65	7.21	6.22	5.67	5.32	5.07	4.89	4.74	4.63	4.54	4.40	4.25	4.10	4.02	3.94	3.86	3.78	3.69	3.60
12	9.33	6.93	5.95	5.41	5.06	4.82	4.64	4.50	4.39	4.30	4.16	4.01	3.86	3.78	3.70	3.62	3.54	3.45	3.36
13	9.07	6.70	5.74	5.21	4.86	4.62	4.44	4.30	4.19	4.10	3.96	3.82	3.66	3.59	3.51	3.43	3.34	3.25	3.17
14	8.86	6.51	5.56	5.04	4.69	4.46	4.28	4.14	4.03	3.94	3.80	3.66	3.51	3.43	3.35	3.27	3.18	3.09	3.00
15	8.68	6.36	5.42	4.89	4.56	4.32	4.14	4.00	3.89	3.80	3.67	3.52	3.37	3.29	3.21	3.13	3.05	2.96	2.87
16	8.53	6.23	5.29	4.77	4.44	4.20	4.03	3.89	3.78	3.69	3.55	3.41	3.26	3.18	3.10	3.02	2.93	2.84	2.75
17	8.40	6.11	5.18	4.67	4.34	4.10	3.93	3.79	3.68	3.59	3.46	3.31	3.16	3.08	3.00	2.92	2.83	2.75	2.65
18	8.29	6.01	5.09	4.58	4.25	4.01	3.84	3.71	3.60	3.51	3.37	3.23	3.08	3.00	2.92	2.84	2.75	2.66	2.57
19	8.18	5.93	5.01	4.50	4.17	3.94	3.77	3.63	3.52	3.43	3.30	3.15	3.00	2.92	2.84	2.76	2.67	2.58	2.49
20	8.10	5.85	4.94	4.43	4.10	3.87	3.70	3.56	3.46	3.37	3.23	3.09	2.94	2.86	2.78	2.69	2.61	2.52	2.42
21	8.02	5.78	4.87	4.37	4.04	3.81	3.64	3.51	3.40	3.31	3.17	3.03	2.88	2.80	2.72	2.64	2.55	2.46	2.36
22	7.95	5.72	4.82	4.31	3.99	3.76	3.59	3.45	3.35	3.26	3.12	2.98	2.83	2.75	2.67	2.58	2.50	2.40	2.31
23	7.88	5.66	4.76	4.26	3.94	3.71	3.54	3.41	3.30	3.21	3.07	2.93	2.78	2.70	2.62	2.54	2.45	2.35	2.26
24	7.82	5.61	4.72	4.22	3.90	3.67	3.50	3.36	3.26	3.17	3.03	2.89	2.74	2.66	2.58	2.49	2.40	2.31	2.21
25	7.77	5.57	4.68	4.18	3.85	3.63	3.46	3.32	3.22	3.13	2.99	2.85	2.70	2.62	2.54	2.45	2.36	2.27	2.17
26	7.72	5.53	4.64	4.14	3.82	3.59	3.42	3.29	3.18	3.09	2.96	2.81	2.66	2.58	2.50	2.42	2.33	2.23	2.13
27	7.68	5.49	4.60	4.11	3.78	3.56	3.39	3.26	3.15	3.06	2.93	2.78	2.63	2.55	2.47	2.38	2.29	2.20	2.10
28	7.64	5.45	4.57	4.07	3.75	3.53	3.36	3.23	3.12	3.03	2.90	2.75	2.60	2.52	2.44	2.35	2.26	2.17	2.06
29	7.60	5.42	4.54	4.04	3.73	3.50	3.33	3.20	3.09	3.00	2.87	2.73	2.57	2.49	2.41	2.33	2.23	2.14	2.03
30	7.56	5.39	4.51	4.02	3.70	3.47	3.30	3.17	3.07	2.98	2.84	2.70	2.55	2.47	2.39	2.30	2.21	2.11	2.01
40	7.31	5.18	4.31	3.83	3.51	3.29	3.12	2.99	2.89	2.80	2.66	2.52	2.37	2.29	2.20	2.11	2.02	1.92	1.80
60	7.08	4.98	4.13	3.65	3.34	3.12	2.95	2.82	2.72	2.63	2.50	2.35	2.20	2.12	2.03	1.94	1.84	1.73	1.60
120	6.85	4.79	3.95	3.48	3.17	2.96	2.79	2.66	2.56	2.47	2.34	2.19	2.03	1.95	1.86	1.76	1.66	1.53	1.38
∞	6.63	4.61	3.78	3.32	3.02	2.80	2.64	2.51	2.41	2.32	2.18	2.04	1.88	1.79	1.70	1.59	1.47	1.32	1.00

Table F *(Cont.) Critical Values of the F-distribution (α = 0.001)*

df numerator

df denominator	1	2	3	4	5	6	7	8	9	10	12	15	20	24	30	40	60	120	∞
1	4053*	5000*	5404*	5625*	5764*	5859*	5929*	5981*	6023*	6056*	6107*	6158*	6209*	6235*	6261*	6287*	6313*	6340*	6366*
2	998.5	999.0	999.2	999.2	999.3	999.3	999.4	999.4	999.4	999.4	999.4	999.4	999.4	999.5	999.5	999.5	999.5	999.5	999.5
3	167.0	148.5	141.1	137.1	134.6	132.8	131.6	130.6	129.9	129.2	128.3	127.4	126.4	125.9	125.4	125.0	124.5	124.0	123.5
4	74.14	61.25	56.18	53.44	51.71	50.53	49.66	49.00	48.47	48.05	47.41	46.76	46.10	45.77	45.43	45.09	44.75	44.40	44.05
5	47.18	37.12	33.20	31.09	29.75	28.84	28.16	27.64	27.24	26.92	26.42	25.91	25.39	25.14	24.87	24.60	24.33	24.06	23.79
6	35.51	27.00	23.70	21.92	20.81	20.03	19.46	19.03	18.69	18.41	17.99	17.56	17.12	16.89	16.67	16.44	16.21	15.99	15.75
7	29.25	21.69	18.77	17.19	16.21	15.52	15.02	14.63	14.33	14.08	13.71	13.32	12.93	12.73	12.53	12.33	12.12	11.91	11.70
8	25.42	18.49	15.83	14.39	13.49	12.86	12.40	12.04	11.77	11.54	11.19	10.84	10.48	10.30	10.11	9.92	9.73	9.53	9.33
9	22.86	16.39	13.90	12.56	11.71	11.13	10.70	10.37	10.11	9.89	9.57	9.24	8.90	8.72	8.55	8.37	8.19	8.00	7.81
10	21.04	14.91	12.55	11.28	10.48	9.92	9.52	9.20	8.96	8.75	8.45	8.13	7.80	7.64	7.47	7.30	7.12	6.94	6.76
11	19.69	13.81	11.56	10.35	9.58	9.05	8.66	8.35	8.12	7.92	7.63	7.32	7.01	6.85	6.68	6.52	6.35	6.17	6.00
12	18.64	12.97	10.80	9.63	8.89	8.38	8.00	7.71	7.48	7.29	7.00	6.71	6.40	6.25	6.09	5.93	5.76	5.59	5.42
13	17.81	12.31	10.21	9.07	8.35	7.86	7.49	7.21	6.98	6.80	6.52	6.23	5.93	5.78	5.63	5.47	5.30	5.14	4.97
14	17.14	11.78	9.73	8.62	7.92	7.43	7.08	6.80	6.58	6.40	6.13	5.85	5.56	5.41	5.25	5.10	4.94	4.77	4.60
15	16.59	11.34	9.34	8.25	7.57	7.09	6.74	6.47	6.26	6.08	5.81	5.54	5.25	5.10	4.95	4.80	4.64	4.47	4.31
16	16.12	10.97	9.00	7.94	7.27	6.81	6.46	6.19	5.98	5.81	5.55	5.27	4.99	4.85	4.70	4.54	4.39	4.23	4.06
17	15.72	10.66	8.73	7.68	7.02	6.56	6.22	5.96	5.75	5.58	5.32	5.05	4.78	4.63	4.48	4.33	4.18	4.02	3.85
18	15.38	10.39	8.49	7.46	6.81	6.35	6.02	5.76	5.56	5.39	5.13	4.87	4.59	4.45	4.30	4.15	4.00	3.84	3.67
19	15.08	10.16	8.28	7.26	6.62	6.18	5.85	5.59	5.39	5.22	4.97	4.70	4.43	4.29	4.14	3.99	3.84	3.68	3.51
20	14.82	9.95	8.10	7.10	6.46	6.02	5.69	5.44	5.24	5.08	4.82	4.56	4.29	4.15	4.00	3.86	3.70	3.54	3.38
21	14.59	9.77	7.94	6.95	6.32	5.88	5.56	5.31	5.11	4.95	4.70	4.44	4.17	4.03	3.88	3.74	3.58	3.42	3.26
22	14.38	9.61	7.80	6.81	6.19	5.76	5.44	5.19	4.99	4.83	4.58	4.33	4.06	3.92	3.78	3.63	3.48	3.32	3.15
23	14.19	9.47	7.67	6.69	6.08	5.65	5.33	5.09	4.89	4.73	4.48	4.23	3.96	3.82	3.68	3.53	3.38	3.22	3.05
24	14.03	9.34	7.55	6.59	5.98	5.55	5.23	4.99	4.80	4.64	4.39	4.14	3.87	3.74	3.59	3.45	3.29	3.14	2.97
25	13.88	9.22	7.45	6.49	5.88	5.46	5.15	4.91	4.71	4.56	4.31	4.06	3.79	3.66	3.52	3.37	3.22	3.06	2.89
26	13.74	9.12	7.36	6.41	5.80	5.38	5.07	4.83	4.64	4.48	4.24	3.99	3.72	3.59	3.44	3.30	3.15	2.99	2.82
27	13.61	9.02	7.27	6.33	5.73	5.31	5.00	4.76	4.57	4.41	4.17	3.92	3.66	3.52	3.38	3.23	3.08	2.92	2.75
28	13.50	8.93	7.19	6.25	5.66	5.24	4.93	4.69	4.50	4.35	4.11	3.86	3.60	3.46	3.32	3.18	3.02	2.86	2.69
29	13.39	8.85	7.12	6.19	5.59	5.18	4.87	4.64	4.45	4.29	4.05	3.80	3.54	3.41	3.27	3.12	2.97	2.81	2.64
30	13.29	8.77	7.05	6.12	5.53	5.12	4.82	4.58	4.39	4.24	4.00	3.75	3.49	3.36	3.22	3.07	2.92	2.76	2.59
40	12.61	8.25	6.60	5.70	5.13	4.73	4.44	4.21	4.02	3.87	3.64	3.40	3.15	3.01	2.87	2.73	2.57	2.41	2.23
60	11.97	7.76	6.17	5.31	4.76	4.37	4.09	3.87	3.69	3.54	3.31	3.08	2.83	2.69	2.55	2.41	2.25	2.08	1.89
120	11.38	7.32	5.79	4.95	4.42	4.04	3.77	3.55	3.38	3.24	3.02	2.78	2.53	2.40	2.26	2.11	1.95	1.76	1.54
∞	10.83	6.91	5.42	4.62	4.10	3.74	3.47	3.27	3.10	2.96	2.74	2.51	2.27	2.13	1.99	1.84	1.66	1.45	1.00

*Multiply these entries by 100.

Table G   *Critical Values of the Hartley F_{max} Statistic* ($\alpha = 0.05$)

k = number of groups in experiment

$n-1$	2	3	4	5	6	7	8	9	10	11	12
2	39.0	87.5	142	202	266	333	403	475	550	626	704
3	14.4	27.8	39.2	50.7	62.0	72.9	83.5	93.9	104	114	124
4	9.60	15.5	20.6	25.2	29.5	33.6	37.5	41.1	44.6	48.0	51.4
5	7.15	10.8	13.7	16.3	18.7	20.8	22.9	24.7	26.5	28.2	29.9
6	5.82	8.38	10.4	12.1	13.7	15.0	16.3	17.5	18.6	19.7	20.7
7	4.99	6.94	8.44	9.70	10.8	11.8	12.7	13.5	14.3	15.1	15.8
8	4.43	6.00	7.18	8.12	9.03	9.78	10.5	11.1	11.7	12.2	12.7
9	4.03	5.34	6.31	7.11	7.80	8.41	8.95	9.45	9.91	10.3	10.7
10	3.72	4.85	5.67	6.34	6.92	7.42	7.87	8.28	8.66	9.01	9.34
12	3.28	4.16	4.79	5.30	5.72	6.09	6.42	6.72	7.00	7.25	7.48
15	2.86	3.54	4.01	4.37	4.68	4.95	5.19	5.40	5.59	5.77	5.93
20	2.46	2.95	3.29	3.54	3.76	3.94	4.10	4.24	4.37	4.49	4.59
30	2.07	2.40	2.61	2.78	2.91	3.02	3.12	3.21	3.29	3.36	3.39
60	1.67	1.85	1.96	2.04	2.11	2.17	2.22	2.26	2.30	2.33	2.36
∞	1.00	1.00	1.00	1.00	1.00	1.00	1.00	1.00	1.00	1.00	1.00

Table H *Random Numbers*

10480	15011	01536	02011	81647	91646	69179	14194	62590	36207	20969	99570	91291	90700
22368	46573	25595	85393	30995	89198	27982	53402	93965	34095	52666	19174	39615	99505
24130	48360	22527	97265	76393	64809	15179	24830	49340	32081	30680	19655	63348	58629
42167	93093	06243	61680	07856	16376	39440	53537	71341	57004	00849	74917	97758	16379
37570	39975	81837	16656	06121	91782	60468	81305	49684	60672	14110	06927	01263	54613
77921	06907	11008	42751	27756	53498	18602	70659	90655	15053	21916	81825	44394	42880
99562	72905	56420	69994	98872	31016	71194	18738	44013	48840	63213	21069	10634	12952
96301	91977	05463	07972	18876	20922	94595	56869	69014	60045	18425	84903	42508	32307
89579	14342	63661	10281	17453	18103	57740	84378	25331	12566	58678	44947	05585	56941
85475	36857	43342	53988	53060	59533	38867	62300	08158	17983	16439	11458	18593	64952
28918	69578	88231	33276	70997	79936	56865	05859	90106	31595	01547	85590	91610	78188
63553	40961	48235	03427	49626	69445	18663	72695	52180	20847	12234	90511	33703	90322
09429	93969	52636	92737	88974	33488	36320	17617	30015	08272	84115	27156	30613	74952
10365	61129	87529	85689	48237	52267	67689	93394	01511	26358	85104	20285	29975	89868
07119	97336	71048	08178	77233	13916	47564	81056	97735	85977	29372	74461	28551	90707
51085	12765	51821	51259	77452	16308	60756	92144	49442	53900	70960	63990	75601	40719
02368	21382	52404	60268	89368	19885	55322	44819	01188	65255	64835	44919	05944	55157
01011	54092	33362	94904	31273	04146	18594	29852	71585	85030	51132	01915	92747	64951
52162	53916	46369	58586	23216	14513	83149	98736	23495	64350	94738	17752	35156	35749
07056	97628	33787	09998	42698	06691	76988	13602	51851	46104	88916	19509	25625	58104
48663	91245	85828	14346	09172	30168	90229	04734	59193	22178	30421	61666	99904	32812
54164	58492	22421	74103	47070	25306	76468	26384	58151	06646	21524	15227	96909	44592
32639	32363	05597	24200	13363	38005	94342	28728	35806	06912	17012	64161	18296	22851
29334	27001	87637	87308	58731	00256	45834	15398	46557	41135	10367	07684	36188	18510
02488	33062	28834	07351	19731	92420	60952	62180	50001	67658	32586	86679	50720	94953
81525	72295	04839	96423	24878	82651	66566	14778	76797	14780	13300	87074	79666	95725
29676	20591	68086	26432	46901	20849	89768	81536	86645	12659	92259	57102	80428	25280
00742	57392	39064	66432	84673	40027	32832	61362	98947	96067	64760	64584	96096	98253
05366	04213	25669	26422	44407	44048	37937	63904	45766	66134	75470	66520	34693	90449
91921	26418	64117	94305	26766	25940	39972	22209	71500	64568	91402	42416	07844	69618
00582	04711	87917	77341	42206	35126	74087	99547	81817	42607	43808	76655	62028	76630
00725	69884	62797	56170	86324	88072	76222	36086	84637	93161	76038	65855	77919	88006
69011	65797	95876	55293	18988	27354	26575	08625	40801	59920	29841	80150	12777	48501
25976	57948	29888	88604	67917	48708	18912	82271	65424	69774	33611	54262	85963	03547
09763	83473	73577	12908	30883	18317	28290	35797	05998	41688	34952	37888	38917	88050
91567	42595	27958	30134	04024	86385	29880	99730	55536	84855	29080	09250	79656	73211
17955	56349	90999	49127	20044	59931	06115	20542	18059	02008	73708	83517	36103	42791
46503	18584	18845	49618	02304	51038	20655	58727	28168	15475	56942	53389	20562	87338
92157	89634	94824	78171	84610	82834	09922	25417	44137	48413	25555	21246	35509	20468
14577	62765	35605	81263	39667	47358	56873	56307	61607	49518	89656	20103	77490	18062
98427	07523	33362	64270	01638	92477	66969	98420	04880	45585	46565	04102	46880	45709
34914	63976	88720	82765	34476	17032	87589	40836	32427	70002	70663	88863	77775	69348
70060	28277	39475	46473	23219	53416	94970	25832	69975	94884	19661	72828	00102	66794
53976	54914	06990	67245	68350	82948	11398	42878	80287	88267	47363	46634	06541	97809
76072	29515	40980	07391	58745	25774	22987	80059	39911	96189	41151	14222	60697	59583
90725	52210	83974	29992	65831	38857	50490	83765	55657	14361	31720	57375	56228	41546
64364	67412	33339	31926	14883	24413	59744	92351	97473	89286	35931	04110	23726	51900
08962	00358	31662	25388	61642	34072	81249	35648	56891	69352	48373	45578	78547	81788
95012	68379	93526	70765	10593	04542	76463	54328	02349	17247	28865	14777	62730	92277
15664	10493	20492	38391	91132	21999	59516	81652	27195	48223	46751	22923	32261	85653

Reprinted by permission from W. H. Beyer (ed.), *Handbook of Tables for Probability and Statistics* (Cleveland, Ohio: CRC Press, Inc., 1968). Copyright © 1968 by The Chemical Rubber Co., CRC Press, Inc. Table H is taken from Table XII.4 on pp. 480 and 481.

Table H *Random Numbers (Cont.)*

16408	81899	04153	53381	79401	21438	83035	92350	36693	31238	59649	91754	72772	02338
18629	81953	05520	91962	04739	13092	97662	24822	94730	06496	35090	04822	86772	98289
73115	35101	47498	87637	99016	71060	88824	71013	18735	20286	23153	72924	35165	43040
57491	16703	23167	49323	45021	33132	12544	41035	80780	45393	44812	12515	98931	91202
30405	83946	23792	14422	15059	45799	22716	19792	09983	74353	68668	30429	70735	25499
16631	35006	85900	98275	32388	52390	16815	69298	82732	38480	73817	32523	41961	44437
96773	20206	42559	78985	05300	22164	24369	54224	35083	19687	11052	91491	60383	19746
38935	64202	14349	82674	66523	44133	00697	35552	35970	19124	63318	29686	03387	59846
31624	76384	17403	53363	44167	64486	64758	75366	76554	31601	12614	33072	60332	92325
78919	19474	23632	27889	47914	02584	37680	20801	72152	39339	34806	08930	85001	87820
03931	33309	57047	74211	63445	17361	62825	39908	05607	91284	68833	25570	38818	46920
74426	33278	43972	10119	89917	15665	52872	73823	73144	88662	88970	74492	51805	99378
09066	00903	20795	95452	92648	45454	09552	88815	16553	51125	79375	97596	16296	66092
42238	12426	87025	14267	20979	04508	64535	31355	86064	29472	47689	05974	52468	16834
16153	08002	26504	41744	81959	65642	74240	56302	00033	67107	77510	70625	28725	34191
21457	40742	29820	96783	29400	21840	15035	34537	33310	06116	95240	15957	16572	06004
21581	57802	02050	89728	17937	37621	47075	42080	97403	48626	68995	43805	33386	21597
55612	78095	83197	33732	05810	24813	86902	60397	16489	03264	88525	42786	05269	92532
44657	66999	99324	51281	84463	60563	79312	93454	68876	25471	93911	25650	12682	73572
91340	84979	46949	81973	37949	61023	43997	15263	80644	43942	89203	71795	99533	50501
91227	21199	31935	27022	84067	05462	35216	14486	29891	68607	41867	14951	91696	85065
50001	38140	66321	19924	72163	09538	12151	06878	91903	18749	34405	56087	82790	70925
65390	05224	72958	28609	81406	39147	25549	48542	42627	45233	57202	94617	23772	07896
27504	96131	83944	41575	10573	08619	64482	73923	36152	05184	94142	25299	84387	34925
37169	94851	39117	89632	00959	16487	65536	49071	39782	17095	02330	74301	00275	48280
11508	70225	51111	38351	19444	66499	71945	05422	13442	78675	84081	66938	93654	59894
37449	30362	06694	54690	04052	53115	62757	95348	78662	11163	81651	50245	34971	52924
46515	70331	85922	38329	57015	15765	97161	17869	45349	61796	66345	81073	49106	79860
30986	81223	42416	58353	21532	30502	32305	86482	05174	07901	54339	58861	74818	46942
63798	64995	46583	09765	44160	78128	83991	42865	92520	83531	80377	35909	81250	54238
82486	84846	99254	67632	43218	50076	21361	64816	51202	88124	41870	52689	51275	83556
21885	32906	92431	09060	64297	51674	64126	62570	26123	05155	59194	52799	28225	85762
60336	98782	07408	53458	13564	59089	26445	29789	85205	41001	12535	12133	14645	23541
43937	46891	24010	25560	86355	33941	25786	54990	71899	15475	95434	98227	21824	19585
97656	63175	89303	16275	07100	92063	21942	18611	47348	20203	18534	03862	78095	50136
03299	01221	05418	38982	55758	92237	26759	86367	21216	98442	08303	56613	91511	75928
79626	06486	03574	17668	07785	76020	79924	25651	83325	88428	85076	72811	22717	50585
85636	68335	47539	03129	65651	11977	02510	26113	99447	68645	34327	15152	55230	93448
18039	14367	61337	06177	12143	46609	32989	74014	64708	00533	35598	58408	13261	47908
08362	15656	60627	36478	65648	16764	53412	09013	07832	41574	17639	82163	60859	75567
79556	29068	04142	16268	15387	12856	66227	38358	22478	73373	88732	09443	82558	05250
92608	82674	27072	32534	17075	27698	98204	63863	11951	34648	88022	56148	34925	57031
23982	25835	40055	67006	12293	02753	14827	22235	35071	99704	37543	11601	35503	85171
09915	96306	05908	97901	28395	14186	00821	80703	70426	75647	76310	88717	37890	40129
50937	33300	26695	62247	69927	76123	50842	43834	86654	70959	79725	93872	28117	19233
42488	78077	69882	61657	34136	79180	97526	43092	04098	73571	80799	76536	71255	64239
46764	86273	63003	93017	31204	36692	40202	35275	57306	55543	53203	18098	47625	88684
03237	45430	55417	63282	90816	17349	88298	90183	36600	78406	06216	95787	42579	90730
86591	81482	52667	61583	14972	90053	89534	76036	49199	43716	97548	04379	46370	28672
38534	01715	94964	87288	65680	43772	39560	12918	86537	62738	19636	51132	25739	56947

Appendix 2: The Family of Means

Although the arithmetic mean (discussed in Chapter 3) is by far the most commonly used average, some other means are occasionally used in the behavioral sciences. Here, the family of means is displayed for comparison.

The Arithmetic Mean ($\bar{X}$)

$$\bar{X} = \frac{\sum X}{N} = \frac{X_1 + X_2 + \cdots + X_N}{N}$$

Example:

What is the arithmetic mean of 5, 6, 9, and 11?

$$\bar{X} = \frac{5 + 6 + 9 + 11}{4} = \frac{31}{4} = 7.75$$

The Harmonic Mean (*HM*)

$$HM = \frac{N}{\sum \left(\frac{1}{X}\right)} = \frac{N}{\frac{1}{X_1} + \frac{1}{X_2} + \cdots + \frac{1}{X_N}}$$

Example:

A man runs a mile at 11 mph and another mile at 9 mph. He then walks a mile at 6 mph and another at 5 mph. What is his average speed? The harmonic mean is appropriate for the rates here because the *time* spent moving at each rate is variable, even though the distance at each rate is constant. (Had the time at each rate been constant and the distances variable, the arithmetic mean would have been appropriate.)

$$HM = \frac{4}{\frac{1}{11} + \frac{1}{9} + \frac{1}{6} + \frac{1}{5}} = \frac{4}{(0.5687)} = 7.03$$

The Geometric Mean (*GM*)

$$GM = \sqrt[N]{(X_1)(X_2) \cdots (X_N)}$$

GM cannot be computed when any *X* values are zero or negative. The geometric mean is sometimes used when the observations are thought to represent members of a *geometric series*. (A geometric series is a sequence of numbers in which each number after the first is the product of the preceding term and a constant; for instance, in the geometric series 1, 3, 9, 27, the constant is 3. In the geometric series 10, 5, 2.5, 1.25, the constant is 0.5.) Sometimes the sizes of populations, from year to year, form a geometric series; geometric series also appear in studies of sensory psychology, where stimulus magnitudes follow such a progression.

(The numbers do not have to follow such a progression in order to compute GM, however.)

Example:

What is the GM of 5, 6, 9, 11?

$$GM = \sqrt[4]{(5)(6)(9)(11)} = \sqrt[4]{2970} = 7.38$$

Extracting the Nth root of a number can be difficult if N is greater than 2. The simplest method is to use logarithms. For example:

$$\log_{10} 2970 = 3.472756$$

$$\log_{10} \sqrt[4]{2970} = \frac{3.472756}{4} = 0.868189$$

$$\sqrt[4]{2970} = \text{Antilog of } 0.868189 = 7.38225 = 7.38$$

In general, $HM \leq GM \leq \bar{X}$. The equality sign holds only if all values in the group are equal. Here, for the same data, $HM = 7.03$, $GM = 7.38$, and $\bar{X} = 7.75$.

The Quadratic Mean (QM)

$$QM = \sqrt{\frac{\sum X^2}{N}}$$

The quadratic mean is used when interest is in the absolute value of the numbers to be averaged and the signs are to be disregarded. We have used the quadratic mean already: the standard deviation is the quadratic mean of the deviation scores about the mean.

Example:

What is the QM of 5, 6, 9, 11?

$$QM = \sqrt{\frac{(5)^2 + (6)^2 + (9)^2 + (11)^2}{4}} = \sqrt{65.75} = 8.11$$

The value 8.11 is also the QM of -5, 6, -9, and -11 as well as the QM of 5, -6, -9, 11, and so on.

Appendix 3: Computational Procedures for Percentiles and Percentile Ranks

When the data have been grouped, percentiles and percentile ranks may be estimated either from a cumulative graph or a cumulative table. In either case, the resulting statistics are estimates; the computation will be the more precise estimate.

The computations require a cumulative frequency table that includes a frequency column. Intervals are designated by their *true limits*. In general, the more intervals that are used, the more accurate the estimate. Here, for illustrative purposes, a table with only five intervals is used:

INTERVAL	f	CUM. FREQ.
149.5–174.5	5	50
124.5–149.5	8	45
99.5–124.5	22	37
74.5–99.5	9	15
49.5–74.5	6	6
	$N = 50$	

Values in the cumulative frequency column show the number of observations falling in or below each interval. The lowest interval, for instance, has 6 observations in it and none below it; hence its cumulative frequency of 6. The next higher interval has 9 observations in it and 6 below, for a cumulative frequency of $9 + 6 = 15$. Similarly, the next higher interval has $22 + 15 = 37$; continue in this manner until you reach the cumulative frequency for the uppermost interval (which should equal N).

Before proceeding, one must determine the interval size i. This is the difference between the upper true limit of an interval and the lower true limit, for example, $174.5 - 149.5 = 25$. In this table $i = 25$ for all intervals.

What is the percentile rank of 140? That is, what percentage of the distribution had scores lower than 140? The answer is given by the expression

$$\text{Percentile rank of } 140 = \frac{\text{Cum. freq. of } 140}{N}\,(100).$$

The cumulative frequency of 140 is the number of scores *below* the value 140 in the above distribution. In a grouped frequency distribution this number must be found by interpolation. Notice that the value 140 lies in the interval 124.5–149.5. In order to proceed, one must assume that the eight scores in that interval ($f = 8$ in frequency column) are distributed evenly across the width of that interval ($i = 25$ scale units), as follows:

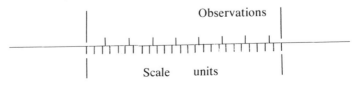

Although the observations probably were not actually distributed this way, it is the assumption we must make once they have been grouped. One must then interpolate to find the number (of those eight) that fall below 140.

$$\text{Cum. freq. of } 140 = \begin{matrix} \text{Cum. freq. below} \\ \text{interval} \\ \text{containing } 140 \end{matrix} + \cfrac{\begin{matrix} \text{Distance between} \\ \text{140 and lower real} \\ \text{limit of interval} \\ \text{containing } 140 \end{matrix}}{\begin{matrix} \text{Total distance} \\ \text{covered by} \\ \text{interval } i \end{matrix}} \begin{pmatrix} \text{Frequency in} \\ \text{interval} \\ \text{containing } 140 \end{pmatrix}$$

For the values in this table,

$$\text{Cum. freq. of } 140 = 37 + \frac{(140 - 124.5)}{25}(8) = 37 + 4.96 = 41.96.$$

This cumulative frequency is then inserted into the formula for percentile rank:

$$\text{Percentile rank of } 140 = \frac{41.96}{50}(100) = 83.92.$$

What score value in the preceding table has 60% of the observations below it? That is, what is the 60th percentile of this distribution? To answer that question, we must find the cumulative frequency of the 60th percentile — the number of scores below the 60th percentile — and then find the value that corresponds to that cumulative frequency.

The cumulative frequency of a given percentile is easy to find. For the 60th percentile, one finds 60% of the total N. Since $N = 50$, 60% of N equals 30. Thus, the cumulative frequency of the 60th percentile is 30. Finding the associated scale value requires interpolation. First find the interval containing the designated cumulative frequency; here, that interval is 99.5–124.5, which contains the 60th percentile. Then,

$$\text{60th percentile} = \begin{matrix} \text{Lower real limit of} \\ \text{interval containing} \\ \text{60th percentile} \end{matrix} + \cfrac{\begin{matrix} \text{Number of scores} \\ \text{below 60th percentile} \\ \text{(cum. freq. of 60th} \\ \text{percentile)} \end{matrix} - \begin{matrix} \text{Number of} \\ \text{scores} \\ \text{needed from} \\ \text{interval to} \\ \text{reach 60th} \\ \text{percentile} \end{matrix}}{\begin{matrix} \text{Number of scores in interval} \\ \text{containing 60th percentile} \end{matrix}} \quad (i)$$

The numerator of the proportion (at the right) is perhaps the most troublesome part of this expression. There are fifteen scores below the interval containing the 60th percentile (this is the cumulative frequency of the interval 74.5–99.5). Since the 60th percentile has thirty scores below it, one needs fifteen more of the scores in the interval 99.5–124.5 to reach the 60th percentile. For the values in this table,

$$\text{60th percentile} = 99.5 + \frac{(30 - 15)}{22}(25) = 99.5 + \frac{15}{22}(25) = 116.54.$$

Appendix 4: Answers to Problems

Listed below are answers to problems for each chapter that have numerical solutions. In most cases, verbal interpretation of these answers is also called for.

Chapter 1

1. a. ratio b. ordinal c. absolute d. nominal e. ratio f. interval g. ordinal h. ratio
2. a. absolute b. ordinal c. absolute
3. a. continuous b. discrete (one cent) c. discrete (one bar press) d. continuous e. continuous
4. a. discrete (1 receptor cell)
   b. continuous (9.5–10.5°)
   c. discrete (1 position)
   d. continuous (52' 0.5"–52' 1.5")
9. a. absolute, discrete
   b. absolute, discrete
   c. ratio, discrete (if one considers "amount of money" as the variable; it is also possible to consider the "number of cents" per year, more accurately described as absolute level measurement)
11. a. 52.5–53.5 b. 3.65–3.75 c. 49.15–49.25 d. 0.2555–0.2565 e. 4.5–5.5
12. a. 47.9 b. 132.8 c. 968.8 d. 0.0 e. 0.2 f. 98.0 g. 2.4 h. 77.5
13. a. 98.906 b. 14.990 c. 175.001 d. 1078.556 e. 232.050 f. 984.656
14. a. 159.5–160.5   b. 421.165–421.175   c. 1.0025–1.0035   d. 146.5–147.5 min.
15. a. either $105,000–115,000 or $107,250–112,500
    b. 7,550–7,650 barrels
    c. 65,500–66,500 people
    d. 11,500,000–12,500,000 light years

Chapter 2

1.

CLASS	f
Senior	5
Junior	8
Sophomore	12
Freshman	5

3.

TRUE LIMITS	MIDPOINT	INTERVAL SIZE
a. 45.5–50.5	48.0	5.0
b. 460.5–500.5	480.5	40.0
c. 0.005–0.105	0.055	0.100
d. 1.005–1.415	1.21	0.41
e. 200.5–300.5	250.5	100.0

4.

MIDPOINT	INTERVAL SIZE
a. 18.00	5.0
b. 150.5	100.0
c. 0.85	0.4
d. 605.5	30.0
e. 8,250.5	2,500.0
f. 0.175	.02

5. There is no single correct answer for these; the answers given represent good choices, but other values may also be appropriate.

INTERVAL SIZE	LOWEST INTERVAL	TRUE LIMITS
a. 4.00	41–44	40.5–44.5
b. 3.00	15–17	14.5–17.5
c. 50.00	251–300	250.5–300.5
d. 10.00	41–50	40.5–50.5
e. 3.00	45–47	44.5–47.5

6. a. 0.3262   c. 0.2238   e. 0.3571
   b. 0.0357   d. 0.0690   f. 0.1595

7.

LIMITS	MIDPOINT	FREQUENCY f	cp
41–50	45.5	4	1.0000
31–40	35.5	11	0.9298
21–30	25.5	26	0.7368
11–20	15.5	15	0.2807
1–10	5.5	1	0.0175

8.

LIMITS	TRUE LIMITS	MIDPOINT	f	cp
160–166	159.5–166.5	163	1	1.0000
153–159	152.5–159.5	156	10	0.9950
146–152	145.5–152.5	149	32	0.9450
139–145	138.5–145.5	142	50	0.7850
132–138	131.5–138.5	135	33	0.5350
125–131	124.5–131.5	128	28	0.3700
118–124	117.5–124.5	121	20	0.2300
111–117	110.5–117.5	114	15	0.1300
104–110	103.5–110.5	107	8	0.0550
97–103	96.5–103.5	100	3	0.0150

9.

LIMITS	MIDPOINT	f	REL. f	cp
82–90	86	7	0.00875	1.00000
73–81	77	31	0.03875	0.99125
64–72	68	42	0.05250	0.95250
55–63	59	65	0.08125	0.90000
46–54	50	117	0.14625	0.81875
37–45	41	132	0.16500	0.67250
28–36	32	150	0.18750	0.50750
19–27	23	149	0.18625	0.32000
10–18	14	72	0.09000	0.13375
1–9	5	35	0.04375	0.04375

10. The distribution in problem 8 is negatively skewed; the distribution in problem 9 is positively skewed. Graphs for these distributions should have points located above interval midpoints.

11. and

12. Frequency distributions and relative frequency distributions should be graphed from the tables below. All graphed points should be located over interval midpoints.

SCORES	MIDPOINT	BUNSEN'S CLASS		ERLENMEYER'S CLASS	
		f	REL. f	f	REL. f
91–100	95.5	6	0.08571	6	0.12000
81–90	85.5	7	0.10000	13	0.26000
71–80	75.5	11	0.15714	12	0.24000
61–70	65.5	11	0.15714	8	0.16000
51–60	55.5	13	0.18571	3	0.06000
41–50	45.5	15	0.21429	5	0.10000
31–40	35.5	7	0.10000	3	0.06000

Bunsen's class is positively skewed, Erlenmeyer's is negatively skewed, suggesting that Bunsen's class found the exam more difficult.

13. a. With eleven intervals, the graph should be made from the following table:

INTERVAL	f
1,120–1,159	2
1,080–1,119	2
1,040–1,079	3
1,000–1,039	6
960–999	14
920–959	31
880–919	12
840–879	6
800–839	8
760–799	11
720–759	5

b. With five intervals, the graph should be made from the following table:

INTERVAL	f
1,100–1,199	3
1,000–1,099	10
900–999	55
800–899	16
700–799	16

c. Using 11 intervals shows the distribution to be slightly bimodal, whereas using 5 intervals shows a unimodal distribution.

14. a. 36.9 b. 29.9 c. 58.0 d. 74.6 e. 99.95 f. 26.8

Chapter 3

1. a. 1,074 b. 542 c. 422 d. 117,072
2. a. 413 b. 44,440 c. 835 d. 27,507 e. 28,376 f. 114,244
3. mean = 107.4; median = 106.5
4. mean = 23.2; median = 22.5; modes = 21 and 26
5. a. mode c. median e. median
   b. mean d. mean f. mean

6.

	MEAN	MEDIAN	MODE
a.	23.0	23.0	24*
b.	66.4	62.0	61*
c.	69.6	72.5	75

*not enough observations in the group for the mode to have meaning

7. Throwing out the highest and lowest values in each group leaves each median's value unchanged.
   a. mean = 22.0 (−1.0 change)   b. mean = 66.6 (+0.2 change)
   c. mean = 70.5 (+0.96 change)

8. a. 32.1   b. 455.1

9. mean = 222.5; median = 213.0

10. problem 8 mean = 134.6; problem 9 mean = 37.9

11. mean = 101.4; median = 101.3. In a symmetric distribution, either value appropriately represents "typical."

12. mean = 37.9; median = 36.1; skew is slightly positive.

13. mean = 1029.1, computed either way.

14. a. grouped data $\bar{X}$ = 72.3   b. raw data $\bar{X}$ = 71.5   c. The raw data mean is the exact value; the grouped data mean is an estimate that may differ from the exact value because some information is lost in the grouping process.

15. median = 75.5; grouped data mean = 72.3. Because the mean is less than the median, we decide that skew is negative.

Chapter 4

1. a. $\bar{X}_1$ = 69.5, $range_1$ = 64; $\bar{X}_2$ = 69.9, range = 64
   b. s_1 = 19.23, s_2 = 12.35

2. $\bar{X}$ = 501.4, s = 103.5

3.

	GROUP I	GROUP II
a.	36.8	37.1
b.	39.0	33.0
c.	1,555.67	1,614.92
d.	129.64	134.58
e.	11.4	11.6

4. $\bar{X}$ = 10.8, s = 1.96

5. If s = 0, all values in the distribution are equal to each other. If the computed SS is negative, a computational error has been made. A negative SS is not possible.

6. a. standard score (or z-score)   b. variation (or SS, or sum of squares)   c. range   d. interquartile range   e. standard deviation   f. mean deviation   g. variance

7. with N = 16, $\bar{X}$ = 220.1, s = 718.52; with N = 15, $\bar{X}$ = 34.7, s = 27.8

8. problem 8 distribution: s = 13.4; interquartile range = 19.0
   problem 9 distribution: s = 18.2; interquartile range = 26.1

9.

	N	SS	MS	s
a.	10	2250	225	15
b.	100	14,400	144	12
c.	15	105	7	2.65
d.	36	176,400	4,900	70

10. a. -0.27 b. 0 (to second decimal place) c. -1.11 d. $+1.63$ e. -3.53

11. a. John: Exam 1 $z = +0.77$, Exam 2 $z = +0.78$; Exam 2 slightly better.
    b. Mary: Exam 1 $z = +1.15$, Exam 2 $z = +0.67$; Mary's mean z of $+0.908$ is better than John's mean z of $+0.775$.

12. John: Exam 1 $T = 57.7$; Exam 2 $T = 57.8$
    Mary: Exam 1 $T = 61.5$; Exam 2 $T = 56.7$

13. a. $+0.22$ c. -2.71 e. -0.67 g. $+2.33$
    b. $+1.04$ d. -1.89 f. -0.11 h. $+2.89$

14. a. 522 c. 229 e. 433 g. 733
    b. 604 d. 311 f. 489 h. 789

15. a. 76.67   d. 95.33
    b. 133.33   e. 110.67
    c. 129.33   f. 82.67

Chapter 5

2. a. $z_{750} = +2.50$, $z_{445} = -0.55$, $z_{620} = +1.20$
    b. $z_{180} = +1.20$, $z_{162} = +0.48$, $z_{115} = -1.40$
    c. $z_{2.532} = +2.20$, $z_{2.504} = -0.60$, $z_{2.499} = -1.10$

3. a. non-normal (positive skew) c. normal
    b. normal   d. non-normal (positive skew)

4.

	AREA ABOVE	AREA BELOW
a.	0.0735	0.9265
b.	0.3121	0.6879
c.	0.5040	0.4960
d.	0.0028	0.9972
e.	0.8389	0.1611

5. a. 76.1 b. 95.6 c. 99.9 d. 1.1 e. 23.9 f. 7.6

6. a. 60 b. 71.5 c. 76.3 d. 48.5

7. a. 97.7 b. 112.2 c. 140.0 d. 162.3 e. 182.3

8. a. 10.39 b. 11.12 c. 7.21 d. 26.36

9. a. 0.1336 b. 0.0796 c. 0.0080

10. a. 0.8664 b. 0.9876 c. 0.9996 d. 0.99994

11. a. 0.1056 b. 0.5328 c. 0.3085 d. 0.6678 e. 0.4013

12. a. 210.1 b. 166.3 c. 195.0 d. 148.7

13. a. 77.3   b. 77.3   c. 54.7   d. 70.7   e. 80.4
14. Using true limits of measurements and interpolated areas:
    a. 0.3638   b. 0.2417   c. 0.1066
15. a. 617.5 or better   b. yes (minimum score = 782.6)   c. 524.7   d. 0.6826
    e. 0.0062

Chapter 6

1. a. positive   b. negative   c. positive   d. negative   e. positive
2. b. $r = +0.88$
3. Scatterplot should show these data to have a moderately high positive correlation.
4. a. $r = -0.23$ b. $r = +0.84$ c. $r = +0.99$ d. $r = -0.83$
5. $r = +0.807$ (based on all 25 pairs of observations)
6. rho $= +0.95$
7. $r = +0.182$ (based on the indicated 8 pairs of observations)
8. d. Pair c shows the least degree of correlation.
9. a. $r = +0.676$ b. $r = +0.903$ c. $r = -0.275$
10. a. Both scatterplots show a positive relationship; the correlation between annual incomes and weekly entertainment expenses is higher than the correlation between annual income and automobile values.
    b. Between annual incomes and weekly entertainment expenses: $r = +0.984$. Between annual incomes and automobile values: $r = +0.634$.
11. Between GPAs and algebra grades: rho $= -0.600$
    Between GPAs and art grades: rho $= +0.928$
12. Between judges 1 and 2, rho $= +0.90$
    Between judges 1 and 3, rho $= +0.38$
    Between judges 2 and 3, rho $= +0.14$
13. $r_{pb} = -0.126$
14. For item #7, $r_{pb} = +0.726$; for #9, $r_{pb} = +0.033$; for #14, $r_{pb} = -0.420$

Chapter 7

1. a and b are linear

2.

SLOPE	INTERCEPT
a. +3	+2
b. −1	+4.5
c. +3	−2
d. +1	+1
e. +0.5	+4

3.

	Y' SLOPE	X' SLOPE
a.	$+10.46$	$+0.0239$
b.	$+1.96$	$+0.49$
c.	-0.00050	-1216.69
d.	-0.00948	-4.22

4. a. Correlation between X and Y is perfect (r = either $+1.00$ or -1.00).
   b. Correlation between X and Y is 0.
   c. Correlation between X and Y is 0.
   d. Correlation between X and Y is perfect.

6. a. 66.7g b. 32.0g c. 48.0g d. 324.0g e. 256.8g

7. Slope = 0.0916, Intercept = -5.608
   a. 3.55 b. 5.84 c. 5.29 d. 2.54

8. Slope = 10.128, Intercept = 64.564
   a. 105.1 b. 85.8 c. 102.0 d. 98.0

9. $Y' = 475$ (satisfactory) b. best estimate = $\bar{Y} = 500$ c. $X' = 185$

10. a. $Y' = 3.02$ b. $Y' = 2.63$ c. $Y' = 3.49$ d. $Y' = 2.83$
    e. Using H.S. GPAs as predictors, $s_{est_y} = 0.373$; using entrance exam scores as predictors, $s_{est_y} = 0.406$.

11. a. 66.04 in. b. 71.17 in. c. 62.97 in. d. 139.77 lb e. 189.59 lb
    f. 117.63 lb

12. a. 3.53 b. 1.16 c. 21.79 d. 0.0149

13. $r^2 = 0.567$; $1 - r^2 = 0.433$

14. a. 0.5625 b. 0.1024 c. 0.1849 d. 0.9604

15. a. $r = 0.894$ b. $r = 0.707$ c. $r = 0.949$ d. $r = 0.316$

Chapter 8

3. $12! = 479,001,600$ possible orders; $15! = 1,307,674,368,000$ possible orders.

4. a. 15 b. 0.0667 c. 0.3333

5. 252 teams of 5 are possible from a group of 10.
   56 teams of 5 are possible from a group of 8.
   p(5 tallest men are selected) = $1/252 = .00397$

6. a. p(Protestant) = .3636; b. p(Republican) = .3788
   c. p(Catholic or Protestant) = 0.7273; d. p(non-Catholic) = 0.6364
   e. p(Catholic and Republican) = 0.0909

7. a. p(Republican and Republican) = 0.1435
   b. p(non-Catholic and non-Catholic) = 0.4050
   c. p[(Protestant and Jewish) or (Jewish and Protestant)] = 0.0771
   d. p[(non-Protestant and non-Republican) and (non-Protestant and non-Republican)] = 0.1931

8. a. p(Female and Democrat) = 0.2727
   b. p(Male and "Other") = 0.0364
   c. p(Male Democrat) = 0.4545

    d. $p(\text{Democrat}|\text{Female}) = 0.4545$

    e. 0.3576

10. a. 0.0625 b. 0.3750 c. 0.9375

11. a. 0.0001 b. 0.3669 c. 0.9536

12. a. 0.5000 b. 0.0215 c. 0.2000 d. 0.3830 e. 0.0026 f. 0.0228

14. a. 0.1788 b. 0.5020 c. 0.2266 d. 0.0104 e. 0.6704

15. a. 0.00052 b. 0.95492

Chapter 9

5. a. $p(z > +0.50) = 0.3085$ b. $p(z > 3.16) = 0.0008$
   c. $p(z > 3.00) = 0.0013$ d. $p(z > +2.00 \text{ or } z < -2.00) = 0.0456$

6. a. $p(z < -2.14) = 0.0162$ b. $p(-1.43 < z < 2.86) = 0.9215$
   c. $p(z > +2.86) = 0.0021$

7. a. 0.1587 b. 0.0062 c. 0.3085

9. a. 0.0041 b. 0.0467 c. 0.8208

10. a. $p(z > 3.06) = 0.0011$ b. $p(z < -2.04) = 0.0207$
    c. $p(z < -1.02) = 0.1539$

Chapter 10

1. a. $s = 1.48$ $\hat{s} = 1.71$
   b. $s = 15.28$ $\hat{s} = 17.08$
   c. $s = 34.95$ $\hat{s} = 38.29$
   d. $s = 2.52$ $\hat{s} = 2.66$

3. a. 8.10 b. 4.16 c. 95.15 d. 0.97 e. 0.306 f. 40.07

4. z-observed = 3.33. Reject H_0 at 0.001 alpha level.

6. t-observed = 1.43, $df = 24$. Do not reject H_0 at 0.05 alpha level.

7. t-observed = -2.50, $df = 99$. Reject H_0 at 0.02 alpha level.

8. t-observed = -0.87, $df = 15$. Do not reject H_0.

9. t-observed = -53.56, $df = 949$. Reject H_0 with alpha lower than 0.001.

10. t-observed = 2.03, $df = 24$. Reject H_0 at 0.05 alpha level (one-tailed test).

12. t-observed = 2.09, $df = 6$. Do not reject H_0 at 0.05 alpha level.

13. a. One-tailed test is appropriate because carrying the passenger could only
       lower fuel efficiency.
    b. t-observed = 2.87, $df = 13$. Reject H_0 at 0.01 alpha level (one-tailed test).

14. a. t-observed = -1.50, $df = 8$. Do not reject H_0 at 0.10 alpha level.
    b. t-observed = -5.00, $df = 99$. Reject H_0 at 0.001 alpha level.

15. t-observed = -0.03, $df = 27$. Do not reject H_0 at 0.05 level.

Chapter 11

3. a. $p(z > +1.00) = 0.1587$ b. $p(z > +2.50) = 0.0062$
   c. $p(z > +2.40) = 0.0082$ d. $p(z > -3.82) = $ about 0.9999

4. a. 53897.4 b. 114.92 c. 0.716 d. 1390.7

5. b. t-observed $= 0.85$. Do not reject H_0.
6. t-observed $= 1.82$, $df = 26$. Reject H_0 at alpha $= 0.05$ level in the one-tailed test.
7. t-observed $= 3.53$, $df = 28$. Reject H_0 at 0.01 level.
9. t-observed $= -4.18$, $df = 33$. Reject H_0 at 0.001 level.
10. t-observed $= -3.30$, $df = 11$. Reject H_0 at 0.01 level.
11. t-observed $= 0.098$, $df = 7$. Do not reject H_0.
12. a. independent samples t-observed $= 1.99$, $df = 8$. Do not reject H_0 at 0.05 level.
    b. related samples t-observed $= 3.38$, $df = 4$. Reject H_0 at 0.05 level.
13. t-observed $= 1.20$, $df = 18$. Do not reject H_0 at 0.05 level.
14. t-observed $= -2.78$, $df = 9$. Reject H_0 at 0.05 level.

Chapter 12

3. a. est $\mu = 304.50$ b. est $\sigma = 218.80$ c. est $P = 0.20$
4. 95 percent confidence interval for $\mu = 202.10$–406.90.
   99 percent confidence interval for $\mu = 164.53$–444.47.
5. a. 226.7   b. 68.9   c. 0.360
6. a. 208.51–244.85   b. 203.09–250.27
7. a. 4.50   b. 44.61   c. 0.612
8. a. 90 percent confidence interval for $\mu = 496.71$–503.29.
   b. 95 percent confidence interval for $\mu = 496.08$–503.92.
   c. 99 percent confidence interval for $\mu = 494.85$–505.15.
   The lower the level of confidence, the narrower the interval.
9. a. 90 percent confidence interval for $\mu = 431.56$–568.44.
   b. 95 percent confidence interval for $\mu = 417.44$–582.56.
   c. 99 percent confidence interval for $\mu = 388.12$–611.88.
   The larger the sample size, the narrower the confidence interval.
10. The error is constructing a confidence interval that does not include the parameter value; the probability of doing so is 1.0 minus the confidence level. At the 95 percent confidence level, for instance, the probability of making such an error is 0.05.

Chapter 13

2. χ^2-observed $= 6.32$, $df = 5$. Do not reject H_0.
3. a. With Yates' correction, χ^2-observed $= 4.08$, $df = 1$. Reject H_0 at 0.05 level.
   b. $p(r \geqslant 10) = 0.0193$; for two-tailed test this is doubled, for an alpha level of 0.0386.
4. χ^2-observed $= 137.87$, $df = 3$. Reject H_0 at 0.01 level.
5. χ^2-observed $=$ about 10.40, $df = 5$ (with normal curve divided into eight intervals). Do not reject H_0 (thus, decision is that curve is normal).
6. χ^2-observed $= 9.70$, $df = 4$. Reject H_0 at 0.05 level.
7. χ^2-observed $= 41.00$, $df = 1$. Reject H_0 at 0.01 level.

8. χ^2-observed = 8.33, df = 2. Reject H_0 at alpha level of 0.02.
9. a. With sign test, difference between groups is not significant at the 0.10 level (two-tailed test).
   b. W = 5. Reject H_0 at the 0.02 level (two-tailed test).
   d. These data would not be appropriate for the t-test because rating scores are probably highly negatively skewed, and they are not interval- or ratio-level data.
11. W = 3. Reject H_0 at the 0.01 level (two-tailed test).
12. R_1 = 131. Reject H_0 at 0.002 level.
13. With sign test, obtained alpha level (two-tailed) is 0.146.
14. W = 8.5. Reject H_0 at 0.02 level.
15. R_1 = 157.5. Reject H_0 at 0.002 level.

Chapter 14

1. a. 70 b. 72 c. 106 d. 351 e. 197 f. 401 g. 53,361 h. 25,627
2. Summary of Analysis

SOURCE OF VARIATION	df	SS	MS	F
Deprivation	3	5192.1	1730.7	21.88
Within Groups	20	1581.8	79.1	
Total	23	6773.9		

Reject H_0 at 0.001 level.

4. F-observed < 1, df = 4, 20. Do not reject H_0.
5. F-observed = 16.44, df = 2, 18. Reject H_0 at 0.001 level.
6. For data in problem 2, F_{max} = 3.36. This does not exceed the tabled critical value (13.7), so one may assume homogeneity of variances. For problem 4, F_{max} = 1.81. Assume homogeneity of variances. For problem 5, F_{max} = 2.37. Assume homogeneity.
7. F-observed = 4.98, df = 2, 12. Reject H_0 at alpha level of 0.05.
8. F-observed = 12.27, df = 2, 27. Reject H_0 at 0.001 level.
9. F-observed = 2.39, df = 3, 16. Do not reject H_0.
10. For ANOVA, F-observed = 38.59, df = 1, 18. Reject H_0 at 0.001 level. For t-test, t-observed = 6.2118, df = 18. Reject H_0 at 0.001 level. Note that $t^2 = F$.

Chapter 15

1. b. *Summary of Analysis*

SOURCE OF VARIATION	*df*	*SS*	*MS*	*F*
DRUG 1	2	74.7	37.3	<1
DRUG 2	2	168.0	84.0	<1
INTERACTION	4	8925.3	2231.3	23.17
Within Group	27	2600.0	96.3	
TOTAL	35	11768.0		

For both main effects, do not reject H_0.
For the interaction effect, reject H_0 at 0.001 level.

3. a.

SOURCE OF VARIATION	*df*	*SS*	*MS*	*F*
Factor A	3	3,413	1137.7	11.79
Factor B	2	758	379	3.93
Interaction	6	2,914	485.67	5.03
Within Groups	108	10,421	96.49	
Total	119	17,506		

Reject H_0 for all three tests; the main effect associated with Factor A and the interaction effect are significant at the 0.001 level, and the main effect associated with Factor B is significant at the 0.05 level.

b.

SOURCE OF VARIATION	*df*	*SS*	*MS*	*F*
Factor A	5	8,765	1,753	1.52
Factor B	2	11,432	5,716	4.97
Interaction	10	610	61	<1
Within Groups	126	144,900	1,150	
Total	143	165,707		

The main effect associated with factor B is significant at the 0.01 level; other effects are not significant.

4. Factor A main effect F-observed = 2.07, df = 2, 27. Do not reject H_0.
Factor B main effect, F-observed = 29.26, df = 2, 27. Reject H_0 at 0.001 level.
Interaction effect F-observed < 1. Do not reject H_0.

5. a. Factor 1 main effect, F-observed = 4.94, df = 2, 12. Reject H_0 at 0.05 level.
Factor 2 main effect, F-observed = 4.34, df = 1, 12 (not significant).
Interaction effect, F-observed < 1 (not significant). (Interaction effect is the same for all cases.)

b. Factor 1 main effect, F-observed = 23.78, df = 2, 2 (not significant).
Factor 2 main effect, F-observed = 20.91, df = 1, 2 (not significant).

   c. Factor 1 main effect, F-observed $= 23.78$, $df = 2, 2$ (not significant).

      Factor 2 main effect, F-observed $= 4.34$, $df = 1, 12$ (not significant).

6. Factor A main effect F-observed $= 1.42$. Do not reject H_0.

  Factor B main effect F-observed < 1, $df = 1, 12$. Do not reject H_0.

  Interaction effect F-observed $= 91.09$, $df = 1, 12$. Reject H_0 at 0.001 level.

7. For all cases, $k = 4$ and $n - 1 = 9$.

   a. $F_{max} = 2.84$. Assume homogeneity.

   b. $F_{max} = 23.23$. Do not assume homogeneity.

   c. $F_{max} = 4.13$. Assume homogeneity.

9. b. Method main effect, F-observed $= 102.61$, $df = 2, 54$. Reject H_0 at 0.001 level.

    Sex main effect, F-observed $= <1$ (not significant).

    Interaction effect, F-observed $= 67.89$, $df = 2, 54$. Reject H_0 at 0.001 level.

Glossary

This glossary includes a short explanation of 241 concepts covered in the text. For a more thorough treatment see the page numbers listed in parentheses at the end of each entry; in the text most of these terms appear in **boldface type** when introduced.

Absolute Scale: Measurements consisting of frequency counts; the level of measurement achieved when the variable of interest is the number of times each observation category occurs. Observing that twenty-seven anchovy pizzas were ordered at a certain restaurant is absolute-scale measurement; observing that a certain pizza is an anchovy pizza is nominal-scale measurement. (27)

Allowance Factor: Used in constructing confidence intervals, it is the distance (on the measurement scale) between the sample statistic and the limits of the interval. One both adds and subtracts the allowance factor to find, respectively, the upper and lower limits of the confidence interval. (329)

Alpha Level: The probability of rejecting a true null hypothesis, that is, the probability of making a Type I error. The value of alpha is chosen by the statistician, usually 0.05 or less. (277)

Alternative Hypothesis: In hypothesis testing, the state of reality that exists if the null hypothesis is not true. Usually, the alternative hypothesis does not specify a particular sampling distribution, but simply that the null hypothesis distribution is not the source of the observed statistic. (275)

Analysis of Variance: A group of hypothesis tests for evaluating experiments with two or more treatment groups. Differences among group means are used to compute a variance; this is compared to a variance derived from differences within groups (the comparison is done by making an F-ratio). If the between-groups variance is large relative to the within-groups variance (i.e., if the F-ratio is large) the conclusion is that at least one significant difference exists among group means. (373)

ANOVA: abbreviation for **Analysis of Variance**

A Priori Probability: The probability of a future event, calculated from prior knowledge of the number of outcomes possible and their relative frequencies. (217)

Arthmetic Average: See **Arithmetic Mean.**

Arithmetic Mean: Also called the **Arithmetic Average,** it is the most commonly used measure of central tendency. The arithmetic mean equals the sum of the values divided by the number of values; it requires at least interval-scale data. (80)

Asymptotic Relative Efficiency: The relative efficiency of one hypothesis test to another, as one approaches infinitely large sample sizes and a vanishingly small difference between means of the H_0 and H_1 distributions (see **Relative Efficiency**). It is used to compare nonparametric tests with their parametric counterparts. (365)

Averages: See **Measures of Central Tendency.**

Axes: Reference lines that delineate the two (or sometimes more) dimensions of a graph (6). See **Horizontal Axis** and **Vertical Axis.**

Bar Graph: Graphical representation of a frequency-distribution table in which each measurement category is represented by a bar that extends to the appropriate distance in the frequency dimension. See also **Histogram.** (49)

Behavioral Variable: A variable that consists of an overt, observable response by a human or animal subject; also called a **Performance Variable.** Any activity of the subject that can be observed and objectively recorded may be described as a behavioral variable. (21)

Bernoulli Experiment: A simple experiment (process) that has only two possible outcomes, such as flipping a coin, picking a person at random and determining the person's sex, and so on. (234)

Biased Sample: A sample selected so that some members of the population are more likely than others to be picked for sample membership. If one wishes to make generalizations about the population based on sample observations, it is usually desirable to avoid biased samples. (251)

Bimodal Distribution: A frequency distribution with two modes. (77)

Binomial Distribution: The set of probabilities for all possible values of a **Binomial Variable,** for a given value of p (probability of a Bernoulli success) and N (number of Bernoulli trials in the **Binomial Experiment**). The sum of these probabilities must be 1.0. (237)

Binomial Experiment: An experiment (process) consisting of a fixed number of identical and independent Bernoulli experiments. A **Bernoulli Experiment** might, for instance, consist of flipping a coin; the probability of a success might be taken as the probability of "Heads," so that $p = 0.50$. Flipping 7 coins creates a binomial experiment, with $N = 7$ and $p = 0.50$. (235)

Binomial Variable: The number of Bernoulli "successes" in a binomial experiment (See **Binomial Experiment**). The value of this variable, often symbolized r, can be any integer from 0 through N, the number of Bernoulli trials in the binomial experiment. There are thus $N + 1$ possible values of r for a given p and N. (235)

Blind Experiment: An experiment using human subjects in which subjects do not know which level of the independent variable is administered to them. (303)

Ceiling Effect: A negatively skewed distribution often results when a variable cannot have a value higher than a certain value. If an exam, for instance, has a score of 100 as the highest possible score, and if the exam is easy, many observed scores will "bunch" at the upper range of possible scores, rather than produce a symmetrical distribution. This is a ceiling effect. (54)

Central Limit Theorem: A mathematical theorem that states, among other things, that the shape of the sampling distribution of the mean will approach that of a normal distribution as larger samples are used, regardless of the shape of the population distribution(s). (261)

Certain Event: An event that will either certainly happen or certainly not happen, contrasted with an **uncertain event** that may or may not occur. A certain event that will happen has a probability of 1.0, and a certain event that will not occur has a probability of 0.0. (212)

Chi-Square Statistic: A test statistic (usually considered to be a **Nonparametric Test**) that allows one to decide whether observed frequencies are essentially equal to or significantly different from frequencies predicted by a theoretical model. The outcome of the test allows one to decide, in turn, whether or not frequencies are distributed equally among categories, whether or not a distribution is normal, or whether or not two variables are independent. (338)

Coefficient of Determination: The proportion of variation in the scatter plot that is explained, that is, the proportion of one variable's variation accounted for by a regression line through the scatter plot of paired observations. The coefficient of determination is equal to r^2, where r is the Pearson r between both variables. (197)

Coefficient of Nondetermination: Equal to $1 - r^2$, the proportion of one variable's variation that is not accounted for by the regression line. (197) See **Coefficient of Determination.**

Combination: (1) The number of different groups of size r that can be selected from a larger group of size N. If, for instance, there are six people available to make a bridge team of four people, there are fifteen different groups of four that can be selected. The procedure for determining how many combinations of size r from a given N one may have are given in Chapter 8. (220). (2) One experimental group in two-factor analysis of variance, which receives one level of both factors (independent variables). A group that receives level 2 of factor 1 and level 5 of factor 2 is referred to as combination 2,5. (399)

Comparable Forms Method: A means of assessing the **reliability** of a test by computing the correlation between people's scores on one version of a test and the same people's scores on a different version of the test. (170)

Conditional Event: An event that requires a prior event (the condition) to occur before it is considered. The symbol $(A \mid B)$ refers to the event A occurring given that B occurs. If, for instance, A represents the event "Grandmother comes to visit us tomorrow," and B represents the event "it rains tomorrow," then the conditional event $(A \mid B)$ is expressed as the event "Grandmother comes to visit tomorrow, given that it rains tomorrow." The probability associated with this event would be expressed — in everyday language — as the chances that Grandmother comes tomorrow if it rains. (230)

Confidence Interval: A range of values within which the statistician is willing to assert with a specified level of confidence that an unknown parameter value lies. Computed from sample statistics, the width of the confidence interval depends on the level of confidence stated, the sample size, and the variability within the sample. (328)

Consistent Estimator: A statistic used to estimate the value of a parameter that has a higher probability of being close to the parameter value when a large sample is used than when a small sample is used. The sample mean, for instance, is a consistent estimator of the population mean, because the sampling distribution of the mean has a consistently smaller standard error as larger and larger sample sizes are used; the smaller standard error of the sampling distribution means that an estimate of μ based on the value of $\overline{X}$ has a higher probability of being correct than when smaller samples are used. (327)

Continuity: A variable's status as either continuous or discrete. (32) See **Continuous Random Variable** and **Discrete Random Variable.**

Continuous Random Variable: A random variable that may (theoretically) assume any value between two points on the measurement scale. It can thus have an infinite number of possible values between those points. Length, for example, is a continuous variable, whereas number of children per family is not. (240) See also **Random Variable.**

Control Group: A group of subjects in an experiment that receives a zero level of the independent variable. (303)

Correlated Variables: Variables whose values are associated; changes in the value of one variable tend to be associated in a systematic way with changes in values of the other. (146) See also **Positive Relation** and **Negative Relation.**

Correlation Coefficient: A single statistic that indicates both the strength and direction of the relation between variables. Most correlation coefficients have values between -1 and $+1$, with positive values indicating positive relations and negative values indicating negative relations, and the absolute value of the coefficient indicating strength of the relation. This book covers the **Pearson Product-Moment Correlation Coefficient,** the **Spearman Rho,** and the **Point Biserial Correlation Coefficient.** (151)

Correlational Analysis: The statistical methods that allow one to discover, describe, and measure the strength of associations among variables. Correlational analysis includes the various techniques of computing correlation coefficients and regression analysis. (146)

Counterbalanced Design: An experimental design in which the experimenter attempts to counteract, or counterbalance, the effects of some extraneous variables on the dependent variable. In a repeated-measures experiment, for example, the order in which two levels of an independent variable are administered might have an effect on the dependent variable; the possible effects of order could be counterbalanced by administering the levels in one order to some of the subjects and in the reverse order to the other subjects. (307)

Covariation: A statistic whose value reflects the degree to which two variables vary together. It is computed by taking a paired observation's X-deviation score and multiplying it by its Y-deviation score, and adding these products for the entire set of paired observations. The **Pearson r** may be computed from the covariation statistic: the covariation may be positive or negative in value, indicating, respectively, positive or negative association between variables. (158)

Critical Value: In hypothesis testing, the value of the statistic that marks the edge of the region(s) of rejection; a value of the observed statistic (or a value of z, t, F, W, R_1 or chi-square associated with that statistic) that occurs with a relative frequency of alpha or less when the null hypothesis is true. Any observed statistic beyond the critical value falls in the region of rejection and leads to a rejected null hypothesis. (278)

Cumulative Proportion Graph: A graph in which one axis represents values of the variable and the other represents the proportion of the distribution that falls below those values (that is, their cumulative proportions). When the data have been grouped, each point on the graph is plotted over the upper true limit of the interval it represents; each point thus represents the proportion of the observations falling in or below that interval. Under certain conditions, the line connecting points on the graph approximates a curve known as an ogive. (58)

Cumulative Proportion Table: A summary table of a group of observations that has one column listing values of the variable and another column indicating the proportion of the distribution that falls at or below each value. When the data have been grouped, the table lists intervals on the measurement scale rather than individual values. (58)

Data: Observed facts, items of information (plural of *datum*). (17)

Decile: A percentile that is an even multiple of 10. The symbol D_4, for instance, stands for the fourth decile, that is, the 40th percentile. Similarly, D_7 stands for the 70th percentile, the seventh decile. (6)

Decision Rule: A statement that the statistician uses (in a hypothesis test) to specify the criteria for deciding between the null and alternative hypotheses. The decision rule indicates the range(s) of values of the observed statistic (or z, t, F, W, R_1 or chi-square) that lead to rejection of the null hypothesis. (278)

Degrees of Freedom: As used here, a characteristic of the sample statistic that determines the appropriate sampling distribution. The specific degrees of freedom (or, df) for each statistic are given with the discussion of the hypothesis test.

Dependent Variable: This is the variable that is *not* manipulated directly by the experimenter. After the different levels of the independent variable have been administered, all subjects are measured, in the same way, on the same dependent variable. In a study of the effects of time spent studying on grade-point averages, where different groups of students are required to spend specified different amounts of time studying, "grade-point average" is the dependent variable. (300)

Descriptive Statistics: Methods for summarizing and describing data in a clear, precise manner. Strictly speaking, descriptive statistics apply only to the subjects actually observed. (9)

Design Matrix: A chart showing the **Factors, Levels,** and **Combinations** involved in a two-factor experiment. Its horizontal dimension represents one factor, and its vertical dimension represents the other; thus, each column represents a level of the first factor, each row represents a level of the second. Individual cells represent specific combinations (experimental groups). (399)

Deviation Score: The difference between the mean of a distribution and an individual element of that distribution. Deviation scores are always found by subtracting the mean from the element; a positive value indicates an element above the mean; a negative value, an element below the mean. (99)

Direct Differences Method: A method of computing t-obtained in the related-samples test for the significance of the difference between two means. (304)

Direction of Relation: The status of the relations between two correlated variables as either positive or negative. (147) See **Positive Relation** and **Negative Relation.**

Directional Test: A hypothesis test with only one region of rejection, that is, a one-tailed test. A directional test is called for only when certain assumptions can be made. Because the region of rejection is located entirely at one end of the distribution in a directional test, fewer extreme values of the observed statistic will lead to rejection of the null hypothesis than in the nondirectional test with the same alpha level. (287)

Discrete Random Variable: A random variable for which there is a probability associated with occurrence of each value of the variable; a discrete random variable may assume only some values between points on the measurement scale. The number of children per family is a discrete variable, for instance. (231) See also **Random Variable.**

Distribution-Free Tests: Another name for nonparametric tests; they are so called because they require either no or very few assumptions about the population distribution (see **Nonparametric Test**). (338)

Double-Blind Experiment: An experiment conducted so that neither the subjects nor the experimenter are aware (during the experiment) of the level of the independent variable administered to individual subjects during the experiment. The double-blind procedure is used so that neither the subjects' nor the experimenter's knowledge of results that are "supposed" to occur can influence the actual outcome of the experiment. (303)

Elementary Event: A term used in applications of probability theory to designate one possible individual outcome of an experimental trial. If an experiment consisted of selecting a college student at random and determining his or her sex, each man and each woman eligible for selection would be an individual elementary event. (218)

Empirical Frequency Distribution: A frequency distribution tabulated from data that have actually been collected (as opposed to a theoretical frequency distribution, which is constructed from theoretical or mathematical considerations). (43)

Error Effect: In an analysis of variance experiment, the component of an observation's value that is not due to the treatment effect; it is assumed that each error effect (there is one for each observation) is produced by a number of different unknown and uncontrollable factors. (379)

Exhaustive Events: The group of outcomes (in a simple experiment) that includes every possible result of a trial. (216)

Expected Frequencies: In the chi-square test, the frequencies of observations in different categories that would be most likely to appear if the null hypothesis is true. (339)

Expected Value of a Statistic: The mean of a statistic's sampling distribution. (260)

Explained Variation: See **Coefficient of Determination.**

Factor: Another name for **Independent Variable** in an experiment evaluated with **Analysis of Variance**. (375)

Factorial: A mathematical operation, symbolized with an exclamation point (!), in which an integer number is multiplied by the next lower integer, and the product is then multiplied by the next lower integer, and so on, until the final product is multiplied by the integer 1. For instance, $4! = (4)(3)(2)(1) = 24.0$. By convention, $0!$ is taken to be equal to 1.0. The number of possible **permutations** possible with n objects is $n!$. (220)

Floor Effect: A positively skewed distribution often results when a variable cannot have a value below a certain value. If an exam, for instance, has a score of 0 as the lowest possible score, and if the exam is quite difficult, many observed scores will "bunch" at the lower range of possible scores, rather than produce a symmetrical distribution. This is a floor effect. (54)

Fixed Factor: In **Analysis of Variance,** an independent variable whose levels represent all levels to which experimental results will be generalized. In other words, conclusions will be drawn only about levels actually used in the experiment. (416)

F-**Ratio:** A statistic used to test the hypothesis that two sample variances are drawn from the same sampling distribution (and hence represent unbiased estimates of the same population parameter variance) against the alternative that the two variances were drawn from different sampling distributions. The test statistic is formed by making a ratio of the two variances; the sampling distribution of the statistic is determined both by the number of degrees of freedom in the numerator variance and the number of degrees of freedom in the denominator variance. (382)

Frequency Distribution: A table or graph that shows the number of times (frequency) with which different values of the variable occur in a group of observations. (43)

Frequency Polygon: Graph of a frequency distribution in which the horizontal axis represents different values of the variable and the vertical axis represents frequencies with which those values occur. In constructing the frequency polygon, a dot is placed over each value of the variable at a height corresponding to the appropriate frequency; the dots are then connected with straight lines to form a polygon figure. When the data have been grouped, each dot is placed over the midpoint of an interval. (49)

Frequency Table: In its simplest form, a two-column table with one column listing values of the variable and the other column listing the frequency with which the different values occur in the group of observations. (43)

F-**Test.** A hypothesis test used in the **analysis of variance** and in testing for **Homogeneity of Variances.** The observed statistic is a ratio made of two variances, each of which is computed from a different component of the total variation in the data. See **Analysis of Variance.** (382)

Gaussian Distribution: Another name for a **normal distribution.** (124n)

Goodness-of-Fit Test: Another name for the **One-Way Chi-Square Test,** used primarily when that test is employed to determine whether or not an empirical distribution may be treated as a normal distribution; in such a case, one is making a decision about the "goodness" of the fit between observations and the theoretical normal model. (344)

Grand Mean: In an analysis of variance experiment, the mean of all observations in the experiment. (378)

Grouped-Frequency Distribution: A frequency distribution table or graph in which frequencies are not listed for each possible value of the variable; rather, a frequency is listed for each of a number of *intervals* on the measurement scale. Each interval is a range of values; all observations falling within the limits of the interval add to the frequency count for that interval. Grouped distributions are used most often when data represent observations on a continuous variable or wherever the number of different observed values precludes the use of all in the table. (46)

Hartley F_{max} Test: A special form of **F-Test,** used to decide whether or not the different treatment groups in analysis of variance or t-test exhibit **Homogeneity of Variances.** (390)

Histogram: A form of **Bar Graph** for illustrating the frequency distribution; in the histogram, the horizontal axis represents values of the variable and the vertical axis represents frequencies with which those values occur. A bar is constructed over each value of the variable (or the midpoint of each interval, if the data are grouped) and extended to the appropriate frequency. The term histogram usually refers to such a graph for interval-, ratio-, or absolute-scale data, while the term bar graph usually refers to such a graph for nominal- or ordinal-scale data. (49)

Homogeneity of Variances: A condition existing when one assumes that two or more group variances represent unbiased estimates of the same population variance. The decision as to whether or not a given set of experimental data represent homogeneity of variances may be based on the **Hartley F_{max} Test;** assumption of this condition is one requirement that should be met before conducting the F-test in analysis of variance or the independent samples t-test. (390)

Homoscedasticity: In the scatter-plot diagram representing paired observations on an X and Y variable, the condition that exists when the variance of observed Y values located over one X value is not too different from the variance of Y values located above another X value (or vice versa). When the condition of homoscedasticity is not approximated, utility of the Pearson r and linear-regression techniques is reduced. (155)

Horizontal Axis: As the name suggests, the horizontal dimension of a two-dimensional graph. Sometimes called the X axis or the abscissa, it usually represents values of the variable in frequency distributions. When experimental results are graphed, it usually represents values of the independent variable. (49)

Hypothesis Testing: A technique in inferential statistics in which one makes a decision about the state of reality in the population. The decision consists of either accepting the state of reality proposed by a null hypothesis or rejecting the null hypothesis in favor of an alternative hypothesis; the null usually postulates a very specific set of conditions, the alternative simply that the null is untrue. The decision to reject the null follows the appearance of observed statistics that are unlikely, given that the null is true. In evaluating experiments, the decision to reject a null hypothesis is usually equivalent to deciding that the independent variable produced an effect on the dependent variable. (272)

Independence: Two variables are independent of each other, if changes in the value of one variable are not related in any systematic way with changes in the value of the other variable. (346)

Independent Events: The situation in probability applications in which the occurrence of one event has no influence on the occurrence of another. When two events are independent, the probability that both occur is equal to the product of their individual probabilities, that is, if A and B are independent events, then $p(A \text{ and } B) = p(A)p(B)$. (229)

Independent-Samples Design: An experiment in which subjects are assigned to different experimental groups on a completely random basis. (301)

Independent Variable: The variable manipulated by the experimenter; the experimenter determines which values of the independent variable are received by subjects. In a study of the effects of time spent studying on grade-point averages, where different groups of students are required to spend specified different amounts of time studying, "time spent studying" would be the independent variable, grade-point averages would be the dependent variable. In analysis of variance, another name for independent variable is **Factor**. (300)

Inferential Statistics: Statistical methods that make it possible to draw conclusions about the population based on observations of a sample (selected from that population) and furthermore make a probability statement about those conclusions to aid in their evaluation. Inferential methods include sampling theory, hypothesis testing, and parameter estimation. (9)

Interaction Effect: In a two-factor analysis of variance, an interaction effect exists when the effect of one independent variable (on the dependent variable) depends, at least to some extent, on which level of the other variable is considered. The presence or absence of an interaction effect is ascertained with an F-test. (403)

Intercept: The value at which a graphed line crosses an axis of the graph. When a regression line has been constructed through a scatter-plot diagram, for instance, the Y axis intercept is the point where the regression line crosses the Y axis (the value of Y-predicted corresponding to $X = 0$). A straight line is completely defined in such a graph if the slope and intercept are known. (185)

Interquartile Range: A statistic sometimes used as a measure of variability, the distance between the 75th and 25th percentiles. The interquartile range is more stable than the simple range and can be used with ordinal-level data; it does not, however, reflect the value of every observation in the group (as does the standard deviation). The median and interquartile range are often used together to describe a group, since both are based on percentiles. (97)

Interval Estimate: A Confidence Interval. (329)

Interval Scale: Sometimes called the equal-interval scale, the level of measurement achieved when (a) each observation falls in one, and only one, measurement category, (b) measurement categories can be ordered, and (c) it can be assumed that the intervals (or distances on the measurement scale) between adjacent categories on the scale are equal. It is thus higher than ordinal-level measurement and lower than ratio-level measurement. The value zero on the scale need not represent zero quantity of the variable; the Fahrenheit temperature scale, for instance, represents interval-level measurement. (25)

Joint Event: In probability applications, an event composed of two other events. The joint event (A and B) occurs only when both A and B occur. If A is the event "Grandmother comes for a visit tomorrow," and B is the event "it rains tomorrow," the joint event (A and B) occurs if Grandmother comes tomorrow *and* it rains. (223)

Least-Squares Criterion: The line through a scatter-plot diagram that minimizes the sum of the squared vertical distances between line and data points; all other lines through the same scatter plot would produce a larger sum; the best-fit line through data points. In the regression situation, the two regression lines Y-predicted and X-predicted are the straight-line least-squares solutions for, respectively, the X and Y dimensions. (187)

Level: One value, or instance, of the independent variable used in an experiment. All subjects within a treatment group receive the same level of the independent variable. (301)

Level of Confidence: A term used in constructing confidence-interval estimates of parameter values to specify the statistician's confidence that the interval includes the parameter value. Using procedures for constructing a 95 percent confidence interval, for instance, the statistician would enclose the true parameter value within its limits with 95 percent of the samples drawn from the population. The higher the level of confidence, the wider the interval. (328)

Level of Measurement: Refers to the degree that characteristics numbers describe characteristics of the variable of interest. The higher the level of measurement, the more statistical methods are applicable (22). See also **Nominal Scale, Ordinal Scale, Interval Scale, Ratio Scale,** and **Absolute Scale.**

Level of Significance: See **Alpha Level.**

Limits-of-Confidence Intervals: The upper and lower values at the two ends of a confidence interval. In a symmetrical confidence interval the limits are located one allowance factor above and below the sample statistic. (329)

Linear Relation: A relation between two variables X and Y of the form

$Y = AX + B$, where A and B are constant values. The transformations of raw scores into z-scores, and z-scores into T-scores, are permissible and do not change the shape of the frequency distributions because they are linear transformations; the transformation process makes use of a linear relation. The graph of a linear relation is a straight line. (185)

Linearly Scaled Axis: A dimension of a graph that is scaled so that equal distances on different parts of the axis represent equal intervals on the measurement scale. It is sometimes called an arithmetically scaled axis. (62)

Lower Limit of an Interval: In frequency distributions where data have been grouped, the lower boundary of an interval on the measurement scale. (47)

Main Effect: In a two-factor experiment, each factor (independent variable) is associated with a main effect; the main effect is the effect that factor has on the dependent variable, considered independently of the other factor's main effect or the interaction effect. The two-factor ANOVA includes a separate F-test for the significance of each main effect. (402)

Matched-Pairs Design: In the two-sample experiment, a procedure in which the entire subject pool is arranged in matched pairs, where pair members are similar (matched) on important characteristics. One member of each pair is then assigned to each experimental group. Thus, individual differences among subjects are less likely to account for differences between experimental groups on the dependent variable. (304)

Mean: A term shared by several measures of central tendency (arithmetic mean, harmonic mean, geometric mean, and quadratic mean), which all reflect the value of every observation in the group. In common use, the term usually refers to the **Arithmetic Mean.** (80, Appendix 2)

Mean Deviation: A little-used measure of variability that is literally the mean of the (absolute value of the) deviations about the mean. (98)

Mean Square: Another name for **Variance,** symbolized MS. (105)

Measurement: In the most general sense, the assignment of labels to observations according to a rule or system. In statistics, measurement systems are classified according to level of measurement and usually produce data that can be represented in numerical form. (18, 20)

Measures of Central Tendency: A family of statistics whose purpose is to convey a picture of the typical observation. The most commonly used measure of central tendency is the arithmetic mean, but the median and mode are also useful in certain situations. Measures of central tendency are also called averages. (74)

Measures of Variability: A family of statistics whose purpose is to convey a picture of the dispersion, or spread of a distribution. The most commonly used measures of variability are the standard deviation and interquartile range. (94) See also **Variability.**

Median: A measure of central tendency defined as the point on the measurement scale above which and below which 50 percent of the observations fall; it is thus the 50th percentile. The median is useful in skewed distributions, since it is not as sensitive as the mean to the presence of a few extremely high or low values. It requires at least ordinal-level data. (78)

Midpoint of an Interval: The value located halfway between upper and lower limits of an interval, found by adding upper and lower limits and dividing by 2. When graphing or computing statistics from grouped data, the midpoint of each interval is sometimes used to represent all observations appearing in that interval. (84)

Mode: A measure of central tendency defined as the most frequently occurring observation category in the data (or in grouped distributions, the midpoint of the interval with the highest frequency). The mode is sometimes defined as the most frequently occurring value *in its vicinity* of the measurement scale; in this case, there may be more than one mode in a distribution. (75)

Mutually Exclusive Events: In applications of probability theory, two or more events that cannot both happen on a single trial. On a single flip of a coin, for instance, the events "heads" and "tails" are mutually exclusive. (216)

Negative Relation: The situation in correlational analysis when high values of one variable tend to be associated with low values of another, and vice versa. The price of gasoline and the number of miles an individual drives each week for pleasure are probably negatively related variables. Negative relations are indicated by a negative-value correlation coefficient. (147)

Negative Skew: A descriptive term applied to frequency distributions with many high values and few extremely low values. On a frequency polygon graph, negative skew produces a "tail" in the direction of low values. (53)

Nominal Scale: The level of measurement whose only requirement is that each observation fall in one, and only one, measurement category. Also called categorical measurement, it is the lowest level of measurement. Designating this pizza as "anchovy" and that pizza as "mushroom" constitutes nominal-scale measurement. (23)

Nondirectional Test: A hypothesis test with two regions of rejection, that is, a two-tailed test. The area under the sampling distribution curve equal to alpha is divided into two equal parts at each end of the distribution, creating two regions of rejection; an observed statistic in either region leads to rejection of the null hypothesis. (287)

Nonindependent Events: Two or more events that have a mutual influence on each other. If the occurrence of one event affects the probability that another event will occur, the two events are nonindependent. (228)

Nonparametric Test: A term referring to a large family of statistical tests that, in general, do not require assumptions about the shape of the population distribution (hence their other name, **Distribution-Free Tests**) and which do not make hypotheses in terms of parameters. Nonparametric tests presented in this book are the chi-square tests, the sign test, the Wilcoxon matched-pairs, signed-ranks test, and the Wilcoxon rank-sum test. Because these tests use ranks or frequencies as data, they do not usually require interval- or ratio-level measurement. (338)

Normal Distribution: A mathematically defined "curve" (relation between X and Y variables). Under certain conditions, it is permissible to treat frequency distributions of actual variables as close approximations of the normal distribution (see Chapter 5). This is desirable because relative frequencies of different values in one special normal distribution (the unit normal distribution) have been tabulated. (122)

Null Hypothesis: Symbolized H_0, one of the two proposed possible states of reality in the hypothesis test. The null hypothesis specifies a sampling distribution of the test statistic; the null is rejected in favor of the alternative hypothesis when an observed statistic appears that is unlikely under the null hypothesis. (275)

Observation: An objectively recorded fact or item of information. Statistics are usually applied to collections of observations. (18)

One-Tailed Test: See **Directional Test.**

One-Way Chi-Square Test: An application of the chi-square test in which observation categories differ only with respect to one variable (hence, one way). Usually, the test is concerned with the question of whether the observed frequencies are distributed in observation categories by chance, or because of the influence of some variable other than chance. (338)

Ordinal Scale: The level of measurement above the nominal scale but below the interval scale. The data represent at least ordinal-scale measurement if (a) each observation falls into one, and only one, category, and (b) observation categories can be ordered. Visiting a military base and recording the rank of each soldier met constitutes ordinal-scale measurement. (24)

Outcome: In probability applications, the result of an experimental trial. (216)

Paired Observation: An observation on two variables, where the intent is to examine the relation between variables. Paired observations form the raw material of correlational analysis. Recording both a person's height and weight, and keeping both of those measurements associated with the same person, constitutes collection of a paired observation. (148)

Parameter: A characteristic of a population determined from observations on every member of the population; population parameters of interest to statisticians include the mean, range, median, standard deviation, and many others. A parameter is also a characteristic of a mathematical relation whose value must be specified before the expression can be evaluated. (255, 124)

Parameter Estimates: Attempts to estimate the values of population parameters (for instance, the mean) from statistics computed on a sample selected from the population. Estimates may consist of a single value (a point estimate) or a range of values (confidence interval). (327)

Parametric Tests: A hypothesis test in which the null and alternative hypotheses are stated in terms of population parameter values. Parametric tests covered in this book are the z-tests, t-tests and F-tests. Because these tests involve means and standard deviations, which require interval-level data or better, applications of parametric tests are limited to situations producing interval-level measurement or better. (275)

Partition of Variation: Division of the total variation (SS) around a mean into two or more components that, when added together, equal the total SS. In regression analysis, partitioning of the variation is done to compare variation due to the regression line with variation around the regression line; in analysis of variance, partitioning of the variation is done to allow assessment of the main effect in a single-factor experiment, or each main effect and the interaction effect in a two-factor experiment. (381)

Pearson Product-Moment Correlation Coefficient: A **Correlation Coefficient** that specifies the degree and direction of relation between two interval-level variables (or variables representing higher levels of measurement). Also called the Pearson r, or simply, r, it is the most commonly used statistic in correlational analysis. (151)

Pearson r: See **Pearson Product-Moment Correlation Coefficient.**

Percentile: A point on the measurement scale below which a specified percentage of the group's observations fall. The 20th percentile, for instance, is the value which has 20 percent of the observations below it. (57). (Also see Appendix 3).

Percentile Rank: The percentile associated with a given score or observation. (56) (Also see Appendix 3.)

Perfect Relation: A relation between two variables such that the value of one is completely determined if the value of the other is specified. The relation between height in inches (Variable 1) and height in feet (Variable 2) is a perfect relation; knowing the value of one variable determines exactly the value of the other. (147)

Performance Variable: See **Behavioral Variable.**

Permutation: One order in which a number of objects, people, or events may be arranged in. See also **Factorial**. (220)

Placebo: In drug studies, a placebo is a pill or injection that is supposed to have no physical effect. If the intent is to study the effects of a certain drug, a placebo may be administered to control group subjects so that they will be blind with respect to the level of independent variable administered. (303)

Point Biserial Correlation Coefficient: A correlation coefficient expressing the degree and direction of relation between a dichotomous variable (one that may have only two values) and a continuous variable (or, a discrete variable that may have many values). (167)

Point Estimate: A single value, produced by application of inferential methods to observations on sample members, which is the statistician's best guess of a parameter value. (327)

Pooled Variances: Symbolized s_p^2, it is the weighted mean of two sample variances. Computing the pooled variance is permissible when the condition of homogeneity of variances can be assumed; when the two samples are assumed to come from the same population, s_p^2 constitutes a better estimate of σ^2 than either sample variance alone. Hence, it is used in the estimation of the standard error of the difference between two means. (311)

Population: The complete group of potential observations. (251)

Population Distribution: A frequency distribution for the variable under study, which includes every member of the population. (254)

Population Parameters: See **Parameter.**

Positive Relation: The situation in correlational analysis that exists when high values of one variable tend to be associated with high values of the other, and low values of one tend to appear with low values of the other. Heights and weights of human beings, for instance, tend to represent a positive relation; positive relations are indicated with positive-value correlation coefficients. (147)

Positive Skew: A descriptive term applied to frequency distributions with many low values and a few extremely high values. On a frequency-polygon graph, positive skew produces a "tail" in the direction of the positive values. (53)

Power: The probability of rejecting a false null hypothesis. Or, equivalently, the area under the sampling distribution created by a true alternative hypothesis that lies in the region of rejection for the null hypothesis. Symbolized $1 - \beta$, the exact power is usually unknown; the statistician can, however, increase power by using larger samples (which produces a smaller standard error in the sampling distributions involved, hence less overlap of sampling distributions specified by null and alternative hypotheses). (292)

Probability Density Function: A mathematically specified distribution used to find the probability that a continuous random variable falls within different intervals on the measurement scale. Although the probability density function is used much like a relative-frequency distribution, the vertical axis is not properly labeled "relative frequency," because there are, theoretically, an infinite number of different values that the variable could assume. (240)

Probability Distribution: For discrete random variables, the probability distribution is a relative-frequency distribution; the relative frequencies associated with values indicate the probabilities that they occur. (For a continuous random variable, the probability density function serves as a probability distribution.) (232)

Probability Statement: A statement that describes an event and indicates the probability associated with its occurrence. "The probability of rain tomorrow is 70 percent and "$p(A) = 0.60$" are both probability statements, since they each designate an event and a probability. (212)

Quartile: A **Percentile** that is an even multiple of 25. The 25th percentile is the first quartile, the 50th percentile is the second quartile (it is also the **Median**), and the 75th percentile is the third quartile. The symbol Q_3, for instance, represents the third quartile. (61)

Random Factor: In **Analysis of Variance,** an independent variable whose levels do not represent all levels to which experimental results will be generalized. In other words, conclusions will be extended to include levels not actually used in the experiment. It is assumed that levels actually chosen for use in the experiment were selected randomly from all possible levels that might have been used. (416)

Random Sample: A sample selected from the population so that every member of the population has an equal chance of being selected for the sample. Although random samples are desirable (the assumption of random sampling underlies some inferential methods), they are, strictly speaking, usually difficult to achieve in practice. (252)

Random Variable: A variable that can assume different values; there is a probability associated with occurrence of different values of the variable, and these probabilities constitute a probability distribution. (231)

Range: The distance on the measurement scale between the highest and lowest values in a distribution (more accurately, the distance between the upper true limit of the highest value and the lower true limit of the lowest value). The range is sometimes used as a measure of variability; as such, it is a relatively unstable statistic. Determination of the range requires at least ordinal-level data. (94)

Ratio Scale: The highest level of measurement (together with absolute-scale measurement). Ratio-level measurement is reached when (a) each observation falls in one, and only one, category, (b) observation categories can be ordered, (c) there are equal intervals between adjacent categories on the measurement scale, and (d) a value of zero represents zero quantity of the variable. (26)

Raw Score: A numerical value assigned to an observation which is expressed in the original units of measurement.

Region of Rejection: The area under the sampling distribution specified by the null hypothesis that covers those values of the observed statistic that lead to rejection of H_0, areas under the H_0 sampling distribution beyond the critical values. In a directional test there is one region of rejection; in a nondirectional test there are two regions of rejection. (277)

Regression Analysis: A branch of correlational analysis that makes possible prediction of the value of one variable from observations on another variable. These predictions are based on a collection of previously made paired observations on both variables. Regression analysis, as discussed here, requires that the two variables be correlated and that the relation between them approximate a linear one. (184)

Regression Line: One of two least-squares lines through a scatter plot of paired observations. Each regression line constitutes the collection of predicted values for one of the variables. (187)

Related-Samples Design: In the two-sample experiment, either a matched-pairs design or a repeated-measures design. Both kinds of related-samples design are evaluated with the same related-samples t-test for the significance of the difference between two means (or z-test), which differs from the test applied to independent-samples designs. In related-samples designs the two experimental groups are deliberately matched on some characteristics; thus, differences between groups on the dependent variable are more likely due to differences produced by the independent variable and less likely due to individual differences among subjects. (301)

Relative Efficiency: A ratio made by taking the number of observations involved in one test, and dividing it by the number of observations required in another test that would produce equivalent **Power**. The more efficient test is the one requiring fewer observations to produce a given power value. (365)

Relative Frequency: The proportion of observations that fall in one category or interval. In probability applications the relative frequency of an event is the proportion of trials on which the event occurs. (54)

Relative-Frequency Distribution: A table or graph that shows (a) observation categories, and (b) the proportion of the group that falls in each category, that is, the relative frequency of each category. (54)

Reliability: When referring to tests or measurement devices, the degree to which repeated applications of the same test (or repeated measurements with the same device) on the same individual produce the same measurement. A test is reliable if it consistently produces the same score for the same individual. (169)

Repeated-Measures Design: In the two-sample experiment, a situation where one group of subjects actually serves as both samples. This is accomplished by applying one level of the independent variable to the subjects, making observations on the dependent variable, and then repeating the procedure with the other level of the independent variable. (304) See also **Related-Samples Design.**

Research Hypothesis: A suspicion on the part of a researcher that two or more variables may be related; the idea that leads to a research study. The study must be designed, however, so that the research hypothesis can be approached in terms of statistical hypotheses or other statistical methods. (300)

Rounding Off: Dropping extra digits to the right of the decimal point, when their inclusion adds no real information. It may also mean changing numbers such as 498,361 to 500,000, for purposes of clarity. (34)

Sample: Some, but not all, members of the population. In inferential statistics inferences about the population are made on the basis of observations on sample members. (251)

Sample Distribution: The frequency distribution of all observations in a sample. When a number of different samples are selected from one population, each sample will probably have a sample distribution slightly different from other samples. (256)

Sample Statistics: Characteristics of samples; statistics computed from observations on sample members. The mean of a sample is a sample statistic because only members of the sample contribute to its value. The mean of a population is a parameter (rather than a statistic) because it applies to all members of the population. (257)

Sampling Distribution: A theoretical distribution that can be specified for any statistic that can be computed for samples from a population. It is the frequency distribution of that statistic's values that would appear if all possible samples of a specified size N were drawn from the population. It is the foundation of inferential statistics, because it allows one to specify the probability with which different values of the statistic appear; it is assumed that a statistic computed from sample observations is one value from such a distribution. (257)

Sampling Distribution of the Mean: Theoretically, the frequency distribution of sample mean values that would appear if an infinite number of samples of a given size were drawn from a population and the individual means computed. Its expected value (mean) is equal to the population mean, and its standard error (standard deviation) is equal to the population standard deviation divided by the square root of the sample size. Each sample mean used in hypothesis testing or parameter estimation is considered as one value from such a distribution. (258)

Sampling Distribution of the Difference Between Two Means: Theoretically, the frequency distribution of differences between two sample means, where an infinite number of possible pair of samples (of sizes N_1 and N_2) were drawn from the specified population or pair of populations. It is the theoretical basis for evaluation of the two-sample experiment. (307) See also **Standard Error of the Difference Between Two Means.**

Sampling Procedure: A method of selecting members of the population for inclusion in the sample. Strictly speaking, the proper sampling procedures must be used if inferences about the population are to be made from sample statistics; two such procedures covered here are random sampling and stratified random sampling. (251)

Scatter-Plot Diagram: A two-dimensional graph in which each axis represents values of a different variable; paired observations on both variables are represented as dots on the graph. Scatter plots may be used as a preliminary step in correlational analysis, since it is usually possible to obtain a rough idea of the degree and direction of the relation between variables by visually inspecting the plot. In the scatter plot, unlike the frequency polygon, the dots are not connected with lines. (148)

Score: A numerical value assigned to an observation. The word "scores" is sometimes used interchangeably with the word "data."

Score Interval: In a grouped frequency distribution, the range of observed values is divided into a number of score intervals. The frequency-distribution table lists the number of observations that fall into each score interval. (46)

Semi-Interquartile Range Statistic: Half of the interquartile range, sometimes used as a measure of variability. (98) See also **Interquartile Range.**

Sign Test: A nonparametric alternative to the related-samples t-test, used to decide whether differences between two groups (or two treatment levels) are about equally distributed in both directions, or whether differences are primarily in one direction. The test statistic used is the binomial r. (350)

Simple Experiment: In probability applications, any process that leads to one of a group of well-defined outcomes. It need not be conducted in a laboratory setting or even under the control of an experimenter. All that is required is that the possible outcomes of the process be known, and that after the process occurs it is possible to determine which outcome has occurred. (216)

Single-Factor ANOVA: An analysis of variance with only one independent variable (**Factor**), and only one hypothesis test. Rejecting the null hypothesis is equivalent to deciding that there is at least one significant difference among treatment group means. (375)

Skewed Distribution: A distribution in which more observations fall on one side of the mean than the other. (53) See **Positive Skew** and **Negative Skew.**

Slope: In the linear equation $Y = AX + B$, showing the relation between X and Y variables, A is a constant value known as the slope. When the equation is graphed, A is in fact the slope of the resulting line, that is, the number of Y units (vertical units) covered by the line for every X unit (horizontal units). Knowing the slope and intercept of a line completely determines the entire line; in regression analysis, one applies computational methods to find the slope and intercept of the regression line (and thus, the whole line). (186)

Spearman Rank-Difference Correlation Coefficient: A correlation coefficient showing the direction and degree of relation between ranks of two variables in a number of paired observations. It can thus be used when one or both variables produce data at the ordinal level. It is sometimes used as a quickly computed substitute for the Pearson r. (164)

Spearman rho: A shorter name for the **Spearman Rank-Difference** correlation coefficient. (164)

Split-Half Method: A means of assessing a test's reliability, in which items on the test are divided into two sections. Reliability is taken as the correlation between people's scores in one section with their scores in the other section. (170)

Standard Deviation: The most commonly used measure of variability. The standard deviation requires at least interval-level data and reflects the value of every observation in the distribution. Like other measures of variability, it is a single number whose size indicates the spread, or dispersion, of the distribution. (101)

Standard Error: The standard deviation of a sampling distribution. (260)

Standard Error of Estimate: In the regression situation, the standard deviation of observed values around the regression line. Thus, the smaller the standard error of estimate, the more precise future predictions are likely to be. (191)

Standard Error of the Mean: Standard deviation of the sampling distribution of the mean. Symbolized $\sigma_{\bar{X}}$ when determined from a known population σ, and $s_{\bar{X}}$ when estimated from sample statistics, the standard error of the mean depends on both the population standard deviation and the sample size. (260)

Standard Error of the Difference Between Two Means: The standard error (standard deviation) of the sampling distribution of the difference between two means. If it is computed from a known population σ, this standard error is symbolized $\sigma_{\bar{X}_1-\bar{X}_2}$; when estimated from sample statistics, it is symbolized $s_{\bar{X}_1-\bar{X}_2}$. As does the standard error of the mean, its size depends on both population standard deviation and sample sizes. (309)

Standard Score: An individual observation that belongs to a distribution with a mean of zero and a standard deviation of 1. Any distribution of raw scores can be transformed into a distribution of standard scores without changing the shape of the distribution or the relative order or distances between members, because the transformation to standard scores is linear. Also called z-scores. A negative standard score indicates an observation below the group's mean, a positive standard score indicates an observation above the mean. (109)

Statistic: A single number that allows one to summarize, analyze, or evaluate a group of observations. (3)

Statistical Decision: Choosing between states of possible reality on the basis of probability considerations. Hypothesis testing involves a statistical decision in which one either accepts or rejects a null hypothesis. (274) See also **Hypothesis Testing**.

Statistical Power: See **Power**.

Statistics: The area of study that includes methods for producing and interpreting statistics (see **Statistic**). Generally speaking, statistical methods are applied as an attempt to understand large masses of information; to discover and describe characteristics of the data that are not apparent from casual observation; to describe characteristics of a group of observations rather than single observations. (3)

Stratified Random Sampling: A sampling procedure used when the possibility of bias in the sample would otherwise be large. Strata (or subgroups) of the population that have potential meaning for the characteristic under study are identified by the researcher *before* the sample members are selected. The sample is then chosen so that the proportion of the population that falls in each stratum and the proportion of the sample members from that stratum are equal. *Within* each stratum, sample members are randomly applied. (252)

Structural Model: A statement, in symbols, representing an assumption underlying the analysis of variance. The model says that each dependent variable observation is the sum of a number of components, all of which add linearly to produce the score; each component represents a different parameter. (379)

Subject Variable: An independent variable that is not actually manipulated (for individual subjects) by the experimenter because it is a characteristic of the subjects. Sex, for instance, may serve as a subject variable in an ANOVA design; one level of this factor would, of course, be "Male," the other level "Female." A significant effect associated with a subject variable is more properly interpreted as a demonstration of correlation than cause and effect. A subject variable is only manipulated in the sense that the experimenter *assigns* people already having different values of the variable to different groups. (392)

Subscripted Variable: The use of subscripts with a symbol for a variable. Subscripted variables are used so that one may refer to specific observations, or perform calculations on specific parts of the total data set. In ANOVA, for instance, the symbol $X_{2,4}$ represents observation 4 in treatment group 2. (81)

Sufficient Estimator: A sample statistic used to estimate population parameters, which uses all possible information in the sample. Sufficiency is a desirable characteristic of estimators. (327)

Sum of Squares: Another name for variation. Symbolized SS. (105) See also **Variation**.

Summation Notation: The summation symbol Σ directs one to add together a number of individual observations (or add together their squared values, or add some other combination of observations). The summation sign may be used with subscripted-variable notation to indicate clearly which values are to be added; it may be used without subscripts when it is clear which values are to be added. (81)

Symmetrical Distribution: A distribution in which, for every observation on one side of the mean, there is another observation at an equal distance on the other side of the mean. In the graph of a symmetrical distribution, the left half of the polygon (or histogram) is a mirror image of the right half. (53)

t-**Distribution:** The shape of the sampling distribution of the mean. There is a different *t*-distribution for each number of degrees of freedom; when $df \geq 30$, however, the shapes of the *t* and normal distributions are close enough to be considered equal in most cases. Thus, the *t*-distribution is used primarily for small sample situations. Use of the *t*-distribution requires the assumption that the population distribution is normal. (281)

T-**Score:** A standard score that has been transformed to a distribution with a mean of 50 and a standard deviation of 10. (114)

t-**Test:** A hypothesis test that uses the *t*-statistic and the *t*-distribution to arrive at a decision. When small samples are used, and when the population standard deviation is unknown, the hypothesis tests about one mean and the test involving two means are *t*-tests. (280)

Test-Retest Method: A means of assessing a test's **reliability,** in which a group of people take a test and then retake it at some later time. Reliability is taken as the correlation between people's scores on the two test-taking performances. (170)

Transformed Standard Score: A standard score that has been transformed (has had a linear transformation applied to it) so that it now belongs to a distribution with any mean and standard deviation the statistician wishes. Transformed scores are used most often in evaluating test scores. See the text discussion (Chapter 4) of reasons for using standard scores and transformed standard scores. (111)

Treatment Effect: In an analysis of variance experiment, the component of each observation's value that results from receiving one level of the independent variable. It is assumed that every observation under the same level of the independent variable has the same treatment effect. (379)

Trial: In probability applications, conducting a simple experiment once. Sometimes the word trial is used to indicate collection of one datum. (216)

True Limits of an Interval: The distance on the measurement scale actually enclosed by an interval, when grouping data, is the distance between the interval's true limits. The interval's upper *true* limit is halfway between the interval's upper limit and the lower limit of the next-higher interval. The upper true limit of the interval 30–39, for instance, is halfway between the interval's upper limit (39) and the lower limit of 40, that is, 39.5. Similarly, this interval's lower true limit is 29.5. (51)

True Limits of a Number: The upper and lower points on the measurement scale that enclose all values of the variable actually represented by a number. When human weights, for instance, are reported in even 1-lb units, the number 160 has true limits 159.5 and 160.5, since all true weights between these values are reported as 160. (33)

Two-Factor ANOVA: An experiment analyzed with analysis of variance, having two independent variables (**Factors**). In two-factor ANOVA, there are three independent hypothesis tests: there is a test for the main effect associated with factor and for the interaction effect. (398)

Two-Tailed Test: See **Nondirectional Test.**

Type I Error: Rejecting a true null hypothesis. The probability of making a Type I error is equal to the alpha level. (288)

Type II Error: Accepting the null hypothesis when the alternative is true. The probability of making a Type II error is symbolized β, a probability whose exact value is usually unknown. (289)

Unbiased Estimator: A statistic used to estimate parameter values whose sampling distribution has a mean (expected value) equal to the parameter value. $\hat{s}^2$, for instance, is an unbiased estimator of σ^2, while s^2 is a biased estimator of σ^2. (327)

Uncertain Event: An event that may or may not occur in the future, i.e., not a **certain event.** Uncertain events are associated with probabilities *between* 0.0 and 1.0. (212)

Underlying Variable: A characteristic of a person or animal that cannot be observed directly, but which is assumed to exist. Intelligence, emotion, strength of attitudes, and personality traits, for instance, are underlying variables that are assumed to exist, but which cannot be directly observed or measured. (21)

Unexplained Variation: See **Coefficient of Nondetermination.**

Unit Normal Curve: A normal curve that has a mean of zero and a standard deviation of 1. Areas under different parts of the unit normal curve are tabulated in the unit normal table; thus values from other normal distributions must be transformed to values in the unit normal distribution, in order to determine probabilities associated with those values. (131)

Unit Normal Table: A table (like Table A, Appendix 1) that makes it possible to determine areas under specific parts of the unit normal curve. The table values were originally determined by applications of integral calculus to the normal curve equation. Normal curve tables rarely include areas associated with values less than 4 or more than 4 standard deviations from the mean, since almost all values in a normal distribution are located in this range. (132)

Upper Limit of an Interval: In frequency distributions where data have been grouped, the upper boundary of an interval on the measurement scale. (47)

Validity: When referring to tests or measurement devices, the degree to which the test (or device) measures what it is supposed to measure. (169)

Variable: Any characteristic (of a person, object, or situation) that can change value or kind from observation to observation. Performances as well as physical characteristics can be variables. (18)

Variability: Dispersion of a distribution; the extent to which group members differ among themselves. Variability is not the name of a specific statistic, but rather the term applied to the characteristic of dispersion. (94) See also **Measures of Variability.**

Variance: The squared standard deviation, sometimes called the mean square. In some applications, such as the analysis of variance, the variance is useful in its own right as a measure of variability. (105)

Variation: The sum of the squared deviations about the mean sometimes called the sum of squares. In some applications variation is useful in its own right as a measure of variability. (105)

Vertical Axis: As the name suggests, the vertical dimension of a two-dimensional graph. Sometimes called the Y axis or the ordinate, it usually represents frequency in frequency distributions, relative frequency in relative-frequency distributions, and cumulative proportion in cumulative-proportion graphs. When experimental results are graphed, it usually represents values of the dependent variable. (49)

Wilcoxon Matched-Pairs, Signed-Ranks Test: A nonparametric alternative to the related samples t-test for two-sample (or repeated-measures) experiments. Like the sign test, this test allows one to decide whether differences between the two treatment levels are about equally distributed in both directions, or whether differences are primarily in one direction. Unlike the sign test, however, this test also uses the sizes of these differences; thus, it uses more information in the data and has more power than the sign test. (355)

Wilcoxon Rank-Sum Test: A nonparametric alternative to the independent samples t-test for two-sample experiments. The test allows one to decide whether ranks of individual observations are distributed equally between groups, or whether one group receives a significantly greater allotment of high-ranking scores, while the other group receives a significantly greater distribution of low-ranking scores. (359)

Yates's Correction: In computing the obtained chi-square statistic, a correction that must be applied when $df = 1$ and when any expected frequency is small (less than 10). It is a correction for continuity, needed because frequencies are a discrete variable, whereas the chi-square distribution is continuous. (342)

z-Score: A standard score from a distribution with a mean of zero and a standard deviation of 1. (109) See also **Standard Score.**

z'-**Score:** A transformed standard score. (112) See also **Transformed Standard Score.**

z-**Test:** A hypothesis test that uses a *z*-score as the obtained statistic and the normal distribution as the sampling distribution. When population standard deviations are known, the hypothesis tests about one or two means are *z*-tests. (278)

Zero Relation: The situation that exists when values of one variable are not related in any way to values of another. With a zero relation, knowing the value of one variable gives you no indication of the value of the other. Zero relations are represented by correlation coefficients of 0.0. (147)

Bibliography

In the United States alone, there are currently over 100 textbooks intended for exactly the same audience as *Understanding and Using Statistics,* 2nd ed. In addition, there are many hundreds more that are designed for slightly different audiences. Although the latter may feature more mathematical theory, broader coverage, lower or higher level of difficulty, and so on, they nonetheless include some material of interest to readers of this book. Some of the books make good supplementary reading; others do not. Unfortunately, those of you who seek to do additional reading will face a vast and confusing array of titles in your library card catalogue. To help you pursue topics covered in this book, therefore, I recommend beginning with the following readings. Naturally, there are many other books available that are equally good, but I believe you will find what is listed below particularly helpful.

Bayesian Statistics

Novick, M. R. "Introduction to Bayesian Inference." Chapter 16 of Blommers, P. J. and Forsyth, R. A. *Elementary Statistical Methods in Psychology and Education.* 2nd ed. Boston: Houghton Mifflin, 1977.

Hays, W. L. "Some Elementary Bayesian Methods." Chapter 19 of *Statistics for the Social Sciences*, 2nd ed. New York: Holt, Rinehart and Winston, Inc., 1973.

History of Statistics

Dudycha, A. L. and Dudycha, L. W. "Behavioral Statistics: An historical perspective." In Kirk, R. W. (Ed.) *Statistical Issues: A Reader for the Behavioral Sciences.* Belmont, California: Brooks/Cole, 1972.

Haber, A., Runyon, R. P. and Badia, P. (Eds.) *Readings in Statistics.* Reading, Massachusetts: Addison-Wesley, 1970. (Not a history per se, but a collection of historically important statistical papers.)

Walker, H. *Studies in the History of Statistical Method.* Baltimore, Maryland: Williams and Wilkins, 1929.

Introductions in Depth to Statistical Applications

Hays, W. L. *Statistics for the Social Sciences*, 2nd ed. New York: Holt, Rinehart and Winston, Inc. 1973. (Covers a slightly broader range of statistical methods than this book, but with much, much more verbal explanation and more mathematical theory; a very readable book for those with little previous mathematical background.)

Winer, B. J. *Statistical Principles in Experimental Design*, 2nd ed. New York: McGraw-Hill, 1971. (A mathematically rigorous, very complete coverage of analysis of variance and related topics in statistical inference; however, even though substantial mathematical theory is included, the emphasis is on application of statistics to evaluation of experiments.)

Measurement Theory

Adams, E. W., Fagot, R. F. and Robinson, R. E. "A theory of appropriate statistics." *Psychometrika*, 1965, 30, 99–127. (These authors argue that, in effect, statistics are neither appropriate nor inappropriate, per se, for data from any given level of measurement. What is critical is the kind of statements made about those statistics. Difficult reading for the introductory student.)

Baker, B. O., Hardyck, C. D. and Petrinovich, L. F. "Weak measurement vs. strong statistics: An empirical critique of S. S. Stevens' proscriptions on statistics." *Educational and Psychological Measurement*, 1966, 26, 291–309. (These authors conducted an empirical test of the question "Do statistics computed on measures which are inaccurate descriptions of reality distribute differently than the same statistics computed under conditions of perfect measurement?" They conclude that, when doing *t*-tests with equal sample sizes, under most conditions, the answer is a firm "no.")

Stevens, S. S. "Mathematics, Measurement and Psychophysics." In Stevens, S. S. (Ed.), *Handbook of Experimental Psychology*. New York: John Wiley & Sons, Inc., 1951. (A basic reference on Stevens' system of classifying measurement scales.)

Torgerson, W. S. *Theory and Methods of Scaling*. New York: John Wiley & Sons, Inc., 1958. The first three chapters are particularly useful: Chapter 1, "The Importance of Measurement in Science"; Chapter 2, "The Nature of Measurement"; Chapter 3, "Classification of Scaling Methods."

Misuse of Statistics

Huff, D. *How to Lie with Statistics*. New York: W. W. Norton and Company, 1954.

Runyon, R. P. *Winning with Statistics*. Reading, Massachusetts: Addison-Wesley, 1977.

Nonparametric Statistics

Bradley, J. V. *Distribution-Free Statistical Tests*. Englewood Cliffs, New Jersey: Prentice-Hall, 1968.

Siegel, S. *Nonparametric Statistics for the Behavioral Sciences*. New York: McGraw-Hill, 1956.

Problem Workbooks

Runyon, R. P. *Descriptive Statistics: A Contemporary Approach.* Reading, Massachusetts: Addison-Wesley, 1977.

Runyon, R. P. *Inferential Statistics: A Contemporary Approach.* Reading, Massachusetts: Addison-Wesley, 1977.

Runyon, R. P. *Nonparametric Statistics: A Contemporary Approach.* Reading, Massachusetts: Addison-Wesley, 1977.

Statistical Issues

Kirk, R. W. (Ed.) *Statistical Issues: A Reader for the Behavioral Sciences.* Belmont, California: Brooks/Cole, 1972.

Tests and Test Theory

Anastasi, A. *Psychological Testing.* 4th ed. New York: Macmillan, 1976.

Uses of Statistics, Illustrative Applications

Tanur, J. M. et. al. (Eds.) *Statistics: A Guide to the Unknown.* San Francisco: Holden-Day, 1972. (Forty-four short, highly readable accounts of statistical applications in a very wide variety of fields, including opinion polling, cloud seeding, medicine, military leadership, geology, and history, to name a few.)

Index